About Island Press

Island Press is the only nonprofit organization in the United States whose principal purpose is the publication of books on environmental issues and natural resource management. We provide solutions-oriented information to professionals, public officials, business and community leaders, and concerned citizens who are shaping responses to environmental problems.

In 2001, Island Press celebrates its seventeenth anniversary as the leading provider of timely and practical books that take a multidisciplinary approach to critical environmental concerns. Our growing list of titles reflects our commitment to bringing the best of an expanding body of literature to the environmental community throughout North America and the world.

Support for Island Press is provided by The Bullitt Foundation, The Mary Flagler Cary Charitable Trust, The Nathan Cummings Foundation, Geraldine R. Dodge Foundation, Doris Duke Charitable Foundation, The Charles Engelhard Foundation, The Ford Foundation, The George Gund Foundation, The Vira I. Heinz Endowment, The William and Flora Hewlett Foundation, W. Alton Jones Foundation, The John D. and Catherine T. MacArthur Foundation, The Andrew W. Mellon Foundation, The Charles Stewart Mott Foundation, The Curtis and Edith Munson Foundation, National Fish and Wildlife Foundation, The New-Land Foundation, Oak Foundation, The Overbrook Foundation, The David and Lucile Packard Foundation, The Pew Charitable Trusts, Rockefeller Brothers Fund, The Winslow Foundation, and other generous donors.

About The Global Development And Environment Institute

The Global Development And Environment Institute (G-DAE) is a research institute at Tufts University that is dedicated to promoting a better understanding of how societies can pursue their economic goals in an environmentally and socially sustainable manner. G-DAE pursues its mission through original research, policy work, publication projects, curriculum development, conferences, and other activities directed toward the development of academic programs at Tufts University and elsewhere. Founded in 1993 and directed by Dr. Neva Goodwin and Dr. William Moomaw, G-DAE is jointly administered by the Fletcher School of Law & Diplomacy and the Tufts Graduate School of Arts & Sciences.

A SURVEY OF SUSTAINABLE DEVELOPMENT

FRONTIER ISSUES IN ECONOMIC THOUGHT
VOLUME 6
NEVA R. GOODWIN, SERIES EDITOR

Titles published by Island Press
in the Frontier Issues in Economic Thought series include:

R. Krishnan, J. M. Harris, and N. R. Goodwin (eds.),
A Survey of Ecological Economics, 1995

N. R. Goodwin, F. Ackerman, and D. Kiron (eds.),
The Consumer Society, 1997

F. Ackerman, D. Kiron, N. R. Goodwin, J. M. Harris, and K. Gallagher (eds.),
Human Well-Being and Economic Goals, 1997

F. Ackerman, N. R. Goodwin, L. Dougherty, J. M. Harris, and K. Gallagher (eds.),
The Changing Nature of Work, 1998

F. Ackerman, N. R. Goodwin, L. Dougherty, and K. Gallagher (eds.),
The Political Economy of Inequality, 2000

Jonathan M. Harris, Timothy A. Wise, K. Gallagher, and
Neva R. Goodwin (eds.),
A Survey of Sustainable Development: Social and Economic Dimensions

A Survey of Sustainable Development: Social and Economic Dimensions

EDITED BY
JONATHAN M. HARRIS,
TIMOTHY A. WISE,
KEVIN P. GALLAGHER,
AND NEVA R. GOODWIN

The Global Development
And Environment Institute
Tufts University

ISLAND PRESS
WASHINGTON • COVELO • LONDON

Library of Congress Cataloging-in-Publication Data

Sustainable human and economic development / edited by Jonathan M. Harris

p. cm. — (Frontier issues in economic thought ; 6)

Includes bibliographical references and index.

ISBN 1-55963-862-1 — ISBN 1-55963-863-X

1. Sustainable development. I. Harris, Jonathan M. II. Series.

HC79.E5 S8688 2001

338.9'27—dc21

00-012957

British Library Cataloguing in Publication Data available.

Printed on recycled, acid-free paper

Printed in the United States of America

10 9 8 7 6 5 4 3 2 1

To

Paul Streeten
whose work has been central to the
field of human development

and

Herman Daly
who has led the way on issues of
environmental sustainability

Note to the Reader

The articles presented in this volume have been summarized by the editors, and each has been reviewed by the authors of the original article to ensure accuracy. Summarizing makes it possible to present a wide overview of the field, but inevitably a summary cannot convey the full content of the original work. Readers are urged to refer to the originals for greater detail and context, and full references for all articles summarized are provided at the beginning of the summary.

In general, the summaries presented here do not repeat material from the original articles verbatim. In a few instances it has seemed appropriate to include in the summaries direct quotations from the original text ranging from a phrase to a few sentences. Where this has been done, the page reference to the original article is given in square brackets. References to other books or articles appear in the bibliography.

Contents

Part II. Economics of Sustainability: The Social Dimension

Part III. Global Perspectives: The North/South Imbalance

BUILDING BLOCKS OF SUSTAINABILITY

Part IV. Population and Urbanization

Part V. Agriculture and Renewable Resources

Part VI. Materials, Energy, and Climate Change

POLICIES FOR SUSTAINABILITY

Part VII. Globalization and Sustainability

Authors of Original Articles

[First authors and corresponding authors listed here]

Alice H. Amsden Dept. of Urban Studies and Planning, Massachusetts Institute of Technology, Cambridge, Massachusetts

Anil Agarwal Centre for Science and Environment, New Delhi, India

David Barkin Universitat Autonoma Metropolitana, Xochimilco, Mexico

Nancy Birdsall Carnegie Endowment for International Peace, Washington, D.C.

James K. Boyce Dept. of Economics, University of Massachusetts, Amherst, Massachusetts

Robin Broad School of International Service, American University, Washington, D.C.

Robert S. Browne Teaneck, New Jersey

Severyn T. Bruyn Dept. of Sociology, Boston College, Chestnut Hill, Massachusetts

John M. Byrne Center for Energy and Environmental Policy, University of Delaware, Newark, Delaware

John Cavanagh Institute for Policy Studies, Washington, D.C.

Robert Chambers Institute of Development Studies, University of Sussex, Falmer, Brighton, United Kingdom

Michael Clemens Harvard University, Cambridge, Massachusetts

Priscilla Connolly Universitat Autonoma Metropolitana, Azcapotzalco, Tamaulipas, Mexico

Gordon Conway Rockefeller Foundation, New York, New York

Robert Costanza Institute for Ecological Economics, University of Maryland, Solomons, Maryland

James R. Crotty Dept. of Economics, University of Massachusetts, Amherst, Massachusetts

Herman Daly School of Public Affairs, University of Maryland, College Park, Maryland

Paul R. Ehrlich Dept. of Biological Sciences, Stanford University, Stanford, California

Salah El Serafy Economic consultant, Arlington, Virginia

Daniel C. Esty School of Forestry, Yale Law School, New Haven, Connecticut

Peter Evans Dept. of Sociology, University of California, Berkeley, California

Jonathan Fox University of California, Santa Cruz, California

Peter Fox-Penner Brattle Group, Washington, D.C.

Hilary French Worldwatch Institute, Washington, D.C.

Dharam Ghai United Nations Research Institute for Social Development (UNRISD), Geneva, Switzerland

Charles Gore United Nations Conference on Trade and Development (UNCTAD), Geneva, Switzerland

Kirk Hamilton Dept. of Environment, The World Bank, Arlington, Virginia

M. Jeff Hammond Redefining Progress, Washington, D.C.

Patrick Heller International and Public Affairs, Columbia University, New York, New York

Carlos A. Heredia Pueblo, Tampico, Mexico

Crawford S. Holling Dept. of Zoology, University of Florida, Gainesville, Florida

Peter D. Kinder Kinder, Lydenberg and Domini, Boston, Massachusetts

Sai Felicia Krishna-Hensel Center for Business and Economic Development, Auburn University, Montgomery, Alabama

Melissa Leach Institute of Development Studies, University of Sussex, Falmer, Brighton, United Kingdom

Robert C. Lind Johnson Graduate School of Management, Cornell University, Ithaca, New York

Alain Lipietz CERPREMAP, Paris, France

Michael Lipton Centre for the Comparative Study of Culture, Development and the Environment, University of Sussex, Falmer, Brighton, United Kingdom

Anne Platt McGinn Worldwatch Institute, Washington, D.C.

Juan Martinez-Alier Dept. of Economics and Economic History, Universitat Autonoma de Barcelona, Bellaterra, Spain

Linda Mayoux Cambridge University, Cambridge, United Kingdom

Giuseppe Munda Dept. of Economics and Economic History, Universitat Autonoma de Barcelona, Belleterra, Spain

Norman Myers Environment and Development, Headington, Oxford, United Kingdom

Sunita Narain Centre for Science and Environment, New Delhi, India

Molly O'Meara Sheehan Worldwatch Institute, Washington, D.C.

Talbot Page Dept. of Economics, Brown University, Providence, Rhode Island

Per Pinstrup-Andersen International Food Policy Research Institute (IFPRI), Washington, D.C.

Michael E. Porter Harvard Business School, Boston, Massachusetts

Philip W. Porter Institute for Global Studies, University of Minnesota, Minneapolis, Minnesota

A. Atiq Rahman Bangladesh Centre for Advanced Studies, Dhanmondi, Dhaka, Bangladesh

Shahra Razavi United Nations Research Institute for Social Development (UNRISD), Geneva, Switzerland

David Reed World Wildlife Fund, Washington, D.C.

Robert Repetto World Resources Institute, Washington, D.C.

Jane Rissler Union of Concerned Scientists, Washington, D.C.

Wolfgang Sachs Wuppertal Institute for Climate, Environment and Energy, Wuppertal, Germany

Sara J. Scherr Dept. of Agricultural and Resource Economics, University of Maryland, College Park, Maryland

Richard E. Schuler Cornell Institute for Public Affairs, Cornell University, Ithaca, New York

Vandana Shiva Research Foundation for Science, Technology, and Natural Resource Policy, Dehra Dun, India

George A. Steiner University of California, Los Angeles, California

John F. Steiner California State University, Los Angeles, California

David I. Stern Centre for Resources and Environmental Studies, Australian National University, Canberra, Australia

Paul Streeten Consultant, United National Development Programme, Spencertown, New York

Lance Taylor New School for Social Research, New York, New York

Judith Tendler Dept. of Urban Studies and Planning, Massachusetts Institute of Technology, Cambridge, Massachusetts

Alexandre Timoshenko International Legal Instruments Division, United Nations Environment Programme (UNEP), Geneva, Switzerland

Michael A. Toman Energy and Natural Resources Division, Resources for the Future, Washington, D.C.

Mahbub ul Haq Formerly United Nations Development Programme (deceased)

Peter G. Veit World Resources Institute, Washington, D.C.

Iddo K. Wernick Earth Institute, Columbia University, New York, New York

Allen L. White Risk Analysis Group, Tellus Institute for Resource and Environmental Strategies, Boston, Massachusetts

Foreword

Amartya Sen

One of the contingent rewards of agreeing to write a foreword to a book is not only that one gets a free copy of the book, but also that one is forced to read the book even if its size and substance suggest that it may not be altogether light reading. The reward is inescapably contingent, since it is dependent on whether the book is actually worthwhile to read.

This imposing volume of carefully edited essays passes the test handsomely. It is not only an excellent collection of essays on an extremely important subject, but it is also a reader's delight in that the editors provide an informative tour of a vast—and rapidly growing—field of research, giving the reader the opportunity to make intelligent decisions on what he or she would particularly like to read. I feel very privileged to be able to present this volume to what I hope will be a large readership.

Indeed, with the illuminating and user-friendly introduction that the editors themselves have provided, my task is made much simpler, and I shall use the opportunity to comment briefly on the nature of the subject and how a reader may view a volume of this kind.

What, then, is so special about yet another book on sustainable development? This is certainly a rapidly growing field of research and publishing. The understanding that nature and the environment in which we live are deeply vulnerable may be a new thought, but its far-reaching implications have made this a much studied area of investigation and assessment. The frailty of each individual life (including its ultimate cessation) has, of course, been well understood for a very long time, leading to ancient and modern studies of the so-called "human predicament." But that predicament has been typically seen as a plight of the individual, and frequently contrasted with the durability of mankind as a whole. Even Alfred, Lord Tennyson's great "elegy" grumbled about the partiality of nature and contrasted the infirmity of individual life with the security that nature provides for our group future:

> So careful of the type she seems,
> So careless of the single life.

With the growing recognition that it is not merely the single life, but also the "types" (indeed all the known types) that are threatened, and that the lives that can be led may well stand in great danger of being impoverished or obliterated, environmental studies have become inescapably a major area of intense research and investigations.

This may be reason enough for trying a get a well-selected and well-organized compendium of essays and other contributions, but the case for a book of this kind is stronger than such general reasoning may suggest. If, despite several brilliant contributions that have tried to integrate the environmental literature, it still seems rather murky, this is partly due to the fact that both the nature of the questions asked and the content of the answers given admit a variety of different concerns and motivating contexts. Do we view the environmental challenge from the perspective of preserving nature or that of preserving the lives that human beings can lead? The latter is more anthropocentric than the former, and thus much more limited, and yet it is not clear from what perspective any non-anthropocentric conservationism may be assessed. There are also disputes between different anthropocentric approaches. For example, should we be concerned only with those environmental issues that influence the standard of living of human beings, or also with the conservation of those natural objects that people find reason to value (whether or not they contribute directly to what can be seen as their "standard of living")? And again, how are the judgments that people make (or, alternatively, the interests they actually have) to be exactly identified, in an articulated form, and how are they to be related to particular programs of conservation, which may compete with one another for our limited resources, or even for our narrow span of attention and commitment?

In the environmental literature, each individual analysis tends to make specific assumptions, if only implicitly, on these issues, and they respectively opt for particular lines of reasoning, taking distinctive positions on these contentious matters. But the discerning reader, not to mention the activist environmentalist, has reason enough to wonder how to compare and contrast these different approaches, and how to deal with what may or may not be an embarrassment of riches but certainly is an embarrassment of some sort as a prelude to action. We are, thus, inclined to seek a more comprehensive understanding that would allow us to form our own views of these divisive issues, in the light of what each approach has yielded or seems to promise.

This problem of diversity is endemic in the field. A great deal of the environmental literature has focused in recent years on the task of sustainability, but there have been several distinct characterizations of *what* it is to be sustained. As a result, the implications of sustainability have emerged in very diverse lights in different parts of the literature. To take another source of contrast, the choice variables on which environmentalists concentrate as instruments of conservation can vary greatly depending on the focus of the discipline to which the analysts themselves belong or with which they are most familiar. Economists often have quite a different focus on policy variables (concentrating on markets, prices, taxes, property rights, etc.) than what anthropologists choose to discuss (such as values, perceptions, cultures, etc.). Similarly, natural scientists frequently take a somewhat different route (focusing on scientific possibilities or technical variations) from what social scientists end up discussing. There are many discipline-related contrasts of approach that supplement the diversities related to basic ethics and valuational priorities.

Wide variance of contexts and concerns is, thus, a major feature of environmental studies. Even when there is a general agreement that the environmental challenges are important and that they call for some reasoned response, the direction of investigations can be widely divergent. What this informative and stimulating book does is present, in a single volume, a great many—sixty-six, to be exact—essays and book chapters, with a remarkable diversity of approaches and outlooks. It thus meets a very important need, and does so with efficiency and style.

Bearing in mind this motivation, it is perhaps important not to see the table of contents as an integral and indivisible agenda, each item of which must be fully tackled by each participant, but rather as a menu that tells readers what is being offered so that they can decide precisely what they want to read, in what detail. With a book of this kind, it is extremely important to exercise one's discretion, in the light of one's own interests and the guidance that is provided by the editors. We have to know something about each of the extant approaches but may have good reason, too, to spend a lot more time on some approaches than others. Indeed, the reader may end up reacting against particular contributions included here, even when he or she profits greatly from others. This is a choice that has to be exercised in an informed way. In giving us these options, and, in general, providing a very rich menu that covers a wide cross-section of the massive span of the extant environmental and sustainable development literature, the editors have put us greatly in their debt.

Finally, I should note that this book completes the fine series on "frontier issues in economic thought" that Neva Goodwin has been editing for the Global Development and Environment Institute of Tufts University. Having already done a great deal to advance a comprehensive understanding of ecological economics, consumption and the consumer society, economic and social goals, the changing nature of work, and the political economy of inequality, this volume extends the guided tour to the frontiers of environmental studies. It is pleasing that the series is ending with a volume that is particularly important in its own right. This is, thus, an occasion for a double celebration, and I am very happy—and privileged—to be allowed to join the combined festivities.

Amartya Sen
Master's Lodge
Trinity College, Cambridge

Acknowledgments

The preparation of this volume has depended on the support and assistance of numerous people, including colleagues at Tufts University as well as outside reviewers. Tufts University faculty members and research associates who have contributed to our discussions of the book include Molly Anderson, Steven Block, William Moomaw, Patrick Webb, John Hammock, Julian Agyeman, Elizabeth Kline, and Serguei Andriouchtchenko. Expert reviewers have included David Barkin, Cutler Cleveland, Jonathan Fox, Arthur MacEwan, Paul Streeten, Sandra Postel, and Matthias Ruth. Others who have supplied valuable input at various stages include Hilary French, Kilaparti Ramakrishna, Kelly Sims, Barbara Connolly, and Brent Blackwelder.

Staff members at the Tufts University Global Development and Environment Institute (G-DAE) have played a key role in bringing the volume to completion. Graduate research assistants Kimberly Barry, Nicole Palasz, Johanna Meyer, and Dan Allen did much of the essential work in surveying the vast body of literature relevant to sustainable development and preparing articles for summarization. Absolutely invaluable to the project was the work of library assistant Stacie Bowman, whose abilities to manage the extensive requirements of this complex project never ceased to amaze us. Administrative support was provided throughout by program coordinator David Plancon. G-DAE Senior Research Associate Frank Ackerman, who has been the lead editor for the previous three volumes in the series, was responsible for preparation of Part VI on energy and materials, and our extensive editorial discussions for this volume have benefited from his experience and insight.

Institutional support from Tufts University has been crucial to the project. Tufts library staff have responded to our numerous requests with fast and reliable service, often using interlibrary loan to scour the far reaches of the planet for hard-to-obtain volumes. We would also like to acknowledge the support and advice of our editor at Island Press, Todd Baldwin.

We are grateful for the support of funders for the Frontiers in Economic Thought project, without whose contribution this work could not have been done. In addition to providing support for this volume, the John D. and Catherine T. MacArthur Foundation has played an important role in supporting much of the series. We also greatly appreciated the support for this volume by the Andrew W. Mellon Foundation, the Richard Lounsbery Foundation, the Compton Foundation, the John Merck Fund, and Mr. and Mrs. George D. O'Neill. In addition, support from The Ford Foundation has contributed significantly to our ability to edit this series.

Volume Introduction

by Jonathan M. Harris and Neva R. Goodwin

This volume completes the six-volume series *Frontier Issues in Economic Thought,* which has been produced by the Global Development And Environment Institute at Tufts University. The series has explored crucial issues that are essential to economic theory but that also require insights from other perspectives and disciplines. Earlier volumes have covered the topics of ecological economics; consumption and the consumer society; economic and social goals; the changing nature of work; and the political economy of inequality. In a sense this volume represents the capstone of the series. The earlier volumes have all raised challenges to the standard theories, which have been used both to define and to guide economic development. This book explores possibilities for a different kind of development, one that would integrate the goals of economic prosperity, social justice, and healthy ecosystems.

Our title suggests that the kind of development we would wish to sustain can be described under two headings: social and economic. Of these two concepts, economic development is, historically, the more familiar. Its objective has often been defined (in practice more than in formal theory) as the expansion of consumption and Gross National Product. Such objectives are obviously important for the world's poor, but there is a growing consensus that the single-minded pursuit of growth should not dominate development policy. While this most clearly applies to those countries that are already relatively affluent, it is also relevant to nations whose need for greater economic output is matched by their urgent need for social equity and environmental protection.

The social dimension—often reflected in the term *human development*—may be defined as progress toward enabling all human beings to satisfy their essential needs, to achieve a reasonable level of comfort, to live lives of meaning and interest, and to share fairly in opportunities for health and education. Thus defined, human development is a final goal: an end to which other important pursuits, such as economic development, are the means.

Where, the reader might ask, is the environment in this list of ends and means? There are those who passionately believe that the preservation of a healthy environment is an end in itself; while others, who also consider themselves environmentalists, take a more anthropocentric approach, caring about environmental integrity because it is essential for the achievement of most other goals that go beyond a very short time horizon.

Sustainable development can be pursued without resolving such philosophical issues. The great contribution of the word "sustainable" is that it introduces the issue of time. The growing popularity of the term indicates an increasing awareness that seeds of self-destruction can be contained within short-term achievements in development as it used to be conceived; that is, economic-development-as-GNP-growth. When we pair economic development with human development—giving the latter precedence in defining final goals, and never forgetting that economic development is valuable only as a means to these final ends—then we are less likely to engage in short-sighted activities whose successes will collapse upon themselves. We are also obliged to place a high value on ecosystem health—whether we care about it as an end in itself, or because we recognize that continued human well-being depends on it.

In seeking an alternative approach to economic development it is important to consider not just the views of economists, but also of those who study the social and political aspects of development. It is important not just to accept the recommendations of experts from the global North, some (though not all) of whom may feel sure that the economic successes of the United States and Europe point the way for other nations, but to listen to voices from the South, who have a sharper awareness of global inequities. And it is essential to understand the perspective of ecologists and other natural scientists, who tell us that planetary ecosystems are not simply resources to be exploited but are the complex inheritance of millennia, now endangered by ever-growing human impacts.

This volume aims to synthesize these perspectives. We have drawn from a variety of disciplines, always bearing on economic issues but not always limited to economic theory. We have explored environmental and social perspectives, issues of population and resource use, globalization, corporate power, and strategies for achieving a more sustainable path at the local, national, and global levels. In so doing, we have reviewed an extensive literature on sustainable development, which has proliferated since the World Commission on Environment and Development's 1987 report, *Our Common Future*. The methodology of the Frontiers volumes, summarizing leading articles on each topic, allows us to present the voices and to draw on the expertise of many different authors, while presenting an overview of each topic area in the introductory essays.

We hope that this effort will be of use to scholars working to advance the understanding of the meaning of sustainability, to policy-makers interested in reforming current economic systems, and to students seeking a broad overview of the area, which nonetheless provides significant analytical depth. In this introduction, we seek to define the basic issues of sustainable development and introduce themes that will be developed from a variety of perspectives in the rest of the volume.

Sustainable Development: Defining a New Paradigm

In 1987 the World Commission on Environment and Development sought to address the problem of conflicts between environment and development goals by formulating a definition of sustainable development:

> Sustainable development is development which meets the needs of the present without compromising the ability of future generations to meet their own needs. (WCED 1987)

In the extensive discussion and use of the concept since then, there has been a growing recognition of three essential aspects of sustainable development[1]:

- *Economic*—An economically sustainable system must be able to produce goods and services on a continuing basis, to maintain manageable levels of government and external debt, and to avoid extreme sectoral imbalances that damage agricultural or industrial production.
- *Environmental*—An environmentally sustainable system must maintain a stable resource base, avoiding overexploitation of renewable resource systems or environmental sink functions and depleting nonrenewable resources only to the extent that investment is made in adequate substitutes. This includes maintenance of biodiversity, atmospheric stability, and other ecosystem functions not ordinarily classed as economic resources.
- *Social*—A socially sustainable system must achieve fairness in distribution and opportunity, adequate provision of social services, including health and education, gender equity, and political accountability and participation.

Clearly, these three elements of sustainability introduce many potential complications to the original, simple definition of economic development. The goals expressed or implied are multidimensional, raising the issue of how to balance objectives and how to judge success or failure. For example, what if provision of adequate food and water supplies appears to require changes in land use that will decrease biodiversity? What if nonpolluting energy sources are more expensive, thus increasing the burden on the poor, for whom they represent a larger proportion of daily expenditure? Which goal will take precedence?

Despite these complications, the three principles outlined above do have resonance at a common-sense level. Surely if we could move closer to achieving this tripartite goal, the world would be a better place; equally surely, we frequently fall short in all three respects. Thus there is ample justification for the elucidation of a theory of sustainable development, which must have an interdisciplinary nature. In exploring the nature of such a theory it makes sense to start with the concept of development itself, and then turn to the requirements for sustainability.

Development in Practice—Achievements and Failures

From a historical point of view, the concept of economic development gained broad acceptance only relatively recently. As the historian of economic thought, Roger Backhouse puts it:

> Development economics in its modern form did not exist before the 1940s. The concern of development economics, as the term is now understood, is with countries or regions which are seen to be under or less developed relative to others, and which, it is commonly believed, should, if they are not to become ever poorer relative to the developed countries, be developed in some way. (Backhouse 1991)

This definition immediately points up a significant difference between development economics and much of the rest of economic theory. In neoclassical economic theory, an effort has been made to achieve a positive rather than a normative perspective—that is, to describe what is rather than positing what should be. Development economics, in contrast, is explicitly normative, as Backhouse's description makes clear. As such, it cannot avoid concern with social and political issues, and must focus on goals, ideals, and ends, as well as on economic means.

Beginning after World War II, economists, other social scientists, and policymakers thus adopted a framework of thought that was much more ambitious in its scope than previous formulations of political economy. The announced goal of economic development policy was to raise living standards throughout the world, steadily providing more goods and services to an expanding population. An implicit set of goals coexisted with this official purpose: to reconstruct Europe after World War II, to open markets for Western goods, and to contain communism. Communist economies, of course, had already adopted a model of development through central planning, and part of the impetus of early development theory was to provide an alternative path. W.W. Rostow, for example, gave his seminal work, *The Stages of Economic Growth,* the subtitle: *A Non-Communist Manifesto* (1960). The two sets of goals for development sometimes came into conflict, with the demands of the Cold War distorting the original, more idealistic, perspective.

The international economic institutions set up after World War II, including the International Monetary Fund, the World Bank, and the General Agreement on Tariffs and Trade, were specifically designed to provide a stable framework for world development. The World Bank in particular was intended to provide necessary capital for infrastructure investment in developing nations. This infrastructure, it was hoped, would provide the basis for more productive agriculture and industrialization. However, as the process of development proceeded, the original assumptions underlying this strategy came into question.

One line of critique, which developed in the 1970s, emphasized the need for a specific focus on basic needs (Streeten et al. 1981). Education, nutrition, health, sanitation, and employment for the poor were the central components of this approach—reflecting an acknowledgment that the benefits of development did not necessarily "trickle down" to those who needed them most. The human development perspective has endured and has been expanded in the works of economists such as Amartya Sen (Sen 1981, 1992, 1999; Anand and Sen 1996). While it has remained an undercurrent in mainstream development theory, it has inspired the creation of the United Nations Development Programme's Human Development Index (HDI), which uses health and education measures together with Gross Domestic Product (GDP) to calculate an overall index of development success (UNDP 1990–1999); see Part II of this volume and Part X of Volume 3 in this series (Ackerman et al. 1997) for a detailed treatment of the HDI and associated measures.

A harsher critique came from analysts in the global South, who viewed the process of development as promoting structural relationships that systematically benefited the North at the expense of the South. Theorists including Celso Furtado and Raul Prebisch argued that pervasive inequality and unfavorable terms of trade would lead to a state of persistent "dependency," in which wealth would flow from the "periphery" of less-developed economies to the "core" of dominant nations. This perspective was associated with an advocacy of strong state intervention to promote import-substituting industrialization, using tariffs and state direction of economic development. (See Part III of this volume for further discussion.)

Following the debt crisis of the early 1980s, international lending agencies promoted a strategy of "structural adjustment," including liberalization of trade, eliminating government deficits and overvalued exchange rates, and dismantling parastatal organizations deemed to be inefficient. Structural adjustment was seen as correcting the errors of earlier, government-centered development policies that had led to bloated bureaucracies, unbalanced budgets, and excessive debt. However, critiques of structural adjustment policies have found them at odds with basic-needs priorities (see Part VII of this volume). Market-oriented reforms have often led to greater inequality and hardship for the poor even as aspects of economic efficiency improved. The tension between the basic-needs and market-oriented perspectives on development has thus remained strong.

At the same time that concerns were growing over the harsh social impacts of structural adjustment an environmental critique of development gained prominence. The 1987 report of the World Commission on Environment and Development (the Brundtland Report) focused attention on the relationship between development and its environmental effects, strengthening and advancing concepts that had been introduced at the United Nations Conference on the Human Environment at Stockholm in 1972. Under the leadership of Herman

Daly and Robert Goodland, the World Bank's Environment Department—in sharp contrast to other, more powerful, departments of the Bank—produced numerous reports documenting the adverse environmental effects of unrestrained development and calling for a transition to sustainability (Daly and Goodland 1991; Goodland, Daly, and El Serafy 1992, 1994). The United Nations Environment Programme (UNEP) also produced a series of reports on environmental issues, including economic analysis of the impacts of structural adjustment (UNEP 1996) and most recently the comprehensive report *Global Environment Outlook 2000* (UNEP 1999).

At the turn of the century, what is the 50-year record of the broad-reaching, and historically young, effort at global development? The concept has been widely accepted by countries of varied political structure. There have been remarkable successes—notably in East Asia—and worldwide progress both in standard GDP measures and in measures of human development such as life expectancy and education. There have also been areas of slow or negative growth, especially in sub-Saharan Africa, where GDP increase was slow and food production per capita was in decline even before the rapid spread of AIDS devastated many countries and dramatically lowered life expectancies.

Globally, most countries have made significant advances both in GDP and in Human Development Index measures (see Part II of this volume and the summarized article by Streeten in Part VII). But overall the record of development on a world scale is open to two major criticisms:

- The benefits of development have been distributed unevenly, with income inequalities remaining persistent and sometimes increasing over time. The number of extremely poor and malnourished people has remained high, and in some areas has increased, even as a global middle class has achieved relative affluence.
- There have been major negative impacts of development on the environment and on existing social structures. Many traditional societies have been devastated by overexploitation of forests, water systems, and fisheries. Urban areas in developing countries commonly suffer from severe pollution and inadequate transportation, water, and sewer infrastructure. Environmental damage, if unchecked, may undermine the achievements of development and even lead to collapse of essential ecosystems.

These problems are not minor blemishes on an overall record of success. Rather, they appear to be endemic to development as it has taken place over the past half-century. While those who have consistently voiced criticisms of mainstream development policies have generally lacked the power to change those policies, the importance of these issues is being increasingly acknowledged. Former World Bank chief economist Joseph Stiglitz, for example, has called for alternatives to the mainstream "Washington consensus," which sees market-oriented reform and fiscal stringency as the key to development policy (Stiglitz 1997, 1998). New thinking on development policy is clearly in order.

The straightforward view of development as an upward climb, common to all nations but with different countries at different stages, is misleading and certainly inadequate for the twenty-first century. The absolute gaps between rich and poor nations, and between rich and poor groups within individual countries, are widening, not narrowing. And even if we can imagine all nations reaching stable populations and satisfactory levels of GDP by, say, 2050, can we envision the planetary ecosystem surviving the greatly increased demands on its resources and environmental absorption capacity?

A growing awareness of these challenges to traditional development thinking has led to the increasingly wide acceptance of the concept of sustainable development. The World Bank has produced research on indicators of sustainable development, in particular measures of genuine savings: "The true rate of savings in a nation after due account is taken of the depletion of natural resources and the damages caused by pollution" (see World Bank 1997a and the summarized article by Hamilton and Clemens in Part I of this volume). This new attention to a combination of social and environmental factors indicates that lines of thought that were formerly at the fringes of development policy are making their way into the mainstream.

A Synthesis of Perspectives

Drawing on economic, ecological, and social perspectives, we can identify some of the main themes that are integral to the construction of a new paradigm:

- The original idea of development was based on a progression from traditional to modern mass-consumption society. Within this framework, a tension has developed between the promotion of economic growth and the equitable provision of basic needs. Development as it has proceeded over the last half-century has remained inequitable.
- The conservation of ecosystems and natural resources is essential for sustainable economic production and intergenerational equity. From an ecological perspective, both human population and total resource demand must be limited in scale, and the integrity of ecosystems and diversity of species must be maintained. Market mechanisms often do not operate effectively to conserve this natural capital, but tend to deplete and degrade it.
- Social equity, the fulfillment of basic health and educational needs, and participatory democracy are crucial elements of development, and are interrelated with environmental sustainability.

Taken together, these observations suggest new guidelines for the development process. They also require modifications to the goal of economic growth.

Economic growth in some form is required for those who lack essentials, but it must be subject to global limits and should not be the prime objective for countries already at high levels of consumption. As Alan Durning has suggested, a moderate level of consumption, together with strong social institutions and a healthy environment, represents a better ideal than ever-increasing consumption (Durning 1992).

Sustainability is more than limits on population or restraint in consumption—though these are important. It means that in our choice of goods and technologies we must be oriented to the requirements of ecosystem integrity and species diversity. It also implies that the apparent independence of economics from biophysical science is a luxury we can no longer afford.

In pursuing these modified development goals, it will be necessary to recognize the limits of the market mechanism. During the structural adjustment phase of development policy, the virtues of free markets became an article of faith for policy-makers; this dogma will have to be revised, as the World Bank now acknowledges:

> Reducing or diluting the state's role cannot be the end of the reform story. For human welfare to be advanced, the state's capability—defined as the ability to undertake and promote collective actions efficiently—must be increased. (World Bank 1997b)

Although under some conditions markets may excel at achieving economic efficiency, they are often counterproductive in terms of sustainability. Guided markets may often be useful tools for achieving specific environmental goals, and there is an extensive economic literature on "internalizing externalities" so as to reflect environmental costs and benefits in the market.[2] But, in a broader perspective, it is the social and institutional processes of setting social and environmental goals and norms that must guide sustainable development policy.

In this volume, we present voices from different disciplinary perspectives contributing essential elements of the new paradigm of development. In order to cover these wide-ranging issues in a single volume, 66 articles and book chapters have been summarized. In Part I, we emphasize the environmental perspective. The concept of natural capital is more specifically defined and explored, together with efforts to integrate environmental issues more fully into economic theory. Part II focuses on the social dimensions of sustainability, reviews the concept of human development, and discusses the role of social capital and democratic government. Part III gives special attention to the Southern critiques of the Northern-dominated development regime, and investigates what would be required to establish common ground between North and South on a vision of sustainability.

After reviewing these general perspectives we move to more specific analyses in Parts IV, V, and VI. These sections deal with the relationship between human population and resource needs and the planetary ecosystems that supply these needs. Part IV focuses on population growth and its impacts on the environment, balancing the concept of ecological limits with the social and political factors that affect both the rate of population growth and its social and ecological effects. Part V considers agricultural ecosystems, biodiversity, and resource management, looking both at the extent of damage to ecosystems from human exploitation and the prerequisites for more sustainable resource management. In Part VI, we examine the application of the concept of sustainability to industrial systems and energy use, reviewing analyses from the field of industrial ecology as well as the debate over responses to global climate change.

Institutional and policy issues provide the themes for Parts VII to X. Part VII examines the sweep of globalization, especially its effects on social and ecological sustainability. Critiques of globalization raise the issue of how to alter current trends in a more sustainable direction. Efforts to redirect policy and practice in the direction of sustainability are discussed in Part VIII, which looks at efforts to reform corporate policies through stakeholder activism and government policy. Part IX, which deals with government policies at the national and local levels, provides an overview of analytical and practical approaches to promoting sustainability. Part X takes on issues of global institutional reform, reviewing proposals both for reform of existing institutions and for creating new institutions oriented toward global sustainability.

The Relation of this Book to the Rest of the Frontiers Series

Sustainable Human and Economic Development concludes the series of six topics that we selected for treatment in the Frontiers series, beginning our conceptualization and research in 1993. We will briefly summarize how the first five volumes have led up to Volume 6.

Ecological economics (the topic of Volume 1) is an important new way of understanding the interactions between human economic behavior and ecological realities. We have drawn on this to support our analysis of the environmental constraints within which economic development must occur.

Much of the thrust of contemporary economic activity goes toward promoting a consumer society (Volume 2). In the present book we probe the environmental and social stresses created by this orientation, considering how it contributes to unsustainability in the economies of the world today.

In Volume 3 we posited that any notion of development or progress must include the idea of improvement in human well-being. We also noted that this central idea is often lacking in, or orthogonal to, standard prescriptions for economic development. Volume 6 assumes the positions suggested in Volume 3 (just as Adam Smith's *Wealth of Nations* assumed the positions reached in his *Theory of Moral Sentiments*—to make a rather immodest comparison!), and seeks an understanding of how we can sustain a kind of development that does progressively enhance human well-being.

Work is a defining characteristic of most human lives, including the worker's experience on the job, and the rewards (income, access to other resources, learning opportunities, respect, etc.) that may accompany it. Any possibility of progressing to a preferred state—and of sustaining progress over time—must take into account the fast-changing nature of work in the modern world. This was the subject of Volume 4.

Another issue that will in crucial ways determine what kinds of progress we can have, and sustain, is the topic of inequality. Volume 5 took the political economy approach of stressing the issues of power and money that, together (and they often operate together), account for a large portion of what human beings feel as inequality. Power inequalities help to determine who gets to decide what kind of world we will strive to achieve. Money inequalities have enormous impacts upon patterns of consumption, which in turn help to determine both the development of human potential (for work, for enjoyment, and for understanding problems and solutions) and also the human impact on the environment. Many of the world's societies today are characterized by extreme inequality in both these areas. A sustainable society, while not necessarily completely egalitarian, must be one in which the degree of inequality is tempered and limited by generally accepted concepts of basic social justice.

We can simplify and summarize by saying that Volumes 2, 4, and 5 of the Frontiers series examine what is known about three critical areas of human life: the fulfillment of basic and other needs and wants; the experience of work; and the sources and effects of inequality. (These may be equated with the three fundamental topics of economics: consumption, production, and distribution.)

Volumes 1 and 3 emphasize the ways in which the field of economics interprets human experience: how our economic behavior interacts with our natural environment, and how our economic theories relate to our human goals. Our special objective in these volumes was to inject into economics a keener awareness of context and of goals, because these are two of the three elements that we believe to be most seriously missing from standard economic theory.

The third critical missing element is time, which is essential to an understanding of economic development. Volume 6 stretches out the time context, looking to the past and the future to examine change over time in economic conditions, and in the associated aspects of human well-being and ecological health. The effects of past development in shaping current social and environmental conditions, and the effects of present development practices on the future, are themes that recur throughout the volume.

The essential question of this book is: What is it that we should develop and sustain in order to make progress toward adequate needs-satisfaction and value-filled lives for all human beings? In searching for answers to that question, we present both the general principles and some of the specific requirements of sustainable development. Development theory, as we have noted, has always been normative as well as positive in its analytical vision. Today we require a new normative vision drawing on strong but sometimes neglected traditions in economics as well as political and social theory and combining traditional wisdom with modern technology. Fortunately, there is now an extensive effort by theorists and practitioners from many disciplines to transform the concept of sustainable development into reality. The goal of this volume is to represent the best of these efforts and to help point the way toward a more realistic vision and more effective practice.

Notes

1. See, for example, Holmberg (1992); Reed (1997).
2. See, for example, Markandya and Richardson (1993).

PERSPECTIVES ON SUSTAINABILITY

PART I

Economics of Sustainability: The Environmental Dimension

Overview Essay

by Jonathan M. Harris

Serious consideration of sustainable development requires a rethinking of major elements of economic theory. The standard economic perspective must be broadened to take into account environmental and social perspectives. In this essay, we address primarily the environmental issues, then move to the social perspective in Part II.

To some extent, principles of environmental sustainability can be expressed in standard economic terms. For example, economic theory provides for the internalization of environmental costs, and natural resource economics recognizes the concept of sustainable yield in natural resource systems. But to consider the full implications of sustainability we need to look beyond these formulations to consider the social and ecological contexts of economic activity. This in turn leads to a re-examination of some of the fundamental theoretical frameworks of economics. Among the concepts that need to be re-examined are capital, valuation, distribution, savings, investment, and economic growth. As a result, both the categories of analysis and the policy implications derived from that analysis change significantly when we take sustainability seriously.

A Broader View of Capital and Production

Standard economic theory treats manufactured capital as the key to development, with some attention also to human capital (skills and knowledge embodied in individuals). Although natural resources are acknowledged as an input to the productive process, they are not an important feature of most economic models. In addition, the social and institutional arrangements that provide the basis for economic activity remain very much in the background. Economic theories of sustainability are based on a more expansive concept of capital. Viewing capital broadly as any stock that produces or contributes to a flow of output, we can identify four kinds of capital: manufactured, natural, human, and social.

Manufactured capital is what is ordinarily referred to simply as "capital" in most economic theory. Natural capital corresponds to what standard economic theory traditionally defined as "land"; the reason for the different terminology is to emphasize the essential productive role of this factor, and to broaden the concept to include all environmental functions. Human capital refers to education and skills possessed by individuals. Social capital is used to refer to knowledge and rules embedded in culture and institutions, such as the legal system or the concept of property rights. All four kinds of capital and essential to economic activity, although standard economic theory emphasizes primarily manufactured and human capital.

Manufactured capital is maintained and accumulated through investment. This process, of course, is central to standard models of economic growth. There is a clear and reciprocal relationship between manufactured capital and economic production: more capital makes it possible to expand economic production, and the devotion of a share of production to investment makes it possible to accumulate more capital. The relationship can be defined in mathematical terms, using concepts such as the savings rate and the capital/output ratio, and is easily amenable to measurement and econometric analysis.

Human capital is also fairly well represented in economic models of the labor market and plays an important role in modern growth models. The dynamics of natural and social capital present more problems, and these kinds of capital have received less recognition and attention from economists. Yet for true sustainability to be achieved, all four kinds of capital must be maintained at levels that allow both human well-being and healthy ecosystems.

Manufactured and Natural Capital

In the first article summarized here, **Robert Costanza and Herman Daly** set forth the basic conditions for the maintenance of natural as well as manufactured capital. They point out that neoclassical economics usually treats natural and manufactured capital as fully substitutable. If this approach is accepted, there is no particular reason to conserve natural capital so long as manufactured capital is augmented by a value equal to or greater than the depletion of natural capital. For example, it would be acceptable for a country to cut down its forests if the economic proceeds from the timber sales are used for investment in industrial development.

Even in the neoclassical perspective, however, the principle of *weak sustainability* is appropriate. A well-known principle derived from work by Solow and Hartwick (the "Hartwick rule") states that consumption may remain constant, or increase, with declining nonrenewable resources provided that the rents from these resources are reinvested in reproducible capital (Hartwick 1977; Solow 1986). In this approach, sustainability requires that the *total value* of the two forms of capital remain constant over time. El Serafy has pointed out that in order to assess this value, there must be a full accounting for natural capital depletion (El Serafy, 1993; also see article by El Serafy summarized below).

A *strong sustainability* approach is based on the idea that substitutability between natural and manufactured capital is limited. Rather, the two are seen as *complements*—factors that must be used together to be productive. For example, a fleet of fishing boats is of no use without a stock of fish. In the case of *critical natural capital*—for example, essential water supplies—substitutability is close to zero. While it may be possible, for example, to compensate for some water pollution with purification systems, life and economic activity are essentially impossible without access to water. The strong sustainability approach implies that specific measures distinct from the ordinary market process are necessary for the conservation of natural capital. It also implies limits on macroeconomic scale. The economic system cannot grow beyond the limitations set by the regeneration and waste-absorption capacities of the ecosystem.[1]

Costanza and Daly suggest that a minimum necessary condition for sustainability can be expressed in terms of the conservation of natural capital. This policy goal leads to two decision rules: one for renewable and the other for nonrenewable resources. For renewables, the rule is to limit resource consumption to sustainable yield levels; for nonrenewables, the rule is to reinvest the proceeds from nonrenewable resource exploitation into renewable substitutes. Following these two rules will maintain a constant stock of natural capital. To maintain a constant *per capita* stock of natural capital also requires a stable level of human population, a factor that Daly has emphasized elsewhere (Daly 1991).

The rules suggested by Costanza and Daly for natural capital conservation are rough guides rather than precise theoretical principles. Nicholas Georgescu-Roegen, whose pathbreaking work *The Entropy Law and the Economic Process* outlined the dependence of the economic system on biophysical systems, argued that it is ultimately impossible to maintain a constant stock of natural capital, since all planetary resources will eventually degrade or be used up according to the Second Law of Thermodynamics (Georgescu-Roegen 1971, 1993). But at a more practical level he proposed an approach similar to Costanza and Daly's, reasoning that "the enormous disproportionality between the flow of solar energy and the much more limited stock of terrestrial free energy suggests a bioeconomic program emphasizing such factors as solar energy, organic agriculture, population limitation, product durability, moderate consumption, and international equity" (Georgescu-Roegen 1993; summarized in Krishnan et al. 1995).

The issue of conserving natural capital is part of a broader debate on reconceptualizing economic theory. **Giuseppe Munda** places the issue of natural capital conservation in the context of a contrast between the two economic paradigms of neoclassical and ecological economics. Whereas neoclassical environmental economics seeks to apply the categories of economic theory to the environment, ecological economics attempts to modify the conception of the economic system to acknowledge its role as a subsystem of a broader planetary ecosystem. In this view, standard economic approaches such as valuation in monetary terms can give at best only a partial view of reality. An analytical pluralism that takes into account social and environmental dynamics is essential to the ecological economics paradigm (Norgaard 1989, 1994). For an extensive treatment of the principles of ecological economics, see Volume I in this series, *A Survey of Ecological Economics* (Krishnan et al. 1995). Economic and ecological perspectives on sustainability are discussed in Common and Perrings (1992). A recent evaluation of issues in ecological economics can be found in the articles by Wackernagel, Herendeen, and others in *Ecological Economics* 29 (1999).

The social basis for the conservation of natural capital is explored further in the summarized article by **Talbot Page.** Page suggests that natural capital conservation requires specific recognition in law regarding the "standing" of natural capital as a social asset. He offers an analogy to constitutional law: fundamental social principles must determine the broad outlines of natural resource management. The sphere of "purely" economic analysis is thus limited to more specific decisions within this general framework.

The importance of social capital in natural resource management is evident when "marketization" breaks down traditional social institutions that have governed the use of common property resources such as forests or fisheries. In such cases, it is essential either to find ways to maintain the traditional common property management institutions, or to replace them with effective new institutions. Unfortunately, the development process frequently creates a situation in which neither traditional nor modern forms of social control over resources are effective, resulting in rapid and destructive resource exploitation.

Failures of social capital development are also at the root of many inequities and much human suffering in the development process. This is a central aspect of sustainability that is neglected in standard economic theory and only partly addressed in ecological economics. In Part II of this volume, we address issues of social capital and development in depth.

Intergenerational Equity

Sustainability is sometimes defined as intergenerational equity—ensuring that future generations have an inheritance of natural, social, manufactured, and human capital at least equal to that of the present generation. From the point of view of neoclassical economic theory, sustainability can be defined in terms of the maximization of human welfare over time. Most economists simplify further by identifying the maximization of welfare with the maximization of utility derived from consumption. This formulation certainly includes many important elements of human welfare (food, clothing, housing, transportation, health and education services, etc.) and has the analytical advantage of reducing the problem to a measurable single-dimensional indicator. But it is open to criticism as a serious oversimplification of the nature of human well-being (see, for example, Sen 1999 and Ackerman et al. 1997).

Using the neoclassical definition of welfare maximization, a formal economic analysis raises the question of whether sustainability has any validity as an economic concept. According to standard economic theory, efficient resource allocation should have the effect of maximizing utility from consumption. If we accept the use of time discounting as a method of comparing the economic values of consumption in different time periods, then sustainability appears to mean nothing more than efficient resource allocation—a concept already well established in economics.

One line of criticism of this reductionist approach to sustainability centers on the use of a discount rate to compare present and future costs and benefits. Discounting has been subject to numerous critiques on account of its present bias, especially as the time period under consideration becomes longer (see overview essay for Part VI and summarized article by Lind and Schuler in that section).

A simple example demonstrates the general point. At a discount rate of 10 percent, typically used for cost-benefit analysis, the value of $1 million one hundred years from now is the same as a mere $72 today. Thus it would apparently be justifiable to impose costs of up to $1 million on people 100 years from now in order to enjoy $72 worth of consumption today. By this logic, much resource depletion and environmental damage could be considered acceptable, and even optimal, according to a criterion of economic efficiency.

The problem is that by accepting the use of a discount rate we have implicitly imposed a specific pattern of preferences regarding the relative welfare of present and future generations.[2] Howarth and Norgaard have argued that the use of a discount rate is appropriate for the efficient allocation of this generation's resources but is inappropriate when the rights of future generations are at issue (Howarth and Norgaard 1993). Use of a current market discount rate gives undue weight to the preferences of current consumers. This creates a strong bias against sustainability in the context of issues such as soil erosion or atmospheric buildup of greenhouse gases, where the most damaging impacts are felt over decades or generations.

To achieve intergenerational equity, we need some kind of sustainability rule regarding resource use and environmental impacts. The solution of what Norgaard and Howarth (1991) refer to as "the conservationist's dilemma" is not easy. By imposing a low discount rate for decision-making, we can place a higher value on the future. For example, with a low discount rate future costs associated with soil degradation or global climate change would be weighed more heavily. But at the same time, the use of a low discount rate encourages excessive current investment in manufactured capital (for example, the construction of large dams or nuclear power plants), to the likely detriment of natural capital. It is, of course, possible to use different discount rates for different planning purposes. But such an apparently arbitrary choice has little theoretical justification.

Michael Toman suggests that the issue may be resolved by recognizing that some issues can be appropriately dealt with through neoclassical market efficiency, while others require the application of a "safe minimum standard" approach to protect essential resources and environmental functions. He suggests that the criteria of possible severity and irreversibility of ecological damages should be used to decide which theoretical framework is more appropriate. Others have referred to this approach as the use of a "precautionary principle," which should supersede economic analysis when there is uncertainty about possible outcomes and large potential ecological damage is at issue (Perrings 1991).

The adoption of this reasonable suggestion would have far-reaching implications for economic theory and policy. Note the essential role of "moral imperatives," "public decision making," and "the formation of social values" in Toman's suggested decision framework. None of these appear in the neoclassical economic model, where markets are presumed to be the best resource allocators, and the occasional correction of a "market imperfection" the only appropriate role for government. Thus Toman is in effect asserting the importance of sustainability as a concept independent of standard neoclassical economic analysis, one that requires an explicitly normative and socially determined process of decision-making.

This represents a fundamental shift in the economic paradigm. Much as the Keynesian revolution validated the concept of government intervention to achieve macroeconomic balance, the acceptance of sustainability as a valid social goal places a new complexion on all policy issues concerning the relationship between human economic activity and the environment. Markets may be valuable and essential means, but they cannot determine the ends, which must be arrived at by a social decision process informed by different disciplinary viewpoints. This will require an unaccustomed humility on the part of economists, and a willingness to work together with other social and natural scientists.

Issues of Distribution and Valuation

The advocacy of intergenerational equity also has implications for equity and property rights in the current generation. It makes little sense to talk about equity between generations without acknowledging the great current inequalities of wealth and income. In terms of natural capital and environmental assets, many questions arise as to private and social property "rights" over these assets. Who, for example, has rights to genetic resources, fish stocks, forests, or water? These issues are interlinked with distributional questions. A more equitable distribution of income would have major implications for the use and conservation of water resources, since clean water is a major unmet need in many areas of the developing world. (In neoclassical terms, the "demand" for water is not an "effective demand" if people lack income to back it up.)

Juan Martinez-Alier addresses the question of the relationship between environmental sustainability and social equity. He rejects the idea that the environment can be treated as a luxury good, something that the rich can afford to care about while the poor must focus only on immediate needs. The poor often depend heavily on common property resources, and suffer the most serious consequences of pollution and environmental damage. Thus the battle for a more equitable society often involves resistance to the abuse of natural capital by powerful market actors, including corporations and rich individuals. While there may be cases where the needs of the poor for greater material consumption may require environmental trade-offs, it is the much larger demands of the affluent that directly or indirectly threaten the natural resource base.

Martinez-Alier also deals with some of the paradoxes of economic valuation. He sees this not as a neutral analytic exercise, but as a function of economic power. The valuation of environmental damage may depend strongly on whom that damage affects. Negative externalities suffered by the poor are typically undervalued, both in the market and by economic analysts. Like Talbot Page, Martinez-Alier points out that a rights-based legal analysis may give a very different weighting to environmental factors as they affect individuals and communities. Market-based valuation is only one measuring rod.

This is consistent with the point made by Guiseppe Munda regarding the *incommensurability* of economic and environmental values. Decisions made on such issues ultimately reflect social values, and these in turn are—and should be—shaped by social movements, not just by market outcomes. Sen (1999, 76–85) has similarly emphasized the function of social choice in deciding what weights to give to individual preferences, arguing that the economist's "preference for market-based price evaluation" must be balanced with a public discussion of appropriate social goals. His argument is more general, not limited to environmental issues, but the essential point is similar.

One of the trickiest issues in this area is whether it is possible, or advisable, to provide an effective monetary measure of environmental values. There has been an extensive debate on the issue of the valuation of ecosystem services.[3] Gouldner and Kennedy (1997) argue that the issue is as much a philosophical debate on the basis of value as it is an issue of economic techniques. Costanza and Folke (1997) suggest that valuation should take into account social fairness, economic efficiency, and environmental sustainability. However, they acknowledge the difficulty of integrating these dimensions.

Costanza and colleagues (1998) have put forward an estimate of $16–54 trillion for the value of the world's ecosystem services, an extraordinarily ambitious undertaking that relies heavily on fairly standard economic valuation techniques. Their work has been subject to criticism for a reductionist methodology and for failing to capture the complexity and interdependence of natural functions (Rees 1998; Toman 1998b). The size of their estimate is also heavily dependent on a very high estimate for the value of oceanic nutrient cycling. A research team of biologists from Cornell have suggested a significantly lower estimate of $2.9 trillion for global ecosystem services (Pimentel et al. 1997). This large variation in estimates demonstrates the inherent uncertainty in such calculations. At the same time, some of the critics acknowledge the importance of the "benchmark for environmental discourse" provided by admittedly imprecise figures indicating, in rough terms, the value of ecosystem services to human well-being (Norgaard et al. 1998; for a survey of responses to the ecosystem valuation by Costanza et al., see also Herendeen [1998] and other contributions to the 1998 special issue of *Ecological Economics* [25(1)] on ecosystem services).

At a more microeconomic level, Pearce and Moran (1997) offer methodologies for economic valuation of biodiversity that can be applied to a wide variety of areas, including nontimber forest products, natural genetic resources, watershed services, and wildlife conservation. The best justification for these valuation techniques is that, absent their use, the market system will assign an effective value of zero to many environmental functions. This logic has led other economists to attempt to integrate estimates of environmental and resource values into macroeconomic measures, thereby altering our understanding of concepts such as gross national product and net investment, and creating a new focus on "greening" national accounts.

Green Accounting and Genuine Saving

If economics cannot offer all the answers, it may at least be possible to reform some basic economic measures to give a better fit with environmental and social realities. **Salah El Serafy** discusses "green accounting," which seeks to modify national income statistics to take account of the environment. El Serafy points out that it is very difficult to get a fully "greened" measure of GDP due to the many judgment calls required to value complex environmental impacts, some of which (like species loss) are almost impossible to monetize. For this reason, green accounting has developed along two different lines.

One is the attempt to correct existing national accounts for *natural capital depreciation* using a "weak sustainability" principle (natural capital may be depleted, but this depletion should be accounted for). This is clearly desirable, since the principle of depreciation has long been accepted for manufactured capital. However, it is limited to those areas, like minerals or timber stocks, where valuation is relatively easy, or to quantifiable pollution damages.

The second direction for green accounting is *satellite accounts,* which measure environmental stocks and functions in physical terms, without necessarily attempting valuation (Lange and Duchin 1993). El Serafy suggests that both approaches have their uses. Introducing natural capital depreciation into accounting may have significant trade and macroeconomic policy implications, especially for countries that are heavily dependent on natural resource exports. Satellite accounts, on the other hand, can give a more complete picture of the natural resource base and the state of the environment.

As a step toward integrating economic, environmental, and social policy analysis, the World Bank's Environment Department has developed a measure of *genuine saving* that includes natural capital depreciation. This measure, particularly appropriate for developing countries, indicates that what may appear to be a development "success story" can conceal serious natural capital depletion and in some cases even a net negative genuine savings rate. The summarized article by World Bank researchers **Kirk Hamilton and Michael Clemens** in this section explains the basis of the Bank's calculations and presents specific calculations of genuine savings for developing regions.

To also take account of human capital, Hamilton and Clemens introduce a measure of *extended national investment,* which counts educational expenditures as an integral part of national investment. The combination of this with natural capital depreciation gives a revised measure that indicates the importance of public investment in education for improving the genuine savings picture in many developing nations.

Although it is tempting to seek a single socially and environmentally sound measure of economic activity, it appears that methodological problems make this impossible. (For a broad overview and critique of efforts to create alternatives to standard national income accounts, see England and Harris 1997 or England 1997.) However, specific measures such as those developed by Hamilton and Clemens are clearly of great importance in evaluating the benefits as well as the costs of economic growth. As El Serafy points out, these adjusted measures can have significant implications for macroeconomic and trade policy. Bearing this in mind, it is worthwhile to examine some other perspectives on the environmental impacts of economic growth.

Economic Growth and the Environmental Kuznets Curve

Do environmental conditions get better or worse with economic growth? There is now an extensive literature on this topic, with widely differing results depending on which environmental factors are studied and sometimes on econometric techniques. **David Stern** reviews this literature, evaluating the Environmental Kuznets Curve (EKC) hypothesis according to which environmental conditions worsen during the early stages of industrial development, then improve as income levels rise.[4]

Stern finds that the EKC logic applies only in a limited number of cases, and that several other factors are important. Much depends on which pollutant is considered. While some pollutants, such as sulfur dioxide and particulates, typically rise in the initial stages of development and then decline at higher income levels, others such as urban wastes and carbon dioxide seem to increase continually with income. There is also some evidence that after a downward turning point there may be another upward turn, as expanded consumption outruns stronger environmental protection. For example, nitrogen oxide levels have remained stubbornly high in developed countries as increased automobile traffic offsets improved pollution controls.

Although initial research seemed to identify a "turning point" of around $5,000 per capita income for key pollutants, other studies have found significantly higher turning points, implying that global pollution will continue to increase for decades unless significant policy changes occur (Selden and Song 1994). Critics of the EKC hypothesis argue that it ignores environmental limits, implying that sufficient economic growth can solve any environmental problem (Arrow et al. 1995). Some have pointed out that while higher per capita income can be associated with reduced pollution, the effect depends on political power, with literacy, political rights, and civil liberties having strong effects on environmental quality in developing nations (Torras and Boyce 1998). Others argue that the simple EKC hypothesis ignores the effects of trade, through which higher-income consumers may displace the environmental impacts of their consumption onto others (Rothman 1998; Suri and Chapman 1998). In general, the simple logic that "wealthier means cleaner" is not supported by the empirical evidence, which gives a much more complex picture, especially for global pollutants and ecosystem damage.

Conclusion

Introducing principles of sustainability into economics leads to the re-examination of fundamental economic concepts. In the new and broader perspective that emerges, development is no longer seen in the same light. As Munda makes clear in the article summarized here, the standard economic view of development as a straight-line process leading to industrialization and mass consumption no longer applies. A more nuanced understanding of development as an interaction of social, economic, and environmental factors—what Richard Norgaard has called *coevolution*—is needed (Norgaard 1994). Along similar lines, Martinez-Alier points to the emergence of a new field of *political ecology,* integrating issues of distribution and equity with environmental and economic perspectives.

As we pursue the varied topics of this volume, we will find that this new interpretation of development is indeed rich in theoretical and empirical insights. While economic evidence and analysis remains a central element, a multidisciplinary approach is essential. Before turning to specific topics in sustainable development in Parts IV–VI, we seek to deepen our understanding of the social and international dimensions of development in Parts II and III. In Parts VII–X, we will return to many of the institutional issues that affect policies for sustainable development.

Notes

1. The distinction between weak and strong sustainability is outlined in Daly (1994). Daly (1991) deals with the issue of limits on macroeconomic scale. See also Pearce and Warford (1993, chapter 2) on substitutability between manufactures and natural capital, as well as the concept of critical natural capital. For a critique of the concept of sustainability, see Beckerman (1994). A defense of weak sustainability is offered by El Serafy (1996), and applicability of the concept is discussed in Gowdy and O'Hara (1997). Strong sustainability is defended by Daly (1995) and criticized by Beckerman (1995), while Common (1996) argues that the distinction between weak and strong sustainability is invalid.
2. Neoclassical economists acknowledge that the choice of a discount rate affects intergenerational distribution, but generally do not regard this as a reason to replace discounting with other criteria for intertemporal resource allocation. See Hartwick (1977); Solow (1986). For further discussion of the issue of valuing the future, see the summarized article by Lind and Schuler in Part VI of this volume, and Part IV of Volume III in this series, *Human Well-Being and Economic Goals* (Ackerman et al., eds., 1997).
3. See, for example, the Symposium on Contingent Valuation in *Journal of Economic Perspectives* 8, 4 (1994), which offers a debate between the proponents (Portney 1994; Hanemann 1994) and critics (Diamond and Hausman 1994) of a survey technique for valuation.

4. The original Kuznets Curve hypothesis was that income inequality first increased, then lessened, with economic development (Kuznets 1955). See Volume 5 in this series (Ackerman et al. 2000) for a discussion and critique of this hypothesis on income inequality. Grossman and Krueger (1991, 1995), Shafik and Bandyopadhyay (1992), and Selden and Song (1994) applied a similar principle to pollution levels as related to economic development.

Summary of

Natural Capital and Sustainable Development

by Robert Costanza and Herman E. Daly

[Published in *Conservation Biology* 6, 1 (March 1992), 37–46.]

This article sets forth a minimum necessary condition for sustainability: the maintenance of the total natural capital stock at or above the current level. This condition embodies a precautionary principle: given the extent to which natural capital has already been damaged or depleted, it would be too risky to allow its significant further loss. Methodological issues raised by this basic condition include the measurement and valuation of natural capital, the use of discounting, differentiating between growth and development, and the shortcomings of current macroeconomic measures such as the Gross National Product. Technological issues include the degree to which technical progress can overcome resource constraints on growth. While there are wide disagreements on these issues, this article suggests that a prudent policy for sustainable development must avoid natural capital depletion.

Different Types of Capital

A fundamental definition of capital is "a stock that yields a flow of valuable goods and services over time." Both manufactured capital and natural capital satisfy this definition. While manufactured capital is produced by humans, natural capital is made up of all the natural resources and environmental functions that are essential for human life and economic activity. Natural capital and its products can be viewed simply as physical entities, or they can be given a more economic dimension through valuation. Natural capital can be further differentiated into renewable (or active) and nonrenewable (or inactive) components. Ecosystems are renewable natural capital that actively produce a flow of services; nonrenewable resources generally yield no services until extracted. A third form of capital is human capital: the stock of education, skills, culture, and knowledge stored in human beings themselves.[1]

A sustainable system must prevent depletion of its capital stock. Manufactured, natural, and human capital all require continual maintenance. Excessive harvest of ecosystem products can reduce the capacity of renewable natural capital to produce services and to maintain itself. Nonrenewable natural capital requires little or no maintenance but is depleted with use over time. The concept of sustainability is implicit in a Hicksian definition of income;[2] consumption that requires the depletion of natural capital cannot be counted as income.

Economic theory has focused on manufactured and human capital, because natural capital has been implicitly or explicitly viewed as abundant. But we are now entering an era in which natural rather than manufactured or human capital will be the limiting factor on economic activity. The concern of classical economists (including Smith, Malthus, and Ricardo) with the constraints of natural resource on economic growth gains new relevance in a period when the scale of human activities has an impact that can significantly reduce the flow of ecosystem goods and services. The economy must be viewed as a subsystem of the larger ecological system.

The failure of standard neoclassical economics to account for natural capital distorts analysis at all levels from project evaluation to the health of the entire ecological/economic system. Analyses that have attempted to account for natural capital, and to measure sustainable economic welfare, show that "[i]f we continue to ignore natural capital, we may well push welfare down while we think we are building it up." [40]

Substitutability Between Natural and Human-made Capital

Neoclassical economic theory holds that "reproducible capital is a near-perfect substitute for land and other exhaustible resources."[3] Natural capital is often omitted entirely from growth models, which are based on the two factors of labor and human-made capital. Even if natural capital is included in the production function, mathematical forms such as the Cobb-Douglas function imply high substitutability between natural and human-made capital. But in most cases natural and human-made capital are complements, not substitutes. "It should be obvious that the human-made capital of fishing nets, refineries, saw mills, and the human capital skill to run them does not substitute for, and would in fact be worthless without, the natural capital of fish populations, petroleum deposits, and forests." [41]

Although technological possibilities do exist for substitution among various forms of capital, there are strict limits to these possibilities for many goods. In the production process, natural resources are transformed into products, while the other forms of capital are used to effect this transformation. The object of increasing the stock of human-made capital is to process a larger flow of natural capital, not to make possible a reduced flow. Recycling and efficiency-increasing technical progress may contribute to reducing rather than increasing the flow of "throughput" (inputs processed into outputs), but in general economic growth serves to increase throughput and resource use.

Valuation of Natural Capital

A rational *micro-allocation* of resources can be achieved through balancing the marginal costs and benefits of different resource uses. But the *macro-allocation* of resources between the ecosystem and the economic system is a social decision that is not well handled through the market pricing system. Since the benefits of appropriating resources and ecosystem functions are mostly private, while the costs are largely social, there is an inherent tendency to overexpand the size of the economy relative to the ecosystem.

Because the value of natural capital is not well captured in existing markets, it is necessary to estimate its value. A variety of methods can be used, ranging from willingness-to-pay surveys, which attempt to mimic market behavior, to energy flow analysis, which is not dependent on human preferences. The latter evaluates natural capital in terms of its embodied or captured energy. For example, the value of the endangered coastal wetlands of Louisiana has been estimated in terms of their capacity to capture solar energy for productive use.[4]

A major issue in valuation is the use of discounting to calculate the present value of a stream of benefits or costs over time. While standard economic theory views discounting as rational optimizing behavior, the discount rate is governed by the preferences of current individuals, and gives little weight to future benefits and costs on an intergenerational time-scale. The discount rate for public policy decisions should therefore be significantly lower than the market rate used for private investment decisions. "The government should have greater interest in the future than individuals currently in the market because continued social existence, stability and harmony are public goods for which current individuals may not be willing to fully pay."[5]

Growth, Development, and Sustainability

Development, or qualitative improvement, can occur without growth in the throughput of resources. But it is excessively optimistic to assume, as the Brundtland Commission did, that a five- to ten-fold economic expansion can come from development rather than growth. Thus if the needs of the world's poor (which undoubtedly require some growth) are to be met, population control, consumption limits, and redistribution must be considered along with economic expansion. Other principles for sustainable development include:

- Limit economic scale to the carrying capacity of natural capital.
- Promote efficiency-increasing technological progress.
- Harvesting rates for renewable resources should not exceed regeneration rates.
- Waste emissions should not exceed the renewable assimilative capacity of the environment.
- Nonrenewable resources should be exploited at a rate equal to the creation of renewable substitutes.

To achieve these ends, the burden of taxes should be shifted from income taxes to taxes on energy use and natural capital consumption.[6] This would create powerful incentives for efficiency-improving technological progress. Implementation of this proposal would pose political difficulties, and it would require international agreement and perhaps the use of ecological tariffs to prevent "dumping" of untaxed resource-intensive products. But it may be the most feasible way of providing the economic incentives to achieve sustainability.

Notes

1. Some theorists prefer to divide these stocks between human capital (embodied in individuals) and social capital (embodied in culture and institutions), thus giving a four-capital system. See, for example, Ekins et al. (1992).
2. Sir John Hicks defined income as the maximum amount that a person can consume during a period and still be as well-off at the end of the period as he was at the beginning. See Hicks (1946).
3. William Nordhaus and James Tobin (1972).
4. Costanza et al. (1989).
5. See Kenneth Arrow (1976).
6. See Jeff Hamond, Tax Waste, Not Work (Part IX, this volume).

Summary of

Environmental Economics, Ecological Economics, and the Concept of Sustainable Development

by Giuseppe Munda

[Published in *Environmental Values* 6 (1997), 213–233.]

This article offers an overview of economic approaches to the concept of sustainable development. Two different economic approaches to sustainability are contrasted: neoclassical environmental economics and ecological economics. Key issues that are identified include weak versus strong sustainability, commensurability versus incommensurability, and ethical neutrality versus acceptance of different values.

The Concept of Sustainable Development

Traditional neoclassical economics analyses the process of price formation by considering the economy as a *closed system.* While classical economists like Malthus, Ricardo, Mill, and Marx saw economic activity as bounded by the environment, neoclassical theory essentially ignored this reality until the 1970s, when a debate began on the social and environmental limits to growth. At this point, some economists such as Ayres and Kneese argued that the economy must be seen as an *open system* that must extract resources from the environment and dispose of wastes back into the environment. The extraction of resources and disposal of wastes causes stress in the life-supporting ecosystem. The growing awareness of actual and potential conflicts between the two systems led to the concept of *sustainable development.*

In standard economic theory, "development" implies both a quantitative change (growth in Gross Domestic Product [GDP]) and a qualitative change (transformation from a pre-capitalist economy based on agriculture to a capitalistic industrial economy). Theories of sustainable development involve both a critique of the quantitative GDP measure[1] and a different view of qualitative transformation. The goals of sustainable development include a harmonization of economic and environmental goals. Since it is difficult to conceive of Western-style economic consumption goals being realized on a planetary scale without massive resource depletion and pollution, this view necessarily entails issues of *distributional equity.* Distribution in this context refers not only to distribution of income and consumption levels, but also to distribution of environmental burdens such as polluted air and water or toxic waste.

Neoclassical Environmental Economics

Environmental economics focuses on (1) the problem of environmental externalities, and (2) the efficient management and intergenerational allocation of natural resources.

Neoclassical economists take their inspiration from Newtonian mechanics, generally believing that economics can be value-neutral, objective, and scientific. Rational decisions regarding "optimal" solutions in neoclassical environmental economics depend on calculations in monetary terms. Natural resources are not seen as imposing binding constraints on economic activity, since *technological progress* and *reproducible human-made capital* can substitute for natural resources. This view underlies the concept of *weak sustainability,* according to which an economy can be considered sustainable if it saves more than the combined depreciation of natural and human-made capital.

While some neoclassical economists deny any special standing to natural capital, others recognize an obligation to keep the value (though not necessarily the physical quantity) of natural capital at a constant level.[2] One obvious problem with this approach is the difficulty of assigning prices to all natural resource functions. To achieve sustainability in this analytical framework, *complete monetary commensurability* is required.

Ecological Economics

Ecological economics focuses on environment-economy interactions but recognizes the existence of *incommensurability* between economic and environmental aspects. Rather than the Newtonian scientific paradigm, it adopts a paradigm of *post-normal science.* This involves a recognition that in the area of global environmental issues, "facts are uncertain, values in dispute, stakes high, and decisions urgent."[3] Uncertainties and values conflicts, generally pushed to the sideline in neoclassical economics, are crucial to ecological economics, which does not claim values neutrality nor an indifference to policy consequences.

In this respect ecological economics is similar to institutional economics: both recognize the importance of different values held by various interested parties, which are reflected in institutional arrangements. No single value-neutral perspective is possible, nor can problems be reduced to a single monetary measure. The distribution of property rights is of fundamental importance, and the interests of stakeholders will shape decision-making. Generally, it will not be possible to identify an "optimal" outcome, but the process of decision-making can at least be transparent.

From this point of view, development is not the straight-line process toward a Western industrialized society envisioned in neoclassical economics. Rather, it can be seen as a process of *coevolution* whereby human society adapts to a changing environment while being itself a cause of environmental change. Cultures, values, beliefs, and economic systems coevolve with ecosystems.[4] The apparent temporary independence of modern productive systems from environmental constraints is an illusion, masking a breakdown of sustainable human/environment relationships. There is no unique or optimal development path; both cultural and ecological diversity are of fundamental importance, and their coevolution moves in unpredictable ways.

Economy-Environment Interaction

Taking the broader ecological economics perspective, we must consider the relationship between three systems:

- The economic system, including production, exchange, and consumption.
- The human system, including biological life processes, culture, aesthetics, and morality.
- The natural system, within which both the economic and human systems are included.

The expansion of the economic subsystem is limited by the size of the global ecosystem. The idea that there are limits to the *scale* of the economic systems leads to a concept of *strong sustainability,* according to which some elements of natural capital are considered *critical,* and not readily substitutable by human-made capital. These critical elements of natural capital must be sustained over time in physical, not economic, terms. This is the theoretical basis for *satellite accounts,* which record physical stock or flow indices of important resources and environmental functions.

According to the thermodynamic law of entropy, resources are degraded and energy used up in all physical and life processes. Complete recycling of materials is impossible, and economic systems are dependent on adequate availability of energy. At the same time, the large-scale use of energy causes increased disposal of wastes into the ecosystem. All theories of development must therefore respect these natural limits on planetary economic scale.

Pluralism and Interdisciplinarity

Traditional monetary evaluation methods such as cost-benefit analysis are based on a partial view of reality connected with only one institution: markets. A more inclusive approach should consider actors and institutions different from the narrow class of consumers. The existence of different perspectives and values should be acknowledged, and a conscious pluralism should be adopted as an approach to decision-making. Attempts to use a single-dimensional measure of value can lead to strange and morally questionable results. For example, the economics research team for the Intergovernmental Panel on Climate Change (IPCC) valued the lives of people in rich countries at up to fifteen times higher than those of people in poor countries.[5]

Rather than pursuing the chimera of "value-free science," analysts and policy-makers should seek to integrate a variety of disciplinary insights. "The impossibility of eliminating value conflicts in environmental policy and the call for a plurality of approaches creates a clear need for environmental philosophers and ethicists to play an important role in ecological economics." [229]

Notes

1. See El Serafy, (Part I, this volume).
2. See, for example, David W. Pearce and K.R. Turner (1990).
3. S.O. Funtowicz and J.R. Ravetz (1994).
4. See R.B. Norgaard (1994).
5. See James P. Bruce (1996).

Summary of

On the Problem of Achieving Efficiency and Equity, Intergenerationally

by Talbot Page

[Published in *Land Economics* 73, 4 (November 1997), 580–596.]

The earth belongs in usufruct to the living.
—Thomas Jefferson

There is a long tradition in economics of separating analyses of equity and efficiency. In this view, benefit-cost analysis should be used to evaluate alternative policy actions, with discounting being used to compare costs and benefits across time periods. Then equity considerations should be assessed independently. This can be called the ***separated approach*** to decision-making. This article proposes an alternative, ***integrated approach,*** in which equity and efficiency are interrelated, and the principle of intergenerational equity provides the basis for sustainability.

Potential Conflicts Between Equity and Efficiency

A distribution system may be efficient but inequitable, for example, if 100 pounds of flour is divided between two people, with one receiving 95 pounds and the other 5. Or, it may be equitable but inefficient, as when 20 pounds of flour is spilled during the distribution, and the recipients then get 40 pounds each. It may be both efficient and equitable (50 pounds each) or inefficient and inequitable (20 pounds spilled, one recipient gets 75, the other 5). However, it may be difficult to say what is equitable. An even division may not be equitable if one person has worked harder than the other to produce the flour. We must also consider incentive effects—the prospect of a larger share may induce someone to work harder, thereby producing a larger total amount to the benefit of both.

Efficiency considerations must thus be balanced against equity issues, in sometimes complex ways. When we consider the long time horizons associated with many environmental decisions, the problem of conflict between principles of equity and efficiency becomes especially important.

Problems with the Separated Approach for Intergenerational Allocation

The use of intergenerational discounting for policy purposes has several well-known problems:

- Extreme sensitivity to the discount rate. $1 million of costs 100 years from now, evaluated at a 10 percent discount rate, yields a present value of future harm of only $72. At a 3 percent discount rate, the present value would be $52,033—a more than 70-fold difference. For long-run decision-making, the choice of the discount rate typically has a larger impact than any other element of the analysis.
- Hypothetical markets as the standard of value. Benefit-cost analysis is intended to mimic market valuation. Market decisions, however, depend on the distribution of wealth and income, which implicitly treats the existing distribution as a normatively acceptable basis for the calculation. Yet this basis is not considered acceptable for many social decisions; typically many decisions are made politically ("one person, one vote") or through a judicial process (the rule of law).
- Intergenerational asymmetry of decision power. As a practical reality, there is a "dictatorship of the present." Even if current actors take an interest in the well-being of future generations, it is still the present generation that controls decisions on resource use. However, there are large differences in the way various institutional processes work with regard to preserving resources for future generations. The use of market power may be significantly more present-oriented than the use of legislative, judicial, or other mechanisms for deciding resource allocation.
- Choice of the discount rate. Even if we accept the use of a market logic, we must still choose a specific discount rate to weigh present versus future. The marginal productivity of capital, often suggested by economists, may not correspond to individual or social time preference. The preferences of future generations cannot easily be included. Also, empirical evidence indicates that people choose a lower discount rate when considering a longer time horizon—something that is not taken into account in benefit-cost analyses.

Intergenerational Equity as Sustainability

Economists sometimes dismiss "sustainability" as a vague and ill-defined concept. Yet many key concepts in economics are equally imprecise. How exactly can we define "money," "capital," or "utility"? The economic definition of "income" put forward by Hicks[1] implies that income must be sustainable, thus introducing the need to define sustainability itself. The concept of sustainability suggested here, based on an integrated approach to equity and efficiency, is no more vague that these key economic terms.

Thomas Jefferson's statement that "the earth belongs in usufruct to the living" is a good working definition of sustainability as intergenerational equity. "Usufruct" has a precise legal definition that specifies a right to use something that belongs to another, provided that the thing itself is not altered or damaged. This can be applied to the earth's resource base taken as a whole. This resource base can be viewed as a commons over generational time. Just as a commons can be destroyed by overuse by many individuals, so the earth's resource base can be destroyed through intensive exploitation by successive generations. But just as there can be successful management principles for the commons, so there can be equitable rules for sharing the earth's productivity between generations.

Identifying these rules depends on assumptions about substitutability, technological progress, and human adaptive capability. If we believe these to be unlimited, then the problem of sustainability becomes trivial. On the other hand, if we believe them to be extremely limited, then sustainability is impossible. It is the middle ground that is the most likely case—there is enough flexibility in productive systems and institutions to make sustainability possible, but only if proper principles for intergenerational justice are established. Further, the problem of intergenerational equity becomes more difficult, and more important, as population and economic output grow relative to the resource base.

An Analogy to Constitutional Law

In jurisprudence, certain principles are established as fundamental or constitutional; within these principles laws are made and implemented based on current social preferences and empirical evidence. Laws and their application are continually changing, based on legislative action and judicial decisions. In contrast, the fundamental constitutional principles can be altered only based on supermajorities or special procedures, something that occurs only rarely. Just as the stable basis of constitutional law provides the "environment" for day-to-day decisions and activities, so the stable basis of the natural environment provides a framework for all economic activity. This suggests that rules for use of the natural resource base should be established based on fundamental principles of intergenerational equity.

To implement this approach, the following steps are necessary:

- Identify key components of the resource base that are essential to sustainability.
- Identify the most effective instruments and decision processes for maintaining the resource base intact.
- Once these instruments are in place, allow ordinary decision-making to take place based on economic criteria, within an intergenerationally equitable decision environment.

Conclusion

This integrated two-tier approach differs from the separated approach used for much current decision-making. Unlike the separated approach, it is not sensitive to discount rates or current market preferences and income allocations, nor is it intergenerationally asymmetric. It goes some way toward resolving potential conflicts between equity and efficiency by allowing economically efficient mechanisms to operate subject to basic principles of equity. However, it poses the problem of distinguishing between fundamental principles and ordinary, day-to-day decision-making. This must be done on a case-by-case basis. For example, the Safe Drinking Water Act employs a precautionary principle to safeguard public health, rather than using cost-benefit analysis or discounting.

The use of this two-tier decision system should become more widespread, both nationally and internationally. It will need to be developed in practice by such institutions as Congress, regulatory agencies, the World Bank, and the International Monetary Fund just as case law and constitutional law have been fine-tuned over the years by court decisions. Although this legal process has been operating for two hundred years in the United States, we can hope that "it will take less time to clarify the issues of intergenerational efficiency and equity in environmental management." [596]

Note

1. See Costanza and Daly (Part I, this volume), note 2.

Summary of

Economics and "Sustainability": Balancing Trade-offs and Imperatives

by Michael A. Toman

[Published in *Land Economics* 70, 4 (November 1994), 399–413.]

The discussion of sustainability has been hampered by uncertainty and lack of uniformity regarding the meaning of the term itself. This paper seeks to identify some common ground among economists, ecologists, and environmental ethicists. Central issues include the requirements for intergenerational equity and the degree of substitutability between natural capital and other forms of capital. The concept of "safe minimum standard," which has been recognized in the ecology, philosophy, and economics literatures, is suggested as a defining principle.

Intergenerational Fairness

Theories of distributive justice can be divided into teleological theories (based on achievement of goals or preferences) and deontological theories (based on innate rights and obligations). A further division can be made between *presentist* theories, which emphasize the current generation and its immediate descendants, and other types of theories that place more emphasis on the future. Yet another division exists between individual-oriented theories and *organicist* conceptions. The latter puts greater weight on community interests. The typical economic concept of discounted intertemporal utility maximization is teleological, presentist, and individualist. It has been subject to ethical criticism on these grounds.

The concept of intergenerational economic efficiency, as defined by the Pareto criterion, does not seem problematical—it simply requires that there be no waste. However, the use of discounting without concern for distributional considerations can impart a presentist bias to calculations of economic welfare. Howarth and Norgaard (1993) have shown that intergenerational equity can be viewed as establishing a fair allocation of endowments among generations. The use of a Rawlsian maximin criterion[1] in their context implies that economic growth should be coupled with a requirement that future generations be no worse off than the present. This approach, however, is still focused on maximization of individual welfare.

A *stewardship* perspective, by contrast, is based on deontological and organicist arguments that invoke an obligation to the entire context of future human life, rather than just to future individuals. This perspective "emphasizes the safeguarding of the large-scale ecological processes that support all facets of human life, from biological survival to cultural existence." [403] The organicist position suggests that there are important social values that cannot be captured in individual utility functions. While this gives a clearer basis for extending concepts of fairness to the intergenerational scale, it also raises questions of individual rights, and poses the danger of the supremacy of the group over the individual.

Resource Substitutability

Assuming that we have some responsibility to future generations, what combinations of capital resources (including natural capital) should be left to our descendants? The answer depends on assumptions about the degree of substitutability between different types of capital. Many economists tend to view capital resources as relatively fungible. From this point of view, large-scale damages to ecosystems are not intrinsically unacceptable, provided compensatory investments in other forms of capital are undertaken.

An alternative view, held by many ecologists and some economists, is that the compensatory investment approach is both ethically indefensible and physically infeasible. While there may be substitutes for some nonrenewable resources, there are no practical substitutes for healthy ecosystems. A related issue is the distinction between local and global impacts. It may be possible to compensate for local environmental degradation through trade, diversification, or migration. But on a global scale, such compensation merely shifts environmental damages around, and ultimately leads to degradation of the entire planetary system.

At the risk of oversimplification, three alternative concepts of sustainability can be derived from this discussion of fairness criteria and substitutability:

- Neoclassical presentism. In this view, sustainability has little standing as a concept distinct from efficient resource use. The present-value criterion is used to evaluate intergenerational welfare, and different forms of capital are considered to be substitutable.
- Neoclassical egalitarianism. This view assumes capital substitutability, but assigns a stronger weight to future interests than is implied by the present-value criterion.
- Ecological organicism. This view emphasizes limited substitutability between natural capital and other assets, and extends the concept of intergenerational fairness from individuals to the species as a whole.

An Extended "Safe Minimum Standard"

A conceptual framework based on the "safe minimum standard" promulgated by Ciriancy-Wantrup (1952) and Bishop (1978), and developed by Norton

(1982, 1992), Page (1983, 1991), and Randall (1986), may be useful in balancing the competing claims of neoclassical efficiency and ecological organicism. "In broad outline, the framework is a two-tier system in which standard economic trade-offs (market and non-market) guide resource assessment and management when the potential consequences are small and reversible, but these trade-offs increasingly are complemented or even superseded by socially-determined limits for ecological preservation as the potential consequences become larger and more irreversible." [405]

In this framework, human impacts on the environment are characterized in terms of "cost" and "irreversibility." The irreversibility metric introduces the ecological concept that large-scale damage to ecosystems is much more harmful and harder to reverse than small-scale disturbances. This gives a two-dimensional classification of resource and environmental impacts, as shown in Figure 1.

The safe minimum standard was developed in the context of species preservation, and its advocates suggest that benefit-cost analysis is inadequate when long-term costs are uncertain but possibly very large. In such cases, the presumption should be in favor of environmental preservation. In Figure I.1, the safe minimum standard applies to the area above and to the left of the dividing line. When impacts are higher in cost and, especially, are likely to be irreversible, the safe minimum standard should override standard economic calculations of

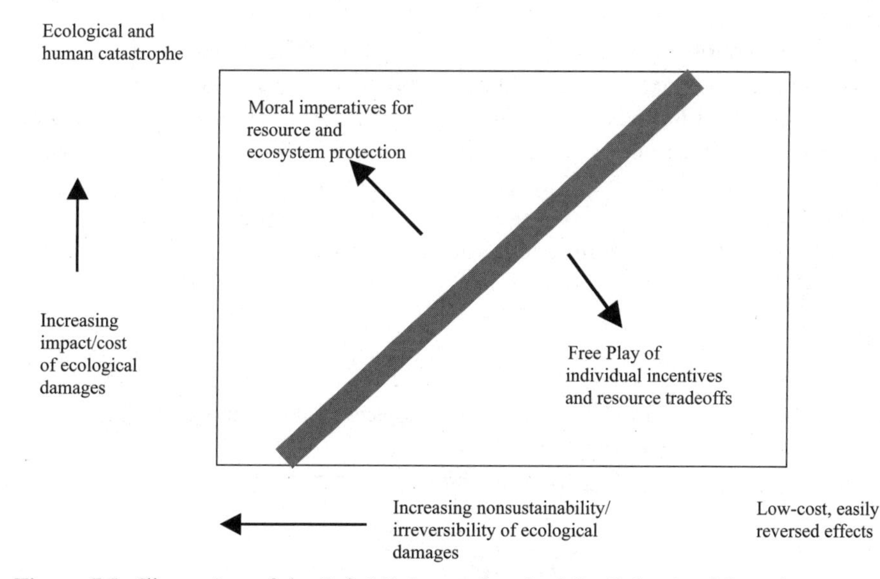

Figure I.1. Illustration of the Safe Minimum Standard for Balancing Natural Resource Trade-Offs and Imperatives for Preservation

cost and benefit. For impacts in the bottom right portion, with modest costs and a high degree of reversibility, individualistic valuations and trade-offs can be relied on. The orientation and placement of the fuzzy demarcation line will be a matter of debate, with ecologists possibly favoring a more vertical line, and neoclassical economists advocating a more horizontal one.

This dualistic approach to decision-making is consistent with the belief that people themselves are dualistic, acting as citizens as well as consumers. In acting as consumers we use individualistic valuations while as citizens we may favor social institutions for environmental management. This view, put forward by Vatn and Bromley (1994), suggests that societies have evolved norms for environmental governance as a way of circumventing the limits of individualistic valuation. This "justifies in particular the imposition of safe minimum standards determined through political discourse and other complex social processes." [409]

Conclusion

Sustainability concerns human values and institutions as well as ecological functions. At the same time, economic analysis without adequate ecological underpinnings is misleading. Both ecologists and economists can contribute to an interdisciplinary understanding of sustainability. Ecologists need to provide information in a form that can be used in economic assessment and also need to recognize the importance of human behavior and incentives. Economists must consider the function and value of ecological systems as a whole and must make greater use of ecological information. While there are difficulties, it may also sometimes be possible to combine economic and ecological perspectives in a single model.

"Despite its continued abuse as a buzz-word in policy debates, the concept of sustainability is becoming better established as a consequence of studies in economics, ecology, philosophy, and other disciplines. With a better understanding of the interdisciplinary theoretical issues, and a better empirical understanding of both ecological conditions and social values, sustainability can evolve to the point of offering more concrete guidance for social policy." [410]

Note

1. A maximin criterion for income distribution choices implies that the best distribution is one that offers the highest minimum income.

Summary of

From Political Economy to Political Ecology

by Juan Martinez-Alier

[Published in *Varieties of Environmentalism: Essays North and South,* ed. Ramachandra Guha and Juan Martinez-Alier (Delhi: Oxford University Press, 1998), Ch. 2, 22–45.]

Ecological economics differs from orthodox economics in its focus on the compatibility between the human economy and ecosystems over the long term. Ecological economists address the issue of translating environmental values into monetary values but are skeptical about expressing future, uncertain, or irreversible externalities into monetary terms. The study of distributional issues in ecological economics constitutes a new field, which is identified in this article as *political ecology.* Like the classic tradition of political economy, political ecology deals with distributional conflicts, but with an added focus on the interests of future generations, other species, and with special attention to nonmarketed natural resources and environmental services.

Rather than seeking to internalize externalities through actual or surrogate markets, ecological economists recognize the incommensurability of many environmental resources and services. This article examines how ecological distribution conflicts are related to allocations of property rights, the distribution of income, and methods of valuing the future.

Ecological and Social Conflicts

Environmentalism is sometimes seen as the product of prosperity, a luxury for those whose material needs are satisfied. This has led many Marxists to scorn environmentalism as an upper-class fad. However, this is profoundly mistaken. In the developing world, the poor must often defend the environment in the interests of their own survival. There are numerous examples in Latin America of poor communities fighting water pollution from mining, defending forests from timber corporations, battling industrial pollution from smelters and factories, and protecting mangrove forests from the shrimp industry. These people might not describe themselves as environmentalists, but they are on the front lines of crucial ecological battles.

Social and environmental conflicts of this type raise far-reaching issues of ecological distribution and control of natural resources. Marxists, while emphasizing class conflict, have neglected such issues since Engels rejected Podolinsky's attempt in 1880 to introduce human ecological energetics into Marxist economics. However, it is essential to combine environmental and social history. Marxists fear "naturalizing" human history, and indeed there have been attempts to do so, ranging from Malthusianism to Social Darwinism to sociobiology. However, introducing human ecology into history does not so much naturalize history as historicize ecology. Human endosomatic energy use is genetically determined, but the exosomatic use of energy and materials is socially driven, depending on economics, politics, and culture. Demography is related to changing social structures, and human migration patterns depend on economics, politics, and law rather than on natural imperatives.

Distribution and Valuing the Future

The economic system lacks a common standard of measurement for environmental externalities. Estimates of environmental values depend on the endowment of property rights, the distribution of income, the strength of environmental movements, and the distribution of power. The issue is further complicated by the difficulty of determining an appropriate discount rate for weighing future costs and benefits.

How can we justify the use of a positive discount rate? The justification for pure time preference is weak. The argument that future generations will be better off, and will therefore have a decreased marginal utility of consumption, is not acceptable from the point of view of ecological economics. Greater consumption today may well leave our descendants with a degraded environment, and therefore worse off. A more reasonable case for a positive discount rate rests on the productivity of capital. But here we must distinguish between genuinely productive investment and investment that is environmentally damaging. Only *sustainable* increases in productive capacity should count. But the assessment of what is sustainable involves a *distributional* issue. If natural capital has a low price, because it belongs to nobody or to poor and powerless people who must sell it cheaply, then the destruction of nature will be undervalued.

Sustainability needs to be assessed through biophysical indicators that incorporate consideration of ecological distribution. Such concepts include the Ecological Footprint, Appropriated Carrying Capacity or Environmental Space, and appropriation of Net Primary Product (NPP). These measures do not translate easily into monetary terms because of *incommensurability*. The monetary values given to externalities by economists are a consequence of political decisions, patterns of property ownership, and the distribution of income. There is thus no reliable common unit of measurement, but this does not mean that we cannot compare alternatives on a rational basis through multicriteria evaluation. Eliminating the spurious logic of monetary valuation opens a broad political space for environmental movements.

Ecological Distribution Conflicts

Ecological distribution refers to social, spatial, and temporal asymmetries or inequalities in the use by humans of environmental resources and services, and in the burdens of pollution. For example, an unequal distribution of land, together with pressure of agricultural exports on limited land resources, may result in degradation by subsistence farmers working on mountain slopes that would not be cultivated so intensively under a more equitable distribution of land. Other examples include the inequalities in per capita energy use and accompanying carbon emissions, territorial asymmetries in sulfur dioxide emissions and the burden of acid rain, and intergenerational inequity between the use of nuclear energy and the burdens of radioactive waste.

The transfers involved in these unequal distributions have no agreed-on monetary values. They have, however, become the subject of political discussion, for example in North/South negotiations over carbon emissions, or in the environmental justice movement over the siting of toxic waste facilities and polluting industries in the United States.

There is also a gender dimension to ecological inequality, as shown by the prominent role of women in local environmental movements in Peru. Women's role in provisioning and care of the household leads to a special concern with such issues as scarcity and pollution of water and lack of firewood. Women often have a smaller share of private property and depend more heavily on common property resources. Also, women often have specific traditional knowledge in agriculture and medicine, which is devalued by intrusion of market resource exploitation or state control.[1]

Some have suggested that environmentalism arises as a result of a change in values away from material consumption toward appreciation of environmental amenities. This may be true for the more affluent, but it fails to describe the "environmentalism of the poor" reflected in numerous grassroots movements against environmental destruction. These include the rubber tappers in Brazil, the Chipko movement and the resistance to the Narmada dam complex in India, and the Ogoni struggle against Shell in Nigeria. Ex-slaves in the Trombetas river region in Brazil have fought hydroelectricity generation and bauxite mining; local Amazon fishermen have defended communal management against the intrusion of commercial fishing; and women in the Brazilian northeast have defended the babassu palm against landowners seeking to clear land. In Peru, villagers have fought against pollution from copper mining and smelters, and in Ecuador against waste dumping, coastal pollution, and destruction of forests and rivers by the oil industry.

International Externalities

Global exploitation of nature raises the issue of the internationalization of externalities. The value of such externalities is clearly related to outcomes of distributional conflicts. What is the true value of a barrel of Texaco oil, a bunch of bananas, or a box of shrimp from Ecuador? The answer depends on the value of the damages caused in production, but this valuation is a product of social institutions and distributional conflicts. Damages of $1.5 billion have been claimed from Texaco in connection with oil extraction in Ecuador as compensations for oil spills, deforestation, and disruption of the life of local communities. The plaintiffs are Indians and other local people. The Ecuadorian government, not a plaintiff in the suit, has tried to arrange an out-of-court settlement for about $15 million—one hundredth of the damages sought. How should the New York court where the case was brought assess damages: according to U.S. values or according to Ecuadorian values? Can poor people be persuaded or coerced into accepting a settlement of much lower value than U.S. citizens would expect?

Similar cases have been brought by unions from Costa Rica and Ecuador in a Texas court against Shell, Dow Chemical, United Fruit, and others, seeking compensation for male sterility in Costa Rican and Ecuadorian workers caused by the pesticide DBCP. How much is a case of male sterility worth? Does the value of this externality depend on the distribution of income? As Lawrence Summers, former chief economist for the World Bank, put it in a now infamous memo, "the measurement of the costs of health-impairing pollution depends on the foregone earnings from increased morbidity and mortality. From this point of view a given amount of health-impairing pollution should be done in the country with the lowest cost, which will be the country with the lowest wages."[2] But courts may not necessarily be bound by the logic of the market. International legal cases such as these provide a practical arena for observing social and institutional influences on the valuation of externalities.

Commercial shrimp cultivation in Ecuador has also caused substantial losses to people who make their living from sustainable use of mangrove forests on the Pacific coast. This has not yet become the subject of a court case, but similar issues of valuation and property rights arise. Throughout the developing world, regions that have been developed on the basis of extractive enterprises are the victims of ecologically unequal exchange. The externalities that they suffer are chronically undervalued by the market. The externalization of social and environmental costs by resource-extracting firms gives rise to new social movements.

As the market system spreads over the world, it generates political responses to the inequities of ecological distribution. Global issues, first raised by scientists, also provide the basis for local or national campaigns, as in Southern demands for compensation for the "ecological debt" created by high Northern carbon emissions. The rich behave as if they were owners of a disproportionate part of the planet's carbon dioxide absorption capacity, and then dump excess carbon into the atmosphere as if they owned that too. In an ecological sense, Southern nations are in a creditor position and can use this to advantage in international negotiations. There is a basis for an alliance between Southern groups seeking to protect the rainforest (a carbon sink) or to oppose oil extraction and Northern environmentalists trying to restrict oil use and carbon emissions.

Issues of ecological distribution thus provide a link between political ecology and political economy in many domestic and international policy areas. These environmental conflicts, and the resistance movements that they engender, provide the research agenda for the evolving field of political ecology.

Notes

1. Bina Agarwal (1992).
2. "Let Them Eat Pollution" (1992). Lawrence Summers, appointed Secretary of the U.S. Treasury in 1999, was chief economist at the World Bank at the time this internal memo was leaked.

Summary of

Green Accounting and Economic Policy

by Salah El Serafy

[Published in *Ecological Economics,* 21 (1997), 217–229.]

The idea of "green," or environmental, accounting has become popular, and a number of proposals have been made for modifying national accounts to include a consideration of resource depletion and environmental deterioration. In this article, the author suggests that certain weaknesses still pervade the new proposals. National accounts are primarily an economic framework and are not suitable for an adequate representation of all environmental changes. There are also serious problems involved with the valuation of resource and environmental stocks. Especially for developed nations where the main environmental concern is with pollution impacts, greening the accounts is of limited value. However, integrated resource accounting is vital for developing nations that have a heavy dependence on natural resources and for which conventional accounting can lead to distorted and destructive macroeconomic and trade policies. This paper suggests that green accounting can help to ensure income (or "weak") sustainability as a step toward a stronger ecological sustainability.

Principles of Green Accounting

Advocates of green accounting have different concerns, including preserving the stock of environmental assets and measuring the effect of environmental changes on welfare. This paper has a more precisely defined goal: the proper measurement of national output and expenditure. "Selling natural assets and including the proceeds in the gross domestic product, GDP, is wrong on both economic and accounting grounds." [218] To get a proper estimate of *net value added*, we must subtract depreciation of assets. This is done for produced assets in the calculation of net domestic product (NDP). Even though NDP is rarely estimated, depreciation of produced assets is fairly small and predictable. Declines in natural assets, on the other hand, may be large and volatile, and are not reflected at all in the estimates of GDP commonly used for macroeconomic analysis.

The concept of green accounting presented here thus involves no value judgment about preserving the environment. It simply embodies a correct accounting principle for estimating sustainable *income*. In accordance with this principle, economic policies will need to be reassessed once national accounts have been adjusted for natural asset losses.

However, greening the national accounts cannot fully capture many aspects of environmental deterioration such as biodiversity loss, nor can they provide a solution for a broad range of environmental problems. For these purposes, physical rather than economic indicators of environmental change are more appropriate.

Definitions of Sustainability

There are different possible definitions of sustainability. *Weak sustainability* has been used to refer to sustainable income, which should include only value added and exclude the proceeds of asset sales. This approach reflects the accounting principle of keeping capital intact for income-estimation purposes. If capital is consumed, then allowance must be made for capital consumption or depreciation. Weak sustainability is a *positive* rather than a *normative* concept, requiring only a correct approach to income estimation. Also, since income accounting is done on a year-by-year basis, this approach does not guarantee long-term sustainability.

Strong sustainability, on the other hand, requires maintaining the *stock* of natural capital intact, including the waste-assimilation services of the environment. Advocates of strong sustainability argue for the existence of a complementary relationship between natural resources and produced capital. This means that damaged or depleted natural capital cannot easily be replaced with manufactured capital. For nonrenewable resources, this principle implies that the equivalent of the user cost[1] of depletable natural resources should be invested in developing renewable substitutes. Strong sustainability is appropriate to a long-run normative approach. While (for income sustainability) it may be appropriate in the short term, or for an individual firm, to deplete natural assets in order to build up produced capital, it would be environmentally irresponsible to assume that this can be done without limit over an extended period of time.

Satellite Accounting Systems

The United Nations Statistical Division has proposed a system of integrated economic and environmental accounting, SEEA.[2] This proposal, the outcome of a process of discussion among different international agencies, including the United Nations Environment Programme (UNEP) and the World Bank, is intended as a compendium of information on points of contact between the environment and the economic system. However, it does not offer any definitive recommendation on reform of macroeconomic accounting systems, and it is referred to by its authors as an "interim version," indicating that the discussion is still in progress.

The SEEA approach appears to focus on accounting for environmental *stocks*. The *flow* accounts are derived from changes in stocks during the accounting period. There are two weaknesses in this approach. First, it is impossible to compile a comprehensive list of all environmental stocks. Second, the valuation of environmental stocks using current prices means that the flow estimates are affected by price volatility during the estimation period. This compromises both the environmental and the economic information provided by such estimates. For economic purposes, a better approach would be to calculate the user cost component of resource declines and either subtract this from GDP as capital consumption or (much better) exclude it from the gross product altogether.

In dealing with pollution, the costs of regulating or cleaning up pollution should be considered as intermediate inputs to be charged against output. Where effective regulations are lacking, pollution costs can be estimated by calculating the theoretical cost of meeting acceptable standards, based on current technology. Like the calculation of user cost, this should not be viewed as a radical new departure but simply as an overdue correction of national accounts that currently value environmental damages incorrectly at zero.

We must distinguish between two different goals for environmental accounting. If the objective is to describe the state of the environment, then physical measures of resource and environmental stocks should be used. This could be done within a system of *satellite accounts,* which must be deliberately separated from the economic accounts. If the objective is to reform the system of national accounts, then a procedure focused on estimating user costs and intermediate environmental costs is needed to achieve realistic macroeconomic measurements. In the latter case, environmental stocks should be kept firmly in the background.

Policy Implications of Greening the National Accounts

If economists accept conventional GDP estimates, then their policy recommendations are likely to be wrong in the case of natural resource dependent economies. Output estimates may be exaggerated by 20 percent or more and true estimates of capital formation may turn out to be nil or negative. Factor productivity estimates are thrown into question when neither the products nor the inputs are measured correctly. Capital/output ratios will be incorrect if they ignore rapid liquidation of natural capital. Sophisticated macroeconomic models based on such data will give highly questionable results for guiding long-term development.

International trade will tend to align domestic with international prices. But international prices are often distorted by agricultural subsidies, political and military interventions, and the failure to internalize externalities. This will encourage the selling of natural resources below full environmental cost, and the situation gets worse in view of the often upward-sloping supply curves of many poorer countries' exports.

The impact of natural capital depletion will be especially large in estimates of national savings and investment. Estimates of "genuine savings" by the World Bank indicate that many countries' net savings and capital formation may in fact be negative, a clear indicator of unsustainability.

The export of natural capital also distorts exchange rates and creates a bias against nonresource-exporting sectors, including manufacturing. This phenomenon is recognized by economists as the "Dutch Disease," but methods used to estimate exchange rate overvaluation will not be reliable when proceeds from the unsustainable export of natural assets finance an import surplus. In this case, an apparent stability of the domestic price level will be illusory, masking significant damage to nonresource exporting sectors that must compete with artificially cheap imports. In the balance of payments accounts, a trade deficit may be concealed, or may appear to be a surplus, since the proceeds of natural capital exports are recorded incorrectly in the current account.

"Greening the national accounts is more important for economic than for environmental policy . . . especially for those countries whose natural resources are rapidly eroding, and the erosion is counted misleadingly in GDP as value added. Once the accounts are greened, macroeconomic policies need to be re-examined along the lines elaborated in this paper." [228]

Notes

1. The user cost can be thought of as the cost imposed on future extraction by using up a resource today. Its calculation depends on the expected lifetime of the resource and on the interest rate realistically to be expected on the new investments.

2. United Nations Department for Economic and Social Information and Policy Analysis (1993).

Summary of

Are We Saving Enough for the Future?

by Kirk Hamilton and Michael Clemens

[Published in *Expanding the Measure of Wealth: Indicators of Environmentally Sustainable Development* (Washington, D.C.: The World Bank, 1997), Ch. 2, 7–18.]

This article applies the concept of environmentally adjusted accounting to measures of savings and investment. Adjusting savings measures to reflect environmental depletion fits well with many of the traditional concerns of development economics, including the savings-investment gap and the importance of investment finance for development.[1] "Developing 'greener' national accounts holds the additional promise of treating environmental problems within a framework that the key economic ministries in any government will understand." [7] This framework may help to bridge the gap between environmental ministries and economic ministries. In addition, broader development concerns can be included in adjusted savings and investment measures by taking account of investment in human capital.

Genuine Saving

Genuine saving is defined as the true rate of saving in a nation after due account is taken of the depletion of natural resources and the damages caused by pollution. The policy implications of such a measurement are straightforward: negative rates of genuine saving must eventually lead to declining well-being. To correct negative or inadequate genuine saving, a variety of interventions are possible, including macroeconomic policy, environmental policy, and human resource policy.

The commonly used measure of wealth accumulation is *gross saving:* GNP minus public and private consumption. Gross saving is equivalent to gross domestic investment less net foreign borrowing. A more accurate measure is *net saving,* or gross saving minus the value of depreciation of produced assets. *Genuine saving* is obtained by subtracting the value of resource depletion and pollution damages from net savings.

Resource depletion is measured as the total rents on resource extraction and harvest. Rent is defined as the difference between the value of production at world prices and the total costs of production. For renewable resources, net depletion (harvest minus new growth) is calculated. For pollution costs, the adjustment represents pollution emissions valued at their marginal social cost. It is often difficult to obtain country-specific estimates for pollution damages, and the general estimates presented here use carbon dioxide emissions as a proxy for other pollutants. Some important ecological values, such as biodiversity and watershed protection, are not included due to difficulty of estimation, and the value of soil erosion is also omitted for the same reason.

To take account of human capital investment, the concept of *extended domestic investment* is introduced. In this formulation, current educational expenditures such as teachers' salaries and textbooks are treated as investment. (This differs from the standard accounts, which include only capital expenditures such as school building construction as investment.) A complete valuation of human capital would be extremely complex; this approach serves as a first approximation.

These two broad categories of adjustments to standard savings accounts move the figures in opposite directions. To examine their separate impacts, regional trends are first discussed using only the depletion and degradation adjustments. Then the calculations are modified by the introduction of extended domestic investment, which gives a revised genuine savings estimate including both environmental depreciation and human resource investment.

Regional Trends in Genuine Saving

An analysis of regional trends in genuine saving (Figure I.2) reveals a remarkable pattern in sub-Saharan Africa: genuine savings rates rarely exceeded 5 percent during the 1970s, and after that plunged into the negative range, where they have remained ever since. These negative rates have been accompanied by persistently low regional indicators of human welfare, including education, nutrition, and medical care.[2] In Latin America and the Caribbean, genuine savings

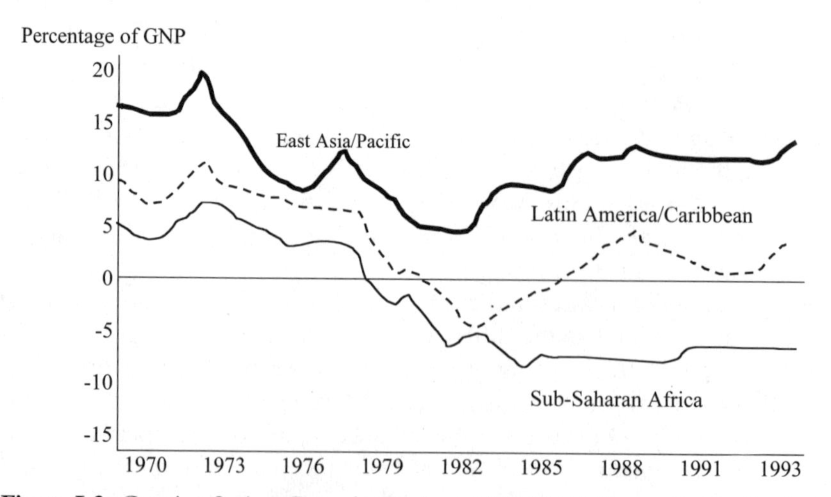

Figure I.2. Genuine Savings Rates by Region, 1970–1993: East Asia/Pacific, Latin America/Caribbean, Sub-Saharan Africa.

declined from 8 to 9 percent of GNP in the 1970s, falling into the negative range during the 1980s debt crisis. They have since recovered into a low positive range, below 5 percent of GNP.

In contrast, the East Asia and Pacific region has shown generally strong rates of genuine savings, recently exceeding 15 percent of GNP. High-growth economies such as China (including Hong Kong), the Republic of Korea, Singapore, Thailand, and Taiwan dominate these regional statistics. Indonesian and Malaysian rates are lower, on a par with some of the higher Latin American rates. Lao People's Democratic Republic, Papua New Guinea, and Vietnam have negative genuine savings but have recently been improving toward the positive range.

In the Middle East and North Africa, genuine savings are consistently negative, indicating that the enormous rents from crude oil exports have been only partly invested, with a large portion being spent on imported food and manufactures (Figure I.3). Modestly positive saving by Algeria, Egypt, Israel, Morocco, and Tunisia has not offset the large negative genuine savings rates of the major oil exporters.

In South Asia, near-zero genuine savings in Bangladesh and Nepal have been offset by positive genuine savings of close to 10 percent in India, giving a net

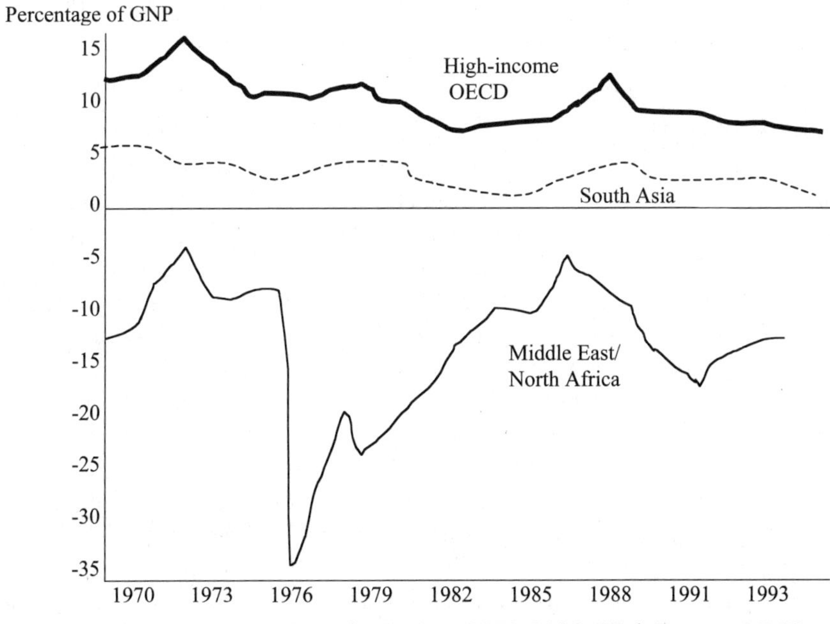

Figure I.3. Genuine Savings Rates by Region, 1970–1993: High Income OECD, South Asia, Middle East/North Africa.

positive rate for the region. High-income OECD (Organization for Economic Cooperation and Development) countries have genuine savings of around 10 percent, buoyed by high investment and lack of natural resource dependence. Western Europe and Japan are the biggest savers at 10 to 15 percent, while more resource-intensive economies such as Australia, Canada, and the United States achieved only 1 to 3 percent rates.

Investing in Human Capital

The world's nations spend trillions of dollars every year on investment in human capital through their education systems. If the notion of wealth is expanded to include human capital, educational spending, including current spending, must be counted as investment. Making this adjustment increases the genuine savings measure for economies such as Chile, where current educational expenditures represent about 3.1 percent of GNP. Using this approach, about one-third of Chile's adjusted genuine savings of around 10 percent can be attributed to education.

Introducing human capital investment into the picture accentuates the differences between high- and low-saving countries. The disparities between low, medium-, and high-income countries become more marked when educational investments are considered (Figure I.4). Adding human capital investment markedly improves regional genuine savings measures; even sub-Saharan African rates, while remaining negative, are closer to zero.

Policy Implications

The main impact of the genuine savings calculation is to reveal the extent to which resource-rich countries have been consuming, as opposed to saving, natural resource rents. This is consistent with economic evidence that growth rates have been weaker in resource-intensive economies.[3] For these countries to improve their performance, changes are needed in both macroeconomic and environmental policies. A key issue is the reinvestment of natural resource rents collected through government royalties. Some of the highest-quality outlets for public investment are in human capital. Ministries of finance and human resources can use genuine savings measures to help guide public investment policy.

For resource ministries, efficient resource extraction and harvest, security of tenure for producers, and capture of rents through resource royalties should be high priorities. For environment ministries, the challenge is to reduce pollution damages. In India, for example, 1991 pollution damages were calculated at roughly 2.5 percent of GNP, lowering genuine savings rates from 10.5 percent to 8 percent.

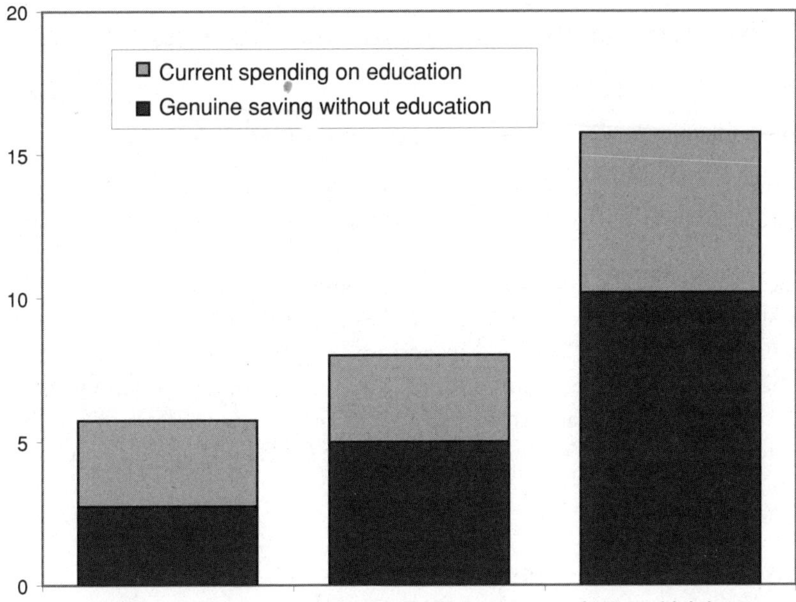

Figure I.4. Contribution of Current Education Expenditures to Genuine Saving.

"The bottom line in the analysis of genuine saving is that policies leading to persistently negative savings rates must entail, eventually, declines in welfare. The intuition, or hope, that greening the national accounts could influence policies for sustainable development has taken some time to be realized, but the analysis of saving and wealth holds out this possibility. This analysis also emphasizes that the key economic ministries, the human resources ministry, and the resource and environmental ministries all have important policy levers at their command if the goal is to achieve sustainable development." [16]

Notes

1. D.W. Pearce and G. Atkinson (1993).
2. World Bank (1996).
3. J.D. Sachs and A.M. Warner (1995).

Summary of

Progress on the Environmental Kuznets Curve?

by David I. Stern

[Published in *Environment and Development Economics,* 3 (1998), 173–196.]

According to the Environmental Kuznets Curve hypothesis (EKC), an inverted U-curve relationship exists between indicators of environmental degradation and levels of income per capita. This in turn has been taken to imply that economic growth will eventually reduce environmental impacts associated with the early stages of economic development. This paper discusses the theoretical underpinnings of the EKC hypothesis and reviews many of the empirical studies that have used econometric methodology to investigate EKC relationships. The available evidence indicates that a number of other factors affect or modify the income-environment relationship, and that the EKC logic applies only in a limited number of cases.

Theoretical Basis of the EKC

The original Kuznets curve asserted an inverse U-curve relationship between income inequality and income levels.[1] Advocates of the EKC hypothesis argue that as development begins rates of land clearance, resource use, and waste generation proceed rapidly. But at higher levels of development better technology, improved environmental awareness and enforcement, and structural economic change favoring services and information-intensive production techniques lead to improved environmental conditions. According to Beckerman (1992), a strong advocate of this logic, "There is clear evidence that, although economic growth usually leads to environmental degradation in the early stages of the process, in the end the best—and probably the only—way to attain a decent environment is to become rich."

Critics of the EKC hypothesis argue that empirical evidence for the relationship is weak and applies only to a subset of indicators. In addition, evidence on existing environmental conditions in poor and rich countries are not good predictors of the dynamic relationships associated with economic growth. And even where EKC-type patterns hold true, global "turning points" toward lower total pollution are decades away.

Theoretical models of economic growth and pollution generate EKC-type relationships under appropriate assumptions about consumer preferences, pollution control policies, and substitutability in production. Empirical evidence is needed to evaluate whether these models are plausible under real-world conditions.

Empirical Evidence on EKC Relationships

The first empirical EKC study, by Grossman and Krueger (1991), estimated EKCs for sulfur dioxide (SO_2), smoke, and suspended particulate matter (SPM) as part of a study of the potential impacts of NAFTA (North American Free Trade Agreement). They identified turning points for SO_2 and smoke at around $4,000–$5,000 of per capita income using a Purchasing Power Parity measure and lower turning points for SPM. However, at income levels over $10,000–$15,000, all three pollutant levels appeared to increase again.

A study used in the 1992 World Development Report estimated EKCs for ten different environmental indicators.[2] The results differed for each indicator. Availability of clean water and urban sanitation improved with higher income, while river quality worsened. SO_2 and SPM conformed to the EKC pattern, with turning points of $3,000–$4,000, while municipal waste and carbon emissions per capita increased unambiguously with rising income. Panayotou also found an EKC pattern for SO_2, SPM, and nitrogen oxides (NO_x), with turning points around $3,000–$5,500 (see Figure I.5).

A study by Selden and Song (1994) found EKC relationships to exist for airborne emissions, but with much higher turning points, in the $6,000–$12,000 range. This is significant because it implies that global pollution levels will continue to increase for decades before mean income reaches this range.

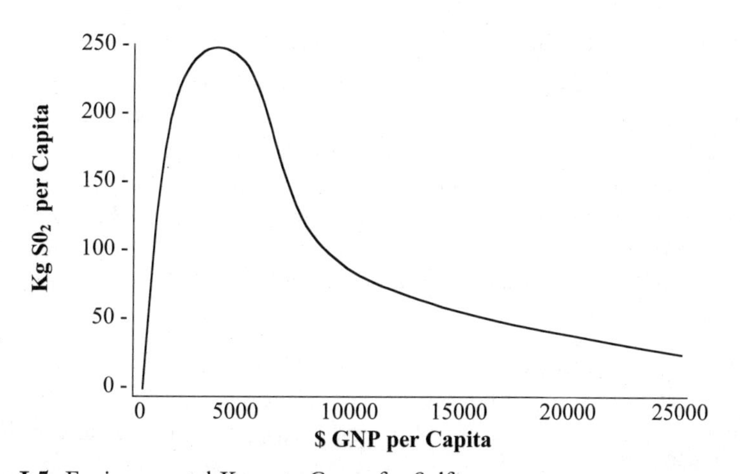

Figure I.5. Environmental Kuznets Curve for Sulfur.
Source: Panaytou, T. (1993) "Empirical Tests and Policy Analysis for Environmental Degradation at Different Levels of Development." Geneva: International Labour Office Working Paper WP238.

Critiques of the EKC

Stern et al. (1996) identify seven major problems with the basic EKC hypothesis:

- *Simultaneity and irreversibility*—Environmental damage may not be easily reversible, and the effects of widespread environmental damage may affect economic growth.
- *Trade effects*—Specialization by developing countries in resource-intensive or pollution-intensive production may increase degradation in these countries while reducing it in developed countries. However, when today's developing countries become rich they will not be able to reduce pollution-intensive production by importing such products from poor countries.
- *Econometric problems*—Different functional forms seem to give different EKC results, indicating that the underlying relationship may not be robust.
- *Ambient concentrations versus emissions*—Some of the basic EKC studies measure ambient pollution levels in urban areas. However, even if these levels decline, total emissions may still be increasing, but spread over a wider area.
- *Asymptotic behavior*—Functional forms that imply that pollution levels could go to zero are in conflict with basic principles of thermodynamics, which specify that use of resources inevitably implies the production of wastes. For this reason, even if strengthened environmental standards lead to a decline in pollution levels, further increases in consumption will raise levels again. Some of the empirical studies seem to indicate this N-shaped pattern, but the econometric evidence is not conclusive.
- *Mean versus median income*—The turning point estimates for a number of studies are around current world mean per capita income, which at first glance might imply that we can expect a decline in global pollution levels as income grows. However, the world income distribution is heavily skewed, so that there are much larger numbers of people below the mean per capita income than above it. It is the much lower median income that is relevant, implying that global pollution levels will increase for decades to come.
- *Aggravation of other environmental problems*—With economic development, levels of some pollutants decline, but others increase. The mix of effluents typically shifts from sulfur and nitrogen oxides to carbon dioxide and solid waste, but total waste per capita may not decline. Greater energy use per capita, which tends to accompany economic growth, can serve as a proxy for multiple environmental impacts, some of which are monotonically increasing.[3]

Other Determining Factors

Recent studies have identified important factors other than income that serve as determinants of levels of environmental degradation. One of these is trade, which can shift environmental impacts. Rothman (1998) suggests that an examination of the environmental impacts generated by consumption within a country, rather than production, shows that impacts increase with higher income levels for all except a few categories of consumption goods.

Other variables that appear significant in affecting environmental quality include degrees of political freedom, spatial intensity of population and economic activity, economic structure, and price effects. Moomaw and Unruh (1997) find that the oil price shocks of the 1970s, not income changes, were the triggering factor leading to changes in per capita CO_2 emission trends.

Conclusions

"There has been progress in understanding the scope and determinants of the EKC in the last few years and some progress in methods of investigation. Evidence continues to accumulate that the inverted-U shape relation applies to only a subset of impacts, and that overall impact, perhaps approximated by per capita energy use, rises throughout the relevant income range." [192] In addition to the importance of structural change, technological progress, and political democracy, there is "increasing evidence that the EKC is partly determined by trade relations. If this is so, the poorest countries of today will find it more difficult than today's developed countries to reduce their environmental impact as income rises." [192]

Notes

1. This relationship has also been the subject of controversy: See Ackerman et al. (2000).
2. N. Shafik and S. Bandyopadhyay (1992).
3. See Suri and Chapman (1998).

PART II

Economics of Sustainability: The Social Dimension

Overview Essay

by Timothy A. Wise

> "Sometimes one gains the impression that the development debate is just a succession of fads and fashions. But the evolution from economic growth, via employment, jobs and justice, redistribution with growth, to basic needs and human development represents a genuine evolution of thinking and is not a comedy of errors, a lurching from one slogan to the next."
> —Paul Streeten, *Thinking About Development* (1995)

Much of the sustainable development discourse, especially when it appears in the economics literature, focuses on environmentally sustainable development. Since the Brundtland Report, however, there has been considerable effort to broaden the concept of sustainability beyond environmental concerns by recognizing the myriad social dimensions of sustainability. The report itself highlighted the ways in which poverty is both a cause and an effect of environmental degradation. But the social dimensions of sustainability extend beyond poverty and its connection to the environment to include a range of issues often ignored in environmental circles.

This section presents some of the research in economics and related disciplines that has attempted to integrate social issues into the conception of sustainable development. It is intended to highlight the ways in which development that fails to meet basic needs and allow democratic participation for all is not desirable and may not be sustainable. These efforts have taken their most coherent form with the evolution of the "human development paradigm," which has found an institutional home in the United Nations Development Programme (UNDP) and an ambitious set of comparative data and analysis in the annual *Human Development Report.*

This essay examines not only the concept itself but also recent attempts to refine it and apply it to contemporary development issues, in particular the "sustainable livelihoods" approach and "social exclusion" theories. It then turns to the poverty-environment nexus, attempting to go beyond the Brundtland Report's implication that the poor are significantly responsible for environmental destruction. After touching on the vast array of research on gender, sustainability, and development, the essay then assesses the usefulness of the concept of "social capital," a term that borrows conceptually from economics and is applied widely, most interestingly as it relates to the effectiveness of government action. It concludes with a discussion of democracy, participation, and empowerment as they relate to issues of sustainability.

The Origins of Human Development

Years before environmentalists began campaigning for an approach to economic development that would take account of the environmental impact of economic life, an equally influential group of economists was working to address the overemphasis on economic growth in prevailing development strategies. Much as environmentalists argued that the economics profession overlooked humanity's increasing consumption of the planet's stock of natural capital, these economists argued that the overemphasis on economic growth overlooked the ways in which that growth improved—or failed to improve—the quality of life for different segments of the population. In an attempt to address neoclassical economics' shortcomings in explaining distributional issues, this approach focused on the persistence of poverty even in high-growth economies

Two distinct schools of thought merged to form what has come to be known as the "human development paradigm." Mahbub ul Haq and Paul Streeten promoted a "basic needs" approach in the late 1970s, arguing that the traditional focus on economic growth needed to be augmented by one that emphasized meeting the basic needs of all members of society. They pointed out that many programs, such as education, nutrition, and health care, represented investments in human capital that were shown to be productive for generalized economic growth. The basic needs approach went beyond efficiency arguments to challenge the prevailing orthodoxy within the development community, calling for major shifts in the power balance within highly skewed societies (Streeten et al. 1981).

The other school of thought underpinning the human development paradigm came out of the work of Indian economist Amartya Sen. Sen argued for a shift in emphasis from incomes to outcomes and from per capita income growth to improved quality-of-life outcomes. The centerpieces of Sen's development theories are his linked notions of capabilities, functionings, endowments, and entitlements. He defined capabilities as the set of choices available to different individuals and groups within society, while functionings refer to the options actually chosen by individuals (Sen 1981, 1992). Both concepts recognize that in a stratified society individuals have differential access to resources and opportunities. Sen also offered the related concepts of entitlements and endowments. The former refers to an individual's ability to exercise effective command over endowments, which Sen defines as an individual's commodities, wealth, and other productive resources, most notably labor. Sen showed that the cause of the 1943 Bengal famine was not a generalized food shortage but rather the poor's inability to "establish their entitlement over an adequate amount of food." (Sen 1999, 162)

These two approaches merged to form the human development paradigm in the 1980s. The first article summarized in this section, by **Mahbub ul Haq,** articulates the principles underlying the human development approach. In 1990, the UNDP published the first of its annual *Human Development Report,* which have advanced human development theories while creating an alternative set of economic measures designed to illuminate the diverse quality-of-life outcomes produced by per capita incomes (UNDP 1990). While those indices now range into areas as far-flung as "social stress and social change," the most enduring contribution of the *Human Development Report* remains its focus on poverty, inequality, and the outcomes of social and economic development.

Poverty and Human Development

Building directly on Sen's outcomes-based approach, UNDP's premises were clear: "Human development is a process of enlarging people's choices. . . . [A]t all levels of development, the three essential ones are for people to lead a long and healthy life, to acquire knowledge and to have access to resources needed for a decent standard of living." (UNDP 1990, 10) The UNDP's Human Development Index (HDI) was constructed to reflect these basic capabilities. Using weighted averages, it adds life expectancy at birth and two measures of educational access to an adjusted real per-capita GDP measure to generate an HDI value. Countries are then ranked by their HDIs, providing an interesting and useful contrast to GDP per capita as a measure of development.

The HDI as a measure is by no means without its flaws. England (1997), in an earlier volume in this series, notes in particular that it is much more useful in comparing developing countries than developed nations, in part because the index gives very little value to per-capita incomes over the world median (about $5,500)—a reflection of UNDP's emphasis on "sufficiency rather than satiety." (UNDP 1994, 91) Still, the HDI has proven to be a helpful way to identify developing countries and regions in which economic growth has failed to produce expected quality-of-life improvements, as well as those in which quality-of-life improvements have been achieved at levels higher than per-capita incomes would suggest. For example, Costa Rica has an HDI nearly equal to that of

Korea despite per capita incomes barely half Korea's. Paraguay and Morocco both have per capita incomes around $3,500, but Paraguay's HDI is much higher, suggesting that the country is more effectively translating growth into human development (UNDP 1997, 20).

The HDI also highlights progress over time, and it is worth noting that several important quality-of-life measures indicate that important progress has been made in many parts of the world since 1970. In both industrial and developing countries there has been dramatic improvement in life expectancy, under-five mortality rates, and adult literacy rates. Interestingly, gains in all categories are shown even for the group of least developed countries, though the figures also highlight how far they still have to go, with life expectancy at only 50 years, under-five mortality over 17 percent, and adult literacy below 50 percent (UNDP 1998, 19).

Issues of global inequality were examined in a previous volume in this series (Ackerman et al. 2000), which addressed some of the problems involved in measuring poverty. Using one of the more accepted measures of absolute poverty—$1 per day in purchasing power parity—we can look at the recent trends. (See Figure II.1) They show that between 1987 and 1998, a period of significant economic growth, the percentage of people in the world in absolute poverty has decreased overall, while the absolute number has risen. This means

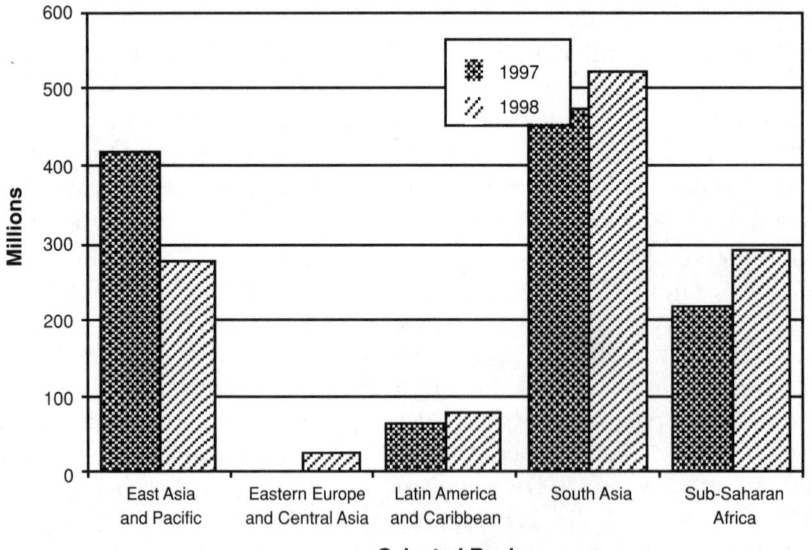

Figure II.1. Number of People Living on Less Than $1 (PPP)/Day 1987 and 1998
Source: The World Bank, "Poverty Trends and Voices of the Poor: Income Poverty—The Latest Global Numbers." http://www.worldbank.org/poverty/data/trends/income.htm.

that some 1.2 billion people, one-fifth of the world's people, live in dire poverty despite strong growth in the global economy. Viewed regionally, we can see that in some areas, such as Latin America, both the percentage and number of people living on less than a dollar a day has grown.

In 1997, UNDP added another wrinkle to its annual report in an attempt to advance beyond income measures of poverty. Their new Human Poverty Index (HPI) was created to recognize that poverty is a state of deprivation and that money income is only a means to provide life's necessities. After all, according to U.N. statistics, 1.3 billion people lack access to safe water, one billion lack adequate shelter, over 800 million are malnourished, a similar number lack access to health services, and 109 million children of primary school age—22 percent of the world's total—are not in school (UNDP 1998, 49).

Instead of looking at income, which can have vastly different impacts depending on the circumstance, UNDP created a composite index for developing countries from three key areas of deprivation for which there is adequate data:

- survival, measured by the percentage of people expected to die before age 40.
- knowledge, measured by adult illiteracy.
- overall economic provisioning, measured by a composite of three variables: the percentage of children under five who are malnourished, the percentage with access to health services, and the percentage with access to safe water.

An interesting feature of the HPI is that, in contrast to the HDI, it recognizes that there are different deprivations relevant to developing and industrial countries. UNDP developed a subsequent HPI-2 for wealthier countries using different indices: survival to age 60, adult illiteracy, income poverty, and long-term unemployment. Given the categories chosen for developing countries, the HPI tends to produce lower poverty rates than income-based measures in countries with relatively developed public sectors and infrastructure. It particularly highlights the importance of education and public health.

For example, Zimbabwe's HPI is much lower than its income-poverty rate—17 percent compared to around 40 percent—due to public services. By contrast, income poverty in Egypt is less than 20 percent but 35 percent are affected by human poverty according to the HPI, reflecting the poor quality of public health and education (UNDP 1997, 21–22). The HPI probably places too great an emphasis on these factors, leading to understatements of poverty in countries with relatively developed public sectors. Indeed, UNDP acknowledges that the HPI should be used in conjunction with income measures of poverty (UNDP 1997, 19). Despite its limitations, the HPI is valuable in reminding us that income poverty and human deprivation do not always move hand in hand.

Refining and Applying Human Development Theory

One of the important areas in which human development theory has been refined is in its treatment of environmental issues. The World Commission on Environment and Development put the issue of poverty and the environment squarely on the policy table, calling poverty "a major cause and effect of global environmental problems." (World Commission on Environment and Development 1987, 3) While this formulation clearly casts the poor as the victims of environmental degradation, it has led to a perception that they are also the perpetrators. As the reports stated, "Those who are poor and hungry will often destroy their immediate environment in order to survive: They will cut down forests; their livestock will overgraze grasslands; they will overuse marginal land; and in growing numbers they will crowd into congested cities." (WCED 1987, 28) This generalized image of the poor as short-term maximizers has led to the unfortunate overgeneralization that poor people cannot at present practice sustainable development. In policy circles, this perspective is compatible with the argument that economic growth, and a rise in per capita incomes, is the only solution to environmental destruction by the poor.

Both the sustainable and human development fields have grappled with this issue in useful and important ways. Robin Broad (1994), using her detailed study of the Philippines, argues that the poor often are a country's leading champions and practitioners of sustainable resource use, as they have a direct interest in preserving the resources on which they depend for their livelihoods. It is important to disaggregate poverty so as to distinguish between the desperate poor and what Sheldon Annis calls the "merely poor." (1992) The latter, he argues, are model resource managers if given secure land tenure and control over natural resources. Broad notes from her Philippines case study that the poor, if well organized to defend their rights, can often become society's strongest advocates for sustainable practices.

Economist Bob Sutcliffe brings the argument back to the issue of overconsumption in the global North, reasserting the necessity for a global redistribution of wealth to achieve both human and sustainable development. "[H]uman development is in danger of being unsustainable unless there is redistribution; and sustainable development is in danger of being anti-human unless it is accompanied by redistribution" (Sutcliffe 1995).

In policy circles, **Robert Chambers** has attempted to put this approach into practice with his "sustainable livelihoods" strategy for rural development. As he explains in the article summarized here, sustainability must begin with an attempt to address basic needs by empowering the poor. Chambers notes that environmental destruction is the result of population growth, migration forced by economic pressure, dispossession of rural livelihoods by "core" interests, and the tendency of businesses, government, and politicians—not the poor—to take a short-term view of resource exploitation. The solution, he argues, is to ensure that the poor have adequate command over resources, rights, and livelihoods. For the rural poor, secure property rights are essential to the stabilization of rural ecosystems.

The sustainable livelihoods approach has been championed by the International Labour Organization (ILO) and other development practitioners as a useful framework for incorporating the employment and livelihood needs of the poor into discussions of sustainable practices. Their detailed case studies have confirmed many of Chambers' hypotheses about the relationship of socioeconomic to environmental sustainability. (See, for example, Ahmed and Doeleman 1995.) These studies particularly stress the importance of addressing livelihood needs with more than simple prescriptions for employment-intensive conservation projects. While such approaches may be beneficial in labor-surplus economies, in other cases labor shortages due to out-migration may make labor-intensive conservation measures impractical.

Social Exclusion, Social Development

The cause of "social development" took center stage at the United Nation's 1995 World Summit for Social Development. This gathering, and the preparatory work that led up to it, produced a great deal of original research, work that is being carried on today by the U.N. Research Institute for Social Development (UNRISD). Among the themes taken up at the conference was that of social integration, one of the theoretical refinements of the human development paradigm. Emerging from European social democratic experiences, the concept of social exclusion has been advanced as a better way to understand the complex interaction between social, economic, cultural, and political systems in meeting the needs of all members of society.

Originally coined in France to refer to members of society excluded from social insurance programs, the term social exclusion is being incorporated into the human development discourse as a way to understand the impact of globalization and economic change on those not clearly benefiting from those processes. In this section we summarize the introduction to an ILO book by **Charles Gore, Jose B. Figueiredo, and Gerry Rodgers** that draws on a wide range of case studies to assess the value of the term in advancing our understanding of rapid economic change. The authors argue that the concept helps us move beyond traditional definitions of poverty, recognizing in particular the rising number of people made permanently superfluous to formal economic activity in the global economy. Gore and his coauthors use their case studies to highlight the limitations in policy prescriptions such as the World Bank's "New Poverty Agenda," which focus too narrowly on labor-intensive growth, health and education investments, and effective safety nets.

Some Southern writers have criticized the social exclusion concept for misidentifying the problem as one of a lack of integration rather than one of marginalization, which they argue is inherent in the international division of labor (Faria 1994). Marshall Wolfe (1995) points out that one of the paradoxes of social integration is that it includes integration into the consumer culture, which is certainly one of the roots of unsustainability. Wolfe repeats a Latin American joke about the population being divided into three groups: those who have credit cards, those who want credit cards, and those who have never heard of credit cards. He notes that globalization's erosion of the size of the third group is a form of integration that undermines efforts toward sustainability. (See Goodwin et al. 1996, the second volume of this series, for a fuller discussion of consumerism.)

Gender in Sustainable Development Theory

The 1995 World Conference on Women in Beijing followed closely on the heels of the World Summit for Social Development, both temporally and conceptually. Held just two months later, the Beijing conference shone a spotlight on the glaring gender inequities that persist to differing degrees in every country, and made efforts to identify, analyze, and address those inequalities.

Within the field of economics, such efforts were not new. There is a long history of attempts to understand gender inequality, touching all branches of economic thought. While it is beyond the scope of this essay to survey these areas,[1] it is important to note some important changes within mainstream development thinking as it relates to gender. In an interesting report on changes in World Bank practices, Myra Buvinic and her coauthors (1996) noted an increased willingness to recognize household inequalities, a stated preference for more participatory lending programs, and an effort to "mainstream" gender issues in Bank programs by replacing stand-alone women's projects with gender components in all aspects of the Bank's work. These changes were part of a renewed focus on global poverty by development institutions, a welcome shift after the 1980s' preoccupation with economic stabilization and growth.

Shahra Razavi, in an article summarized in this section that itself is an overview of a special issue of the journal *Development and Change* on the issue of gender and poverty[2], argues that the relationship between women and poverty has received inadequate study and is fraught with misconceptions. Most research relies on faulty data derived from household poverty surveys, which typically make the assumption that resources are shared equally within the household, inflating the data on women's access to resources. Many studies also attempt to make gender analyses by separating male- from female-headed households. Razavi points out that the latter is a heterogeneous category that includes women who have experienced dramatically different social processes—widowhood, divorce, migration. This leads to evidence that is easy to misinterpret. It also leads to policy errors, which Razavi argues are evident in the World Bank's "New Poverty Agenda" and other development programs.

This critique is echoed throughout the literature on gender and development. Mayoux, for example, suggests the World Bank's focus on "participatory development" may be a way to shift "the costs of development and service provision onto women participants. . . . " (Mayoux 1995, 253) Gita Sen (1999) notes that labor-intensive growth will not significantly reduce female poverty and improve the quality of women's lives if wages remain low and working conditions poor. Palmer-Jones and Jackson (1997) point out that labor-intensive work is also effort-intensive, and there is evidence that increased energy demands on undernourished women can lead to health problems and other decreases in quality of life.

The Concept of Social Capital

The concept of "social capital" has emerged in recent years, and it bears directly on the subject of this section. While lacking an agreed-upon definition, social capital generally refers to the ways in which economic actors interact and organize themselves, magnifying the production resulting from the combination of the three more widely accepted forms of capital: physical, natural, and human. Robert Putnam (1994) is probably the term's foremost popularizer, initially through his studies of development outcomes in different parts of Italy. He broadens the concept to encompass civic associations and other forms of trust-building interaction, and to include political as well as economic impacts.

The concept of social capital seems to be gaining widespread usage, although there is little agreement on describing the phenomenon as a form of capital. The World Bank prominently features social capital in its more recent publications, declaring social capital "the missing link" in development (World Bank 1997a). The Bank's argument in favor of the concept is that, like human capital, social capital is both an input and an output of the development process, both a consumption good and an investment. And, like technology, social capital is more than an input to production; it shifts the entire production function by increasing the productivity of all other inputs. The accumulation of social capital—the levels of trust, cooperation, and institutional coherence in society—increase economic output by decreasing transaction costs.

Social capital, as a concept, has relevance to development discussions precisely because organizations like the World Bank have incorporated it into their theories and strategies and because the term has been widely extended to encompass the effectiveness of state institutions. Proponents argue that high levels of civic involvement—as measured by a seemingly dizzying array of indicators, from church attendance to newspaper readership—are more likely to produce effective government institutions, which are generally seen as likely to be more democratic.

The critiques of social capital theory are widespread. One of the purported advantages of the concept is that it brings together politics and economics in a way that allows new approaches to old themes in political economy. Yet Putzel (1997) and others argue that many applications of the term are curiously devoid of political content, positing but not demonstrating that strong bonds of trust produce desirable democratic outcomes, or any political outcomes at all. More specifically, they note that concepts like trust and cooperation do not reflect the context of stratified societies, where conflict among groups with different interests, within the marketplace and without, are the norm.

Fine (1999) takes the critique further, presenting a broad overview of the literature and concluding that social capital is so ill-defined that it has come to represent in economics anything not reducible to individual exchange relations, with "capital" representing anything other than tangible assets. He considers the issue important because he views the World Bank's appropriation of the term a way to sidestep the failures of its free-market, anti-state development bias. In Fine's view, social capital is a convenient way to overlook societal conflict over economic policy and defuse demands for activist state intervention.

We include in this section summaries of two articles that touch directly on the issue of state intervention, **Peter Evans** from within the social capital discourse, **Judith Tendler** from outside. Both share a common assessment that it is a mistake to view effective government as simply the product of civic involvement. Rather, their studies suggest that government involvement, particularly at the local level, can be a catalyst in promoting civic involvement, which in turn can lead to more effective government. Evans also concludes, based on a series of case studies, that both good government and civic involvement can develop in areas that would be considered to have a relatively low stock of social capital in society. Tendler examines local state institutions to discover ways in which their interconnections with the communities in which they work enhance good governance. She concludes that responsive and effective government institutions are more attainable than we might think, but only if we abandon development strategies that undermine government institutions.

Democracy, Participation, and Empowerment

This brings us to a final social dimension of sustainability: political rights and power. It has become common in development circles to recognize that strategies and projects are unlikely to succeed if they do not involve the willing and informed participation of the intended beneficiaries in both design and implementation. This argument is often extended to an appreciation for the importance of democracy in promoting sustainable development. This widespread acceptance of previously contentious assertions derives from two phenomena, as Robin Sharp (1992) points out. First, there has been a wholesale discrediting of top-down development schemes both by those within the development community and those outside. Second, the demise of many repressive regimes by mobilized citizens has produced a broader appreciation for democracy. Sharp notes that both trends have opened the door to meaningful participation by members of society in their communities and governments.

There has been a good deal of research on the relative benefits of different forms of government for economic growth and sustainable development, much of it contradictory. Lal (1996) provides a good overview of such research, in the process showing that there is little clear correlation between democratic forms of government and economic growth. Such studies are often limited by their focus on forms of government rather than the content of citizen involvement in policies and projects. Singh and Titi (1995) advance and refine the concept of empowerment as an approach that focuses on both the content and process of development, with the poor as both subjects and objects. As Sharp points out, it is important to emphasize not just the form but also the substance of progress toward a pluralist society. If participation is to lead to real empowerment, the key role of government is to guarantee civil and political rights.

It is fitting to conclude by citing Amartya Sen, who fully integrates civil and political rights into his broad-based theories of "development as freedom." He argues that the role of development is to enlarge and enhance the choices—freedoms—available to all. He identifies five linked types of freedom: political freedom, economic facilities, social opportunities, transparency guarantees, and protective security (Sen 1999, 10). Sen sees these freedoms as integrally linked. Perhaps most significant, he sees them in their totality: "Freedoms are not only the primary ends of development, they are also among its principal means." (Sen 1999, 10) Expanding the political, economic, and social choices available to all may be the best route to achieving sustainable human and economic development.

Notes

1. For a useful overview of the state of gender inequality from a human development perspective, see the 1995 *Human Development Report* (UNDP 1995). For a good analysis of the evolution of gender considerations in environment and development theory, see Braidotti et al. (1994) and Kabeer (1994).
2. See special issue of *Development and Change,* 30, 3 (July 1999) on "Gendered Poverty and Well-being."

Summary of

The Human Development Paradigm

by Mahbub ul Haq

[Published in *Reflections on Human Development* (New York: Oxford University Press, 1995), Ch. 2, 13–23.]

Within development theory, the "human development paradigm" claims the broadest vision of a people-centered development process in which economic growth serves to enhance the well-being of the majority. Since 1990, the United Nations Development Programme's (UNDP's) annual *Human Development Report* has developed this vision, elaborated by a team of economists and social scientists headed by the author of this selection, one of the chief architects of the new paradigm.

From Growth to Human Development

Human development is not a new conceptual discovery. The idea that social arrangements, including economic organization, should be judged by the extent to which they promote human good dates back to Aristotle and continues through Immanuel Kant, Adam Smith, Robert Malthus, Karl Marx, and John Stuart Mill. The belated rediscovery of human development has taught us that "the basic purpose of development is to enlarge people's choices." [14]

The human development school distinguishes itself from the economic growth school in that the latter focuses on expanding only one choice—income—while the former seeks to enlarge all human choices—social, economic, cultural, political. It is sometimes argued that expanding income expands these other areas of choice, but that is not the case for a number of reasons.

First, income may not be evenly distributed, limiting choices to those in poverty. Wealth often does not trickle down. More fundamentally, how a society's income is used—the national priorities chosen by a society or its rulers—is just as important as how it is generated. In reality, "there is no automatic link between income and human lives . . . yet there has long been an apparent presumption in economic thought that such an automatic link exists." [14] In addition, wealth may not be necessary at all to fulfill many kinds of human needs, such as democracy, gender equity, and social and cultural support systems.

"The use that people make of their wealth, not the wealth itself, is decisive. And unless societies recognize that their real wealth is their people, an excessive obsession with creating material wealth can obscure the goal of enriching human lives." [15]

The human development paradigm goes beyond the quantity of economic growth to look at the quality and distribution of such growth, recognizing that only public policy can ensure that economic activity produces the desired societal results. We must not reject growth; it is essential to alleviating poverty in poor societies. But we must go beyond growth. This leads us to question the existing structure of power in society, with policies that may vary from one country to the next but that share common threads:

- People move to center stage, with development understood first in terms of its betterment of people's lives.
- Human capabilities are increased through improved health, knowledge and skills, and people have equitable access to opportunity.
- Economic growth is seen not as the end goal of development but as the means to improve lives.
- Political, social, and cultural factors get as much attention as economic factors.
- People are seen as both the means and the ends of development and not regarded narrowly as "human capital" to produce commodities.

Essential Components of Human Development

The four essential elements of human development are equity, sustainability, productivity, and empowerment.

Equity is needed so development does not restrict the choices of many in society. What is important is equity in opportunity, not necessarily in results. But access to political and economic opportunities must be seen as a basic human right. This can involve fundamental restructuring of power. Productive assets such as land may need to be redistributed, as with a land reform. Fiscal policies may be required to achieve greater income equity. Credit systems may need to be reformed to equalize access to credit for those without formal wealth. Political systems may need democratizing to minimize the excessive control of the wealthy. The rights of women, minorities, or other traditionally excluded group must be guaranteed.

Sustainability involves ensuring that human opportunities endure over generations. This means not just sustaining natural capital but physical, human, and financial as well. This should not require preserving every natural resource in its current form. That would be environmental puritanism. We must preserve the capacity to produce human well-being. And we must not preserve present levels of poverty, which are unsustainable in the long run. Indeed, the wide disparities in lifestyles, with a minority leading high-consumption lives, must cease with a redistribution of income and resources from the rich nations to the poorer ones.

Productivity is where economic growth fits in the human development paradigm. This requires investment, both in physical capital and in human capital. It also requires the maintenance of a macroeconomic environment conducive to fulfilling human needs. Raising people's productivity through education and training is an important and productive investment for society, but it should not be seen as simply a means to achieve growth.

Empowerment includes political democracy, freedom from excessive economic controls and regulations, decentralization of power so people can participate meaningfully, and the involvement of all members of civil society—particularly nongovernmental organizations—in making and implementing decisions. Empowerment takes the human development paradigm beyond the human needs approach by incorporating political, social, and cultural rights.

A Holistic Concept

The human development paradigm is therefore a holistic approach to development that incorporates economic growth as one, but only one, feature. Some people mistakenly assert that human development is anti-growth and concerned only with social development. Economic growth is essential for human development, but it must be properly managed. There are four key ways to create the desirable links between economic growth and human development:

- Invest in the education, health, and skills of the people, an approach adopted by many countries, including China, Hong Kong, Japan, Malaysia, Singapore, Thailand.
- Promote the equitable distribution of income and assets, which can produce human development when initial conditions are favorable and where growth is strong (as in China), where they are unfavorable but correctable through public policy and high growth (as in Malaysia), or where they are unfavorable with low growth, in which case public policies can meet basic needs but cannot sustain them (as in Jamaica).
- Structure social expenditures to promote human development even in the absence of strong growth or good distribution. These cases—Cuba, Jamaica, Sri Lanka, Zimbabwe—are generally not sustainable unless the economic base eventually expands.
- Empower people, especially women. This is the best way to ensure that growth will be strong, democratic, participatory, and durable.
- "It is fair to say that the human development paradigm is the most holistic development model that exists today. It embraces every development issue, including economic growth, social investment, people's empowerment, provision of basic needs and social safety nets, political and cultural

freedoms and all other aspects of people's lives. It is neither narrowly technocratic nor overly philosophical. It is a practical reflection of life itself." [23]

Summary of

Sustainable Livelihoods: The Poor's Reconciliation of Environment and Development

by Robert Chambers

[Published in *Real-life Economics: Understanding Wealth-Creation*, ed. Paul Ekins and Manfred Max-Neef (New York: Routledge, 1992), Ch. 7, 214–229.]

One of the practical applications of the human development paradigm is the sustainable livelihoods approach, which proposes a reversal of traditional thinking, arguing that one must start with the needs of the poor if one wants to address problems of environmental destruction.

Starting with the Poor

> [T]he thinking and strategies advocated and adopted with regard to problems of population, resources, environment and development (PRED) have largely perpetuated conventional top-down, centre-outwards thinking, and have largely failed to appreciate how much sustainability depends upon reversals, upon starting with the poorer and enabling them to put their priorities first. [214]

Three main processes stand out in defining the interrelationships among those four areas:

- *Rapid population growth* is common in the South, and it is often the most rapid in fragile rural areas;
- *"Core" invasions and pressures* into Southern rural areas by Northern and/or urban institutions both generate and destroy livelihoods, but for many of the rural poor livelihoods are made less secure.
- *Responses of the rural poor* to population growth and core pressures can involve the unsustainable exploitation of local resources and eventually the migration of significant populations to other areas where livelihoods often remain insecure.

These processes are linked and are not sustainable. The policy goal, then, is to restrain these pressures to enable much larger numbers of rural people to gain secure and sustainable livelihoods.

To create sustainable livelihood security we must overcome "first thinking"—the approach many Northern development professionals take to such problems. "To caricature, the top-down view of 'the rural poor' sees them as an undifferentiated mass of people who live hand-to-mouth and who cannot and will not take anything but a short-term view in resource use. In consequence, it is held, their activities must be regulated and controlled in order to preserve the environment." [216] Though some professionals now recognize that the rural poor are behaving rationally—for example, having large families as a form of old-age security—the poor are still rarely the starting point.

There are overwhelming ethical reasons to put the poor first, but there are also compelling practical reasons. First we must understand what poor people want. While this varies from person to person and from place to place, basic to most is an adequate, secure, and decent livelihood. This includes "security against sickness, against early death, and against becoming poorer, and thus secure command over assets as well as income, and good chances for survival." [217] Poor people want to be able to take the long view.

Putting what they want first integrates PRED by focusing on four goals:

- *Stabilizing population*—It is rational for people who lack secure access to resources and income and who expect some of their children to die to have large families. They are spreading risks and diversifying sources of food and cash, while planning for old-age security. Good health and decent livelihoods are critical to reducing population growth. "[I]n conditions where livelihoods are adequate, secure and sustainable, assets can be passed on to children, children are likely to survive and the benefits of child labor are limited, parents have less reason to want large families." [217]
- *Reducing distress migration*—Poor people are forced to migrate, competing for resources, services, and work in urban areas or increasing pressures on other fragile lands and/or forests. Secure access to resources and livelihoods reduces the pressure to migrate.
- *Fending off core exploitation*—To resist core pressures that would otherwise dispossess them, the poor must be legally, politically, and physically strong, with secure rights to resources. With that security, they are better able to manage and survive.
- *Taking the long view*—Conflicting with conservationist rhetoric, it is core interests—businesses, governments, politicians—that take a short-term view of resource exploitation. "In contrast, poor people with secure ownership of land, trees, livestock and other resources, where confident that they can retain the benefits of good husbandry and pass them on to their children, can be, and often are, tenacious in their retention of assets and far-sighted in their investments." [218] It is the desperate poor, not those who are poor but not desperate, who are more liable to overexploit resources: "Who will plant a tree or invest labor in works of soil conservation

who fears the tree will be stolen, or the land appropriated, or the household itself driven away?" [219]

- "The implication of these four points is that poor people are not the problem but the solution. If conditions are right they can be predisposed to want smaller families, to stay where they are, to resist and repulse short-term exploitation from the cores, and to take a long view in their husbandry of resources. The predisposing conditions for this are that they command resources, rights and livelihoods which are adequate, sustainable and above all secure." [219]

Going Beyond Outmoded Thinking

To achieve such a reversal in traditional approaches to the problems of the rural poor, we need to go beyond "first" thinking. Two expressions of this are environment thinking and development thinking. Environmental thinking takes the long view, values the future more than the present, and emphasizes the negative environmental effects of "development" and of poor people's livelihoods. Development thinking takes the short-to-medium view, discounts future benefits of present actions, sees progress as an increase in production, and views livelihoods simply as labor. In each case, professionals are the critical actors, environmentalists in the first and economists and development professionals in the second.

The necessary step is to move to "livelihood thinking" where the poor are the critical actors and the starting point, and the priority is meeting both their basic short-term needs and their long-term security. Sustainable Livelihood Thinking (SLT) integrates these by focusing on "enabling very poor people to overcome conditions which force them to take the very short view and 'live from hand to mouth.' . . . It seeks to enable them to get above, not a poverty line defined in terms of income or consumption, but a sustainable livelihood line defined to include abilities to save and accumulate, to adapt to changes, to meet contingencies, and to enhance long-term productivity. . . . This will stabilize use of the environment, enhance productivity and establish a dynamic equilibrium . . . of population and resources." [221] This is not an add-on to existing approaches but an alternative.

This immediately leads to a search for potentials and opportunities to help more people gain adequate, secure, and sustainable livelihoods, biologically, economically, and in terms of social organization. The potential is as immense as it has been unrecognized. Bio-economic potential is often tremendous, because changes in land management on degraded land can often unleash remarkable potential for both production and livelihoods. In parts of India, for example, poor people could be growing perennial trees on degraded land, stabilizing the environment and increasing production of many forest products tenfold, in the process supporting more people in a sustainable way.

Practical Implications

This approach generates an agenda for research in which five areas stand out:

- *The nature of secure and sustainable livelihoods*—The traditional view of deprivation has an urban concern with employment and income that does not fit rural conditions: most poor rural people have multiple sources of food as well as cash income. It also neglects vulnerability and the importance of security. SLT shifts attention to assets as well as flows of food and income.
- *Sustainable livelihood-intensity*—This concept can serve as a project criterion in which a development project will be appraised on the basis of the number of people it moves above the sustainable livelihood line.
- *Policies for sustainable livelihood security*—There is a long list of policies that can promote sustainable livelihood security, including peace, equitable and secure rights and access to resources, access to basic services, safety nets, land reform, and many more.
- *Support for the new professionalism*—The challenge here is to change curricula, training methods, the selection process for technical personnel, and much else, to develop a new generation of professionals trained to think and act in a way that breaks with "first" thinking.
- *Appraisal, research, and development by the poor*—The poor, those affected by such policies and projects, must be involved in the process of analyzing problems and designing and implementing solutions.

This provides common ground for professionals and the poor. "For it is precisely secure rights, ownership and access, and people's own appraisal, analysis and creativity, which can integrate what poor people want and need with what those concerned with population, resources, environment and rural development seek." [229]

Summary of

Markets, Citizenship, and Social Exclusion

by Charles Gore,
with contributions from Jose B. Figueiredo and Gerry Rodgers

[Published in *Social Exclusion: Rhetoric, Reality, Responses* (Geneva: International Labor Organization, 1995), Ch. 1, 1–40.]

The term *social exclusion,* which originally was coined in France to refer to those excluded from social insurance programs, is now used as an important analytical concept in understanding the interconnected social, cultural, political, and economic processes that can lead to poverty. In an era of rapid globalization, the term takes on added importance as we seek to understand the impact of rapid economic change on the most vulnerable members of society, particularly in developing and transitional economies. In this introduction to a book prepared by the UNDP and the ILO for the 1995 World Summit for Social Development, the author draws on the book's wide range of case studies—on Mexico, Peru, Cameroon, Tanzania, Tunisia, Yemen, India, Thailand, Russia, and Kazakhstan—to assess the value and policy implications of the concept of social exclusion in broadening and deepening our understanding of the impact of globalization on the development process.

Social Exclusion in a Global Context

Although the concept of social exclusion emerged within a Eurocentric context, it has value in analyzing developing countries as well. The term goes beyond traditional income-based measures of poverty to assess the various processes that contribute to deprivation. In the European context, social exclusion has come to refer to the society's failure to grant meaningful rights and entitlements to all residents. As such, it incorporates considerations of access to legal, political, cultural, and social rights into assessments of access to economic resources. At a time when globalization and neoliberal economic programs are adding to structural poverty, increasing the informalization of labor markets, and reducing social benefits, social exclusion offers a framework for developing alternatives to the European welfare state.

Some have argued that the social exclusion approach adds little more than a European framework to concepts better developed in Latin America, which have focused on structural marginalization. In this view, the problem is not a lack of integration, as the social exclusion concept would suggest, but the marginalization of large groups in society, a process inherent in the social division of labor and grounded in the peripheral integration of developing economies into the world capitalist economy.

The advantages of the social exclusion approach, highlighted in many of the case studies, are descriptive, analytical, and normative. It has descriptive value in that it defines poverty as a state of relative deprivation, going beyond the simple income-based definitions. Analytically, it has value in highlighting the interrelationships between poverty, productive employment, and social integration, all of which are important in understanding the processes associated with globalization. In particular, it helps address the rise in the number of those made permanently superfluous in the global economy; the problems created by increasingly blocked international migration; the backlash by some social groups trying to curb competition; and technologically driven skill polarization and the dualization of labor markets.

To give the concept of social exclusion global relevance, four modifications are worth considering. First, we need to incorporate international relations—including trade, aid, migration, and so forth—and the nature and design of the international regimes that underpin them. Second, where the European focus has been on exclusion from labor markets, in the global context other factor markets are important as well, particularly access to land, credit, and other bases of livelihood. Third, while European analysis has focused on social rights, political and civil rights assume greater importance in societies without developed democratic state institutions or significant welfare provisions. Finally, it is important to go beyond national analyses of exclusion, particularly in societies in which national institutions and rights are not fully developed and in which globalization is creating post-national societies.

One example of the value of the social exclusion concept from the case studies is Partha Dasgupta's theory of labor market exclusion among landless peoples. Dasgupta argues that what those without assets own is not labor power but potential labor power, which they can only convert to actual labor power if they are in adequate health to work. Those with even a small amount of land can meet some of their minimum nutritional requirements from that asset, while the landless must depend entirely on their earnings, which they are less likely to get as they get weaker with declining food intake. Nutritional and health issues have also been shown to be important in situations of surplus labor where day-laborers are selected on the basis of their physical strength and productivity. Illiteracy has also been found to be an important cause of labor market exclusion.

The studies also confirmed the importance of civil and political rights. For example, the Peruvian case showed that 37 percent of peasants lacked legal title to their land in 1984, while 43 percent of shantytown residents in 1991 had no legal title to their urban plots. In addition, "a common, though not surprising, finding of some of the studies is that programs of structural adjustment have undermined the capacity of states to provide health, education and social services." [23] The studies also found vulnerability among indigenous peoples due in part to their unrecognized property rights.

Some Conclusions

In analyzing situations of great inequality, the studies suggest an interplay between four underlying determinants:

- The transnationalization of social and economic life, which is resulting in increasing impacts of developed country policies on less industrialized countries.

- "The changing availability and distribution of assets in situations of increasing scarcity associated with population growth, radical economic transformation and, in recent years, widespread recession and even, in some regions, a disturbing trend of economic decline." [29] These assets are economic (land, finances, skills), political (basic rights), and cultural (social values).
- The social and political structures through which power is exercised and the status and balance of power of societal groups is determined.
- "The nature of the development regime adopted by the national government, including the relative role of state and markets as allocation and accumulation mechanisms; the policies for growth, poverty reduction and structural transformation; and the short- and medium-run programs aimed at economic adjustment and stabilization." [30]

Among the many dimensions of exclusion considered in the case studies, the most important exclusion mechanisms were the organization of markets; the functioning (or nonfunctioning) of governmental institutions; and the presence of discrimination based on gender, caste, ethnicity, and race. The studies also confirmed the particular vulnerabilities of certain groups to exclusion. Children and young people are vulnerable because pressures for early entry into labor markets can undermine future possibilities by limiting educational attainment. Migrants increasingly suffer exclusion, particularly international migrants who face exclusion on the basis of their lack of citizenship rights.

The social exclusion approach has implications for policy, and it highlights some of the limitations of the World Bank's approach to poverty in the 1990s, which was based on labor-intensive growth, increased health and education services, and effective safety nets. Based on the case studies, one can identify several alternative policies:

- Redistributing productive assets (e.g., land) and expanding employment opportunities, both of which receive too little attention.
- Focusing on improving and developing social institutions, including markets, government institutions, and civil society organizations, all of which can be contributors to social exclusion or solutions to it.
- Ensuring civil and political rights, which are critical to securing and sustaining livelihoods.
- Replacing welfare-style targeting of casualties in economic restructuring with a more active attempt at economic integration.

Overall, the studies suggest that the social exclusion concept can contribute a great deal to our understanding of poverty in developing countries. One important research area will be a macro-level analysis of the relationships between capital accumulation, productivity improvements, and social integration.

Summary of

Gendered Poverty and Well-being: Introduction

by Shahra Razavi

[Published in *Development and Change* 30, 3 (July 1999), 409–433.]

Since the early 1990s, mainstream policy institutions have shifted from their preoccupation with stabilization and growth to a renewed focus on global poverty. Multilateral development agencies now follow their "New Poverty Agenda," with its focus on "labor-intensive growth." Meanwhile, discussions of "social exclusion" have been incorporated into analyses of poverty and social disadvantage. While much attention has been paid to women and poverty, the relationship between the two has not received careful study. This article, which is the introduction to a special issue of *Development and Change,* examines some of the shortcomings in the prevailing approaches, including that of the "New Poverty Agenda."

Making Gender Visible: Some Methodological Traps

The relationship between gender disadvantage and poverty at first glance appears clear— for example, in equating female-headed households with poverty. Another common theme is the "win-win"' scenario depicting women's education as a strategy for both reducing fertility and improving welfare. More detailed analyses highlight female disadvantage by disaggregating well-being outcomes. Unfortunately, such "generalizations have tended to replace contextualized social analyses of how poverty is created and reproduced. The gender analysis of poverty also needs to unravel *how* gender differentiates the social processes leading to poverty." [410]

One prevailing methodological bias is the tendency to rely on income and consumption as the best single proxy for poverty, with other aspects of poverty—public goods and services, access to clean air, democracy—reflected in such calculations through the use of "shadow prices," which assign a monetary value to such nonmonetary goods. These are then reflected, for example, in the World Bank's Poverty Assessments (PAs), with the setting of a poverty line. Such data rely too heavily on household surveys, which tend to be narrow, unreliable, and noncomparable. They are also often one-time surveys, making them unsuited to the ongoing monitoring of poverty. Nor is there any consistency in how poverty lines are set. All of this defeats the purpose of collecting quantitative data.

This has profound implications for gender analysis, because household surveys typically ignore the intra-household distribution of income. Most make the assumption that resources are shared equally among all members of the household, inflating women's access to resources. This over-reliance on household data also leads to skewed gender analyses that attempt to make gender visible by separating female- from male-headed households. This presents both methodological and empirical problems. The label "female-headed households" is a heterogeneous category. "By aggregating these distinct categories of households generated through different social processes (e.g. migration, widowhood, divorce), and constructing a simple dualism between male-headed and female-headed households, it becomes impossible to interpret the evidence in a meaningful way." [413]

One constructive alternative to this "money-metric" approach is the analysis of "capabilities and functionings" first developed by economist Amartya Sen. Because this approach measures access to and utilization of resources at an individual level, it is better suited to neoclassical microeconomic analysis. It also allows more useful gender disaggregation of data, which has led to important work documenting female disadvantage in quality of life outcomes.

Outcomes-focused approaches are not free of methodological problems, however. They can lead to problematic comparisons between men and women due to the basic differences in form and function between men's and women's bodies. Many diseases, for example, are gender-specific, as are, of course, reproductive health problems. Nutrition monitoring presents similar problems, as nutritional status can only be assessed and compared once norms and cut-off points are adjusted for gender difference—not an easy process. Social indicators are limited, too, in their ability to generate meaningful causal analysis.

In global comparisons, such outcomes analyses are plagued by data problems. Few developing countries have reliable vital registration systems from which to derive demographic data. Even where such data exists, international agencies often use flawed mathematical models to project current figures. For an indicator as straightforward as literacy, data is old or unreliable; for 19 of 145 countries in the world the most recent adult literacy data is from 1970 or earlier, while in 41 others the figure is from one year in the 1970s.

"Ironically, such over-reliance on simple econometric techniques also marks some of the emerging micro-level feminist research, which uses interview techniques to capture different aspects of female autonomy. . . . [H]ere too the results can be uncontextualized, single-stranded, and difficult to interpret, with a heavy reliance on simple correlations and regressions using a few variables." [417]

It is important to go beyond basic well-being outcomes so as to avoid reducing the issue of female disadvantage to poverty. Prosperity can reduce gender inequalities in such outcomes but still reduce women's autonomy. Basic-needs measures also tend to look at extreme forms of disadvantage such as child mortality, which can obscure other important areas such as women's heavier workloads. Three conclusions can be drawn from an analysis of these issues. First, it is difficult to assess issues of agency and informal power through interview techniques. Second, the extent to which women are constrained by social structures or are able to subvert such structures depends on the issue and context. Third, issues of bodily well-being need to be included within gender analyses of choice and agency.

Other methodological issues emerge as well. The increasing use of iterative processes of participatory rural appraisal (PRA) and participatory poverty assessment (PPA), both of which purport to empower the poor to appraise their own situations and take concrete actions, represent an advance in the use of qualitative analysis in institutions like the World Bank. But such approaches generally rely on data collected in highly public events at which people are under intense community scrutiny. There are also questions about the extent to which subjective perceptions can reveal all we need to know about individual welfare. For example, many PPAs include no analyses of the macroeconomic causes of poverty, an interpretive step often downplayed in favor of simple data collection. Overall, if poverty data is to yield meaningful results, researchers will need to cross-check and "triangulate" their estimates from a variety of sources and methodologies.

Institutions and Entitlements

"The question that has not been explicitly addressed, but which is clearly central to our discussion, is: How can the focus of poverty analyses be sharpened and shifted so that the social, economic, cultural and political processes and the institutions that are implicated in the creation and perpetuation of poverty, become more lucid and central to the enquiry?" [424] The "entitlements" framework developed by Dreze and Sen (1989) made an important contribution to poverty analysis, with entitlements now seen to include not just legal rules defining rights but also socially enforced moral rules, which have an important impact on an individual's access to commodities. Gendered rules of entitlement exist primarily in the household, but also go beyond it.

This makes for a much more complex analysis. For example, an examination of the evidence on the rising incidence of daughter disfavor in parts of India and East Asia during periods of rapid fertility decline suggests a contested set of causal explanations. Similarly, women's lack of secure land rights emerges as an important cause of poverty only in certain contexts, and not, for example, where customary rights give women meaningful usufruct rights to land.

The "New Poverty Agenda"

Many of the problems associated with gender and poverty analyses are evident in the prescriptions of the "New Poverty Agenda," first articulated by the World Bank in its *1990 World Development Report.* Using poverty assessments, this focus has led to policy prescriptions focused on labor-intensive growth, with a three-pronged emphasis on export-led agricultural growth, education, and safety nets.

There are a number of critiques of this approach. First, social and economic relations—the processes that cause and perpetuate poverty—are largely absent. Second, the Bank's orthodox adherence to its three-pronged approach blinds it to the circumstances in any particular country, rendering policies ineffective. Finally, gender issues receive different weight in each of the prongs. Women's issues emerge strongly in safety-net analysis in which women are a disaggregated group. They are often the centerpieces of education analyses. But one rarely finds a gendered analysis of agricultural growth strategies. This reflects the role of micro-economists in the policy arena, where prescriptions call for a reduction in women's "reproductive burden" in order to enable them to more easily switch their labor to tradable goods, the kind of economic activity encouraged by Bank structural adjustment programs.

There are two problems with this approach. First is the assertion that the only reasonable justification for reducing women's work burdens is to enable them to work more, part of the New Poverty Agenda's "policy obsession with extracting work from the poor." [430] Second, the wider policy framework is assumed to be benign, leading to consistent failures to analyze the poverty and gender impacts of, for example, the food-security risks—at both the household and national levels—of increased reliance on export-crop production.

"There are serious doubts about whether agriculture can generate the route out of poverty and destitution in the absence of substantial and sustained developmental support from the government, which in the current policy climate seems difficult to achieve." [431] "The fact that the agenda remains wedded to an abstract theory of labor markets means that it cannot explain the dynamics of female employment. . . . Nor can it explain how labor market arrangements themselves can perpetuate poverty and discrimination." [431]

Ultimately, these analyses demonstrate that one cannot collapse gender analyses of poverty into a welfare agenda, nor can one ignore that gender analyses quickly "move beyond poverty issues into the wider domain of power and subordination. . . . " [432]

Summary of

Government Action, Social Capital, and Development: Reviewing the Evidence on Synergy

by Peter Evans

[Published in *World Development,* 24, 6 (June 1996), 1119–1132.]

Theories of development have tended to downplay or ignore the importance of cooperative, or synergistic, relations between the state and society. At the local, regional, and national levels, such relations are often critical to successful development strategies. Moreover, there is evidence that such relations can be constructed even in societies in which the stock of "social capital" is considered relatively small. This article, drawing on others in a special section on social capital and government in the journal *World Development,* argues that such synergy can be a powerful force for development.

The Structure of Synergy

"'State-society synergy' can be a catalyst for development. Norms of cooperation and networks of civic engagement among ordinary citizens can be promoted by public agencies and used for developmental ends." [1119] Before exploring the social and political conditions that allow such synergy to develop, we need to understand the structure of such relations.

Productive relations between governments and citizens' groups can take many forms. It is important to distinguish between the concepts of complementarity and embeddedness. The former refers to the conventional ways in which a mutually supportive division of labor can exist between the public and private sectors. Where complementarity is strong, the developmental output will be greater than the simple sum of the respective inputs from government and private groups. At its most basic level, this must take the form of the state providing and enforcing the rules and laws that strengthen private organizations and institutions. While traditionally understood to be important in fostering entrepreneurial behavior from economic elites, the rule of law is also important to complement the actions of less privileged groups.

Complementarity also takes the form of the public provision of intangibles such as knowledge through agricultural extension work, which enhances the effectiveness of farmers and peasants. Such intangibles can also include something like broad-based publicity, which can promote the formation of social capital to support government programs. Complementarity can also take a more tangible form, such as government-funded irrigation systems that raise private-sector productivity. Recent research has demonstrated the extent to which such programs, if administered properly, can not only add value to the provision of goods but also increase the stock of social capital by increasing farmers' willingness to work cooperatively.

While the concept of complementarity forces no rethinking of the roles of the public and private sectors, the idea that effective synergy depends on embeddedness challenges mainstream precepts. Embeddedness refers to the "ties that connect citizens and public officials across the public-private divide." [1120] Irrigation is again a useful example. Comparing irrigation programs in Nepal and Taiwan, the latter was found to be much more effective and efficient, in part because of the dense web of ties binding together public officials and farmers at the local level. In contrast to Nepal, where the central government provided the inputs and the local farmers took responsibility for the day-to-day operation of the irrigation systems, in Taiwan the local irrigation officials came from the community, had knowledge of local conditions, and felt significant community pressure to be responsive to the community. "There is a division of labor but it is among a set of tightly connected individuals who work closely together to achieve a common set of goals." [1121]

Embeddedness also refers to government involvement in ventures commonly understood to be the realm of civil society. In Ho Chi Minh City, Vietnam, a successful revolving-credit association was organized by an agency of the city government, with financial support from an international aid organization. The city provided training and technical support, but local village leaders, in and out of government, provided the organizational skills and energy, building on existing kin and friendship ties. This type of synergy "helped transform traditional ties into developmentally effective social capital." [1122] Pervasive examples of embeddedness abound, too, in cases of successful relations between the state and economic elites. The well-known examples of East Asian development highlight the importance of such ties.

Complementarity and embeddedness are not competing conceptions; they are themselves complementary. "(T)he best way to understand synergy is as a set of public/private relations built around the integration of complementarity and embeddedness." [1124]

The Construction of Synergy

Synergy depends to a great extent on existing sociocultural conditions, but it can also be built, even in societies thought to lack such endowments. Obvious examples of endowments that enhance synergy include the stock of social capital within civil society, the effectiveness of government institutions, and even a low degree of inequality within a given society. The absence of such endowments certainly constrains the development of synergistic relations, but it does not preclude it.

Based on the research presented in the prior articles in this issue, it is clear that social capital is crucial to synergy, but the relatively low levels of social capital in many Third World settings are not the key constraining factor in constructing productive state-society interactions. "The limits seem to be set less by the initial density of trust and ties at the micro level and more by the difficulties involved in 'scaling up' micro-level social capital to generate solidarity and social action on a scale that is politically and economically efficacious." [1124] Taiwanese farmers did not start with an exceptional level of solidarity, nor did the Vietnamese communities involved in the revolving-credit program referred to earlier. In a study of rural Mexico, strong community organizations were found to be important but developmentally effective only when they achieved regional scope. Interestingly, this often happened in collaboration with local reformists within national government programs.

Without micro-level social capital, there is little to build on. But such community-level ties are potentially available in many Third World communities. Transforming that social capital into synergy for development requires a competent and responsive set of public institutions. The limits to synergy may rest in government rather than civil society. Studies of East Asian industrialization have highlighted the importance of strong pro-development bureaucracies with close ties to industrial leaders. Similarly, China's relative success in its current economic transition owes partly to the continuous coherence of the government and the resulting ability to restructure local incentive structures to encourage entrepreneurship. This stands in stark contrast to the case of Russia.

One cannot analyze synergy, of course, without addressing issues of politics and interests, which many discussions of social capital fail to discuss. Such discussions often assume the existence of shared interests among a relatively homogenous group of people, who have only to develop trust to achieve collective action. While the idea of synergy presented here takes this a step further to assert that government officials may share the interests of their constituents, social capital is more usefully understood to exist in societies of competing and conflicting interests. How such conflicts are addressed, through political competition or repression, is critical to the emergence of transformative social capital. Political competitiveness seems to have a positive impact on the development of synergy. At the most basic level, it creates an environment in which citizens count. Even in situations of one-party rule at the national level, competing local interests can help generate pressure on local governmental institutions to be responsive to community needs. For such pressure to translate into meaningful action, however, the government must have the administrative capacity and cohesion to produce results.

The nature of the social conflicts underlying political competition is also important. Where inequality is high and conflicts are sharp, it is much more difficult to achieve synergy. "From Taiwan to Kerala, relatively egalitarian social structures are as much of an advantage for synergy as is political competitiveness." [1128] Building synergy for rural development in Taiwan, starting with an agricultural sector dominated by small family-owned farms in a country with one of the lowest Gini indexes in the Third World, is much easier than it is in rural Mexico, where a strong landlord class presides over an excluded peasantry.

Unfortunately, the latter is more typical of Third World societies. "If egalitarian societies with robust public bureaucracies provide the most fertile ground for synergistic state-society relations, most of the Third World offers arid prospect." [1129] Yet synergy can be constructed. Looking at cases where synergy has emerged despite unfavorable conditions, we can identify several factors that contribute. First, social identities and relations are regularly reconstructing themselves in response to changing conditions, and they can be reconstructed to enhance synergy. Second, organizational structures that enhance embeddedness and social capital at both the civic and the governmental levels can make a large difference. Finally, synergy often begins with the redefinition of problems in a way that allows public- and private-sector actors to recognize their respective interests and potential contributions to the development project.

"Synergy is too potent a developmental tool to be ignored by development theories. Like social capital, it magnifies the socially valued output that can be derived from existing tangible assets but requires minimal material resources in its own creation." [1130]

Summary of

Good Government in the Tropics: Introduction

by Judith Tendler

[Published in *Good Government in the Tropics* (Baltimore: Johns Hopkins University Press, 1997), Introduction, 1–20.]

We know much more about bad government in developing countries than we do about good government, thanks to a litany of stories and complaints. These include the familiar charges that public officials are out for private gain, governments overspend and overhire, clientelism trumps merit, public sector workers are poorly trained, and poorly designed programs lead to bribery instead of public service. Economists and social scientists have built their theories based on these richly chronicled behaviors, producing policy prescriptions designed to reduce the size of the public sector. *Good Government in the Tropics,* the introduction of which is summarized here, draws on a detailed look at effective government in one Brazilian state to identify the flaws in mainstream thinking and the elements that can contribute to the development of good government in developing countries.

Flawed Theories, Failing Policies

The mainstream donor community, which includes bilateral and multilateral donor institutions, North American and Western European governments, and aid-giving nongovernmental organizations (NGOs), has a common set of prescriptions for the failings of governments in developing countries. They generally fall into three categories:

- Reducing the size of government through layoffs of "excess" workers, privatization, decentralization, and contracting out for services.
- Ending many of the programs most susceptible to bribery and other forms of malfeasance, such as import and export licensing.
- Putting market-like pressures on government agencies and managers to perform, in part by allowing users to express their preferences and dissatisfactions directly.

The consensus on these issues goes well beyond advocates of neoliberalism. For example, many NGOs concur that government in developing countries is overbearing and that the private sector—including NGOs—could do a better job providing many services. Yet such approaches have led to a consistently flawed set of policies, in a number of respects:

1. Such advice is based on literature that looks principally at poor performance. This has given valuable insights, but not when it comes to understanding what can lead governments to perform well.
2. Where the mainstream development community has looked at "best practices," it has drawn too heavily on imported ideas from industrialized countries (especially Australia, New Zealand, Britain, and the United States) and recently industrialized countries, particularly those of East Asia. Such an approach leads to incorrect interpretations of good performance, as well as the failure to identify examples of good government that do not fit the preconceived mold.
3. The development literature tends to label entire countries or groups of countries as good or bad. While this derives from the 1980s preoccupation with macro-level problems, it leads to little curiosity about the variations between good and bad government within a given country. This leads to advice to bad performers to be more like good performers somewhere else in the world. One example of this is the cottage industry proffering advice to Latin American countries about how to be more like East Asia.
4. "[T]he mainstream development community often filters what it sees through the lens of a strong belief in the superiority of the market mechanism for solving many problems of government, economic stagnation, and poverty." [4] Again, East Asia is a classic example. Until 1993, the donor community wrongly attributed the region's economic success to minimal government intervention, when in fact the contrary was true. The highly interventionist policies that led to such success were considered wrong by the donor community.
5. Many of the prevailing views on the causes of poor performance ignore or contradict the lessons from the emerging literature on industrial perfor-

mance and workplace transformation (IPWT). This field, which originally focused on private firms but now deals with public institutions as well, focuses on innovative practices to increase worker dedication. This involves increasing worker discretion, greater labor-management cooperation, improving trust between workers and their customers, as well as workers and managers. Yet the development community still "starts with the assumption that civil servants are self-interested, rent-seeking, and venal unless proven otherwise." [5] In stark contrast to the IPWT literature, the donor community prescribes reducing the discretion of civil servants, decreasing the size of government without restructuring it to increase worker performance, and eschewing measures to reorganize work in ways that can increase worker commitment.

6. While it is a notable advance for the development community to advocate decentralization, attention to user needs and preferences, and strengthening a civil society capable of demanding accountability, it places excessive faith in such actions. Missing from the prescriptions are measures necessary to build trusting relationships between users and public servants.
7. The development community has shown little interest in the emerging IPWT literature on more flexible labor-management practices despite its consistent complaints that public sector unions are stifling needed reforms. Instead, unions and professional associations are cast as villains in the reform effort, to be undermined or circumvented. "Ironically, this vilification of public-sector employee associations has occurred at a time when the donor community has been celebrating all other forms of associationalism and civil society, including business associations. . . . [W]hile the development community consistently describes public employee associations as the ultimate in *self*-interest, it views all other forms of associationalism in a serious lapse of consistency—as wholesome expressions of the *public* interest." [7] There is some literature that suggests that public sector unions can be key leaders in reform efforts, but there is a strange lack of research into this important area.

Good Government in Northeast Brazil

The purpose of this book is to focus on cases of good government in order to build an argument for thinking differently about public sector reform. I focused on four case studies from the state of Ceara in northeast Brazil, the poorest region of the country and an area known for clientelistic government and poor public administration. Since 1987, reformist governors from the Brazilian Social Democratic Party have received much credit and attention for turning the government around by streamlining a bloated government payroll while introducing innovative and effective new programs. Four of these programs were the subject of this study, including a rural preventive health initiative; a program of business extension and public procurement from informal-sector producers; an employment-creating public works construction and emergency relief program following the 1987 drought; and a program focused on agricultural extension and small farmers.

Five themes emerged from these cases. "Something happened in all of these programs—sometimes unintentionally—that structured the work environment differently from the normal and, in certain cases, from the way experts think such services should be organized." [13]

First, government workers showed unusual dedication to their jobs. Workers reported greater job satisfaction derived from a sense they were more appreciated by their clients and the community. Clients' comments suggested a high degree of trust, which was reminiscent of the IPWT literature on customer service.

Second, the state government created a high level of recognition for these programs and the workers in them through public information campaigns and prizes for good performance. It also created a sense of mission, screening new recruits carefully and providing thorough orientation programs.

Third, workers undertook a wider variety of work than usual, often voluntarily. They were given greater discretion and autonomy, which allowed them to respond to what they thought their clients needed.

Fourth, while it would seem that supervision of such "self-enlarged" jobs would be difficult and could easily lead to rent-seeking activities, this did not happen because, on the one hand, workers wanted to live up to the highly public trust being placed in them and, on the other, their actions were carefully scrutinized by a public armed with new information about its right to public services.

Fifth, contrary to the simple and fashionable models of civil society holding local governments accountable as the central government steps aside, an activist central (state) government devolved some functions to local governments while also taking some traditional powers away from those governments. Moreover, the state government encouraged the organization of civic associations, then worked through them as they turned around and demanded better performance from government. In the end, "both the improvement of municipal government and the strengthening of civil society . . . were in many ways the result of a new activism by central government, rather than of its retreat." [16]

"All of this suggests a path to improved local government that is different, or at least more complex, than the current thinking about decentralization and civil society." [16]

PART III

Global Perspectives: The North/South Imbalance

Overview Essay

by Timothy A. Wise

> The proliferation of . . . international treaties and negotiations has meant an inevitable loss of national sovereignty. Governments across the world must understand that for global cooperation on such issues to continue, this loss of sovereignty has to be shared by all countries. Cooperation cannot be sustained in a situation where some countries are expected to sacrifice their sovereignty, while others use their economic might to preserve their autonomy.
>
> —Anil Agarwal et al., *Statement of Shared Concern*[1]

In recent years, world leaders have engaged in an unprecedented series of international negotiations on a variety of environmental issues, including climate change, hazardous waste, biodiversity, and ozone depletion. Many of these negotiations have broken down over tensions between the industrialized North and the less-developed South. While Northern representatives cite planetary concerns and shared responsibilities, their Southern counterparts respond that they too have the right to develop and that Northern societies whose overconsumption is destroying the environment have to take responsibility for cleaning up the mess.

Tensions are rooted in the historic power imbalance between North and South. This, of course, is strongly related to the history of colonialism, and it is expressed today in wealth disparities between the developed Northern countries and the less developed Southern countries. In seeking paths toward sustainable human and economic development, this imbalance remains a formidable obstacle.

This essay examines the North-South imbalance from a variety of perspectives. After tracing the Southern origins of some prominent schools of development thought, it looks at global inequality, with particular attention paid to the notion that debt of developing countries is an unsustainable South-North drain on resources. The essay then turns to contemporary issues, starting with the most explosive North-South environmental issue to date: the population-consumption debate. After touching on other controversies, such as the issue of biodiversity, the essay concludes with an analysis of global negotiations over environmental issues. It finds reason for hope that pressure from grassroots constituencies in both the North and the South can move government negotiators beyond gridlock.

The "Object" of Development

> Scarcely twenty years were enough to make a billion people define themselves as underdeveloped. (Illich 1981)

> They train you to be paralyzed, then they sell you crutches. (Galeano and Bonaso 1993)

Development theory came into the world with both a subject and an object. As is often the case, the subject was also its creator—the industrialized world. As is also common, the creator theorized a world in its own image, a world in which nonindustrialized countries, many emerging from colonialism, needed only to emulate their former colonial masters in order to industrialize, raise standards of living, and join the ranks of the "developed." Such development would be achieved through the global expansion of the market and through actions by the industrialized countries aimed at those countries deemed underdeveloped. The declarative sentence was complete: The industrialized would develop the underdeveloped.

Not surprisingly, not all "underdeveloped" countries accepted their assigned lot in the development process, and alternative theories of development soon followed. Common to all was a central idea of international economic development theory: that the penetration of market relations and industrialization into noncapitalist or pre-industrial societies will often produce economic processes quite different from those in places where market relations and industrialization are more advanced. Put more simply, being first is a tremendous advantage when it comes to capitalist development, a fact that significantly limits the choices open to those who follow.

While this seems a truism even to the most casual observer of international affairs, it is a contentious assertion within the field of economics. Mainstream economics has stood fast by policies that assume that countries and regions will develop and prosper only if they allow market forces to determine their comparative advantages in the international marketplace and if they resist the temptations to steer their economies in other directions or to regulate the development process through state intervention. Standard economic theories, in fact, would predict that poor countries grow faster than rich ones (World Bank 2000, 14).

There can be little doubt that the vast benefits of economic growth have accrued mainly to the North. World income inequalities have been rising steadily for nearly two centuries; by one estimate the ratio of per capita GDP for the richest and poorest countries in the world grew from 3:1 in 1820 to 72:1 in 1992 (UNDP 1999, 38). While there are countries like Japan and South Korea that have significantly caught up through their development processes, many more have lagged behind, and the poorest countries have stagnated. Since 1970 the wealthiest third of the world's countries showed dramatic growth while the middle and lower thirds exhibited little or no income growth (see Figure III.1), leaving the poorest third with per capita incomes just 1.9 percent of that of the wealthiest third (World Bank 2000, 14).

Mainstream development theory rests on a number of inaccuracies, particularly in relation to the role of the state. The history of capitalist development in what are now called the developed countries was characterized not by free markets but generally by active state intervention, protection of nascent industries through tariffs, and by military action, where necessary, to protect or expand market access for national firms. Moreover, there is ample evidence to suggest that today's developing countries require even more active state intervention, including protectionist measures, if they are to clear any of the additional hurdles placed in their paths by development processes that have come before. (For a brief review of this literature, see the Overview Essay by Tim Wise in Part IX and "A Theory of Government Intervention in Late Industrialization" by Alice H. Amsden, also in Part IX.)

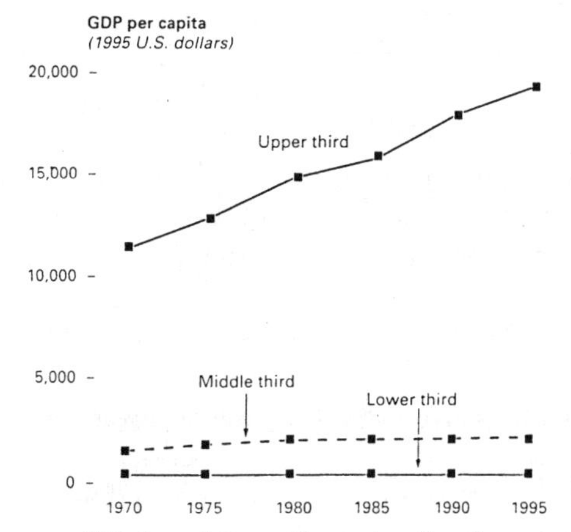

Figure III.1. Incomes of Rich and Poor Countries Continue to Diverge.
Source: The World Bank, *World Development Indicators,* 1999.

When the decolonization that followed World War II produced not a convergence in economic development but instead greater disparities between industrialized countries and the developing world, theorists from the global South emerged to assert alternative theories. **Philip Porter and Eric Sheppard,** in an article summarized in this section, trace the evolution of those theories. They identify two distinct phases. The first, led by Raul Prebisch, Celso Furtado, Andre Gunder Frank, Samir Amin, Dadabhai Naoroji, and others, came to be known as "dependency theory." It held that countries and peoples on the "periphery" of world capitalism were being actively "underdeveloped" by their structural relationship to "core" interests, which extracted resources for their own enrichment. The second, grounded in post-modernism, rejected the entire concept of development as a First World construct. Its leading theorists, such as Arturo Escobar, V. Y. Mudimbe, Edward Said, and Vandana Shiva, likened First World development theory to destructive logging practices: the First World development theory eliminates diversity of indigenous thought and practice just as logging exterminates species of tropical hardwood.

While Porter and Sheppard present a remarkably comprehensive overview of dissident theories, they can offer only a schematic picture of the wide range of current development thinking. Bhalla (1995), for example, examines the concept of "uneven development," with particular reference to India and China, studying unevenness in the form of unbalanced growth in the composition of industries, the urban and industrial bias in relation to rural agriculture, and unequal international terms of trade. Bruton (1998) calls for a reconsideration of what he sees as an oversimplified choice between the failures of import substitution and the promise of outward economic orientation.

Third World Debt: A Lien on Progress

"If the Amazon is the lungs of the world, then the debt is its pneumonia,"observed Luis Agnacio da Silva, a Brazilian labor leader and presidential candidate (George 1992). The Third World's debt crisis seemed to largely fade from view in the 1990s, reflecting not an amelioration of the debt problems facing many Third World countries but the resolution of the crisis facing First World banks. Creditors were temporarily appeased by the International Monetary Fund (IMF) and other international lenders, who restructured debt repayment schedules and offered new loans that brought some of the nonperforming loans out of arrears.

But these loan packages involved little actual forgiveness of debt; they merely stretched out the debt burden over a longer period of time. In fact, between 1990 and 1997 total external debt in low- and middle-income countries increased over 50 percent, from $1.5 trillion to $2.3 trillion (World Bank 2000).

During that same period, developing countries paid out more in debt service than they received in new loans, a $77 billion transfer of wealth from South to North.[2]

Only at the century's end did debt come back onto the public radar screen, this time because of an unprecedented international campaign for debt relief under the banner "Jubilee 2000." Citing the biblical principle of jubilee—debt cancellation to allow new beginnings—the campaign called for massive debt forgiveness for the world's poorest countries. The campaign posed the issue as a matter of sustainable development: "The debt burden inhibits the social and economic development that is needed to lift people out of poverty." (Jubilee 2000/USA 1998) The campaign called for cancellation of external debt for fifty-two highly indebted poor countries, which in 1996 had $370 billion in outstanding debt—$377 per person in countries with per capita incomes of just $425.[3]

There has been remarkably little academic research on the debt issue in recent years, with the exception of occasional case studies. This is unfortunate, given the enormity of the burden on many impoverished countries. Figure III.2 shows the extent of that burden on seven heavily indebted countries. As the figures show, debt service payments, which often do little to reduce outstanding principal, continue to sap remarkably large portions of government budgets. Because public investment is needed to address both environmental and social issues, the debt burden in many countries may represent the largest obstacle to the adoption of sustainable practices. In many countries, governments devote a larger share of their budgets to debt service payments than they do to education

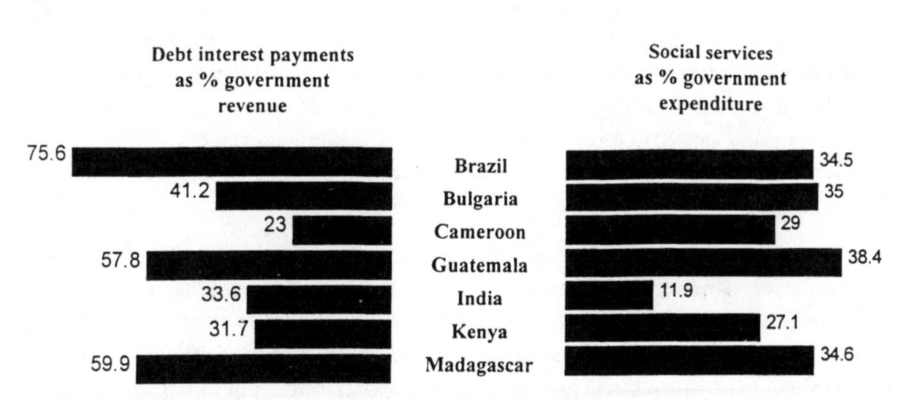

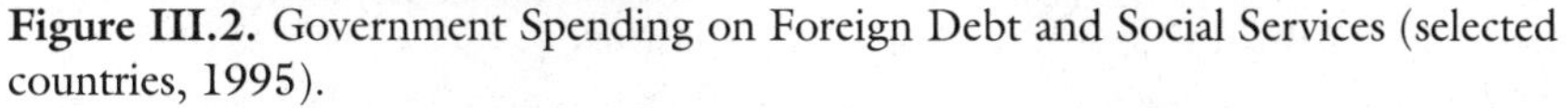

Figure III.2. Government Spending on Foreign Debt and Social Services (selected countries, 1995).
Source: New Internationalist, May 1999, based on data from the World Bank, World Development Report 1998–99.

or health care (Jubilee 2000/USA 1998, 2). Because debt repayments must be in hard currency, they also reduce countries' capacity to import needed goods.

There are additional ways in which debt is unsustainable. Debt restructuring generally comes with a package of structural reforms mandated by international creditors, which contribute to poverty and reduce the government's ability to respond to the needs and desires of its citizens. Susan George (1992) has linked high debt burdens to rates of deforestation, as structural adjustment programs put added pressure on fragile ecosystems by encouraging destructive extractive industries and forcing impoverished farmers onto more marginal lands. (See "Impacts of Structural Adjustment on the Sustainability of Developing Countries" by David Reed in Part VII for a discussion of structural adjustment programs and sustainability). High levels of indebtedness have also been shown to deter foreign direct investment. Over the long term, debt represents a significant form of wealth extraction from the South by the North.

From the creditors' perspective, the main issue seems to be not sustainable development but sustained loan payments. (A popular Brazilian refrain of the 1980s stated that Brazil had borrowed $100 billion, repaid $100 billion, and still owed $100 billion.) The concrete proposals for debt forgiveness, espoused by major international lending institutions such as the Highly Indebted Poor Country Initiative, call not for a new beginning but for the reduction of debt to a "sustainable" level. Interestingly, their definition of sustainable debt is 20–25 percent of export earnings—a figure more than double the ratio applied to Germany after World War II.[4]

The Population-Consumption Controversy

> Consumption clearly contributes to human development when it enlarges the capabilities of people without adversely affecting the well-being of others, when it is as fair to future generations as to the present ones, when it respects the carrying capacity of the planet and when it encourages the emergence of lively and creative communities. (United Nations Development Programme 1998, 38)

No issue suffuses North-South debates on the environment more than the controversy over which constitutes the greater threat to the global environment: overconsumption in the North or high population growth rates in the South. Although the North and the South acknowledge that both issues need attention if they are to move toward sustainability, their differences often undercut needed consensus during international negotiations.

On one side, Northern negotiators warn that the underlying cause of many urgent environmental problems—deforestation, toxic waste, climate change, soil depletion—is the growing pressure of large and rising Southern populations on scarce resources. Though they often acknowledge the role of Northern consumption in environmental degradation, they argue that any constraints on growth or consumption—such as limitations on greenhouse gas emissions—must be shared by those in the South, whose numbers are far greater.

Southerners often respond by invoking the "polluter pays" principle—that those who made the mess are responsible for cleaning it up. They note that Northern consumption and production have generated the vast majority of lasting damage to the planet. And they insist on the "right to development": the right of their populations, long excluded from the benefits of growth, to many of the same privileges Northerners take for granted. Why should China give up tapping its massive reserves of coal for electrification simply because the United States has already done so, in the process creating the environmental problem all are now supposed to solve? (For an overview of the relatively recent assertion of the "right to development," see Nanda 1993.) While they too will acknowledge that population control is a worthy goal, they assert that it can only come with economic development.

Part IV of this book considers the population issue in more detail. Here we seek to examine the North-South consumption imbalance and its implications. The starting point must be an acknowledgment that there are two distinct consumption problems: overconsumption, largely in the North, and underconsumption, largely in the South. While those in the global North are consuming the lion's share of the earth's resources, over one-fifth of the world's people are in desperate need of increased consumption to survive. As Part II showed, 1.2 billion people live on less than one dollar per day; half the world's people live on less than two dollars a day. Figure III.3 illustrates the disparities in world consumption between the rich and the poor. The richest 20 percent account for 86 percent of global consumption, while the poorest 20 percent account for only 1.3 percent. Only in the case of cereals, a food less desired in the diets of the well-off, is there anything close to parity in direct consumption levels, and even this is misleading, because Northern meat consumption indirectly consumes large quantities of cereals in the form of feed grains (Harris 1996).

Massoud Karshenas (1994) goes so far as to introduce a new concept of sustainable development that distinguishes between environmental problems associated with high incomes and advanced technology and those related to underdevelopment and technological stagnation. He notes that in the latter case, increasing consumption can often reduce environmental destruction. Others stress the importance of distinguishing between those in the South who can consume at will—generally the small minority—and those who are largely excluded from consumption—often the overwhelming majority (Camacho 1998). Parikh (1996) concurs, arguing that Northern consumption is clearly "the driving force of economic stress" and that there will be significant environmental benefits to increasing consumption among the world's poorest peo-

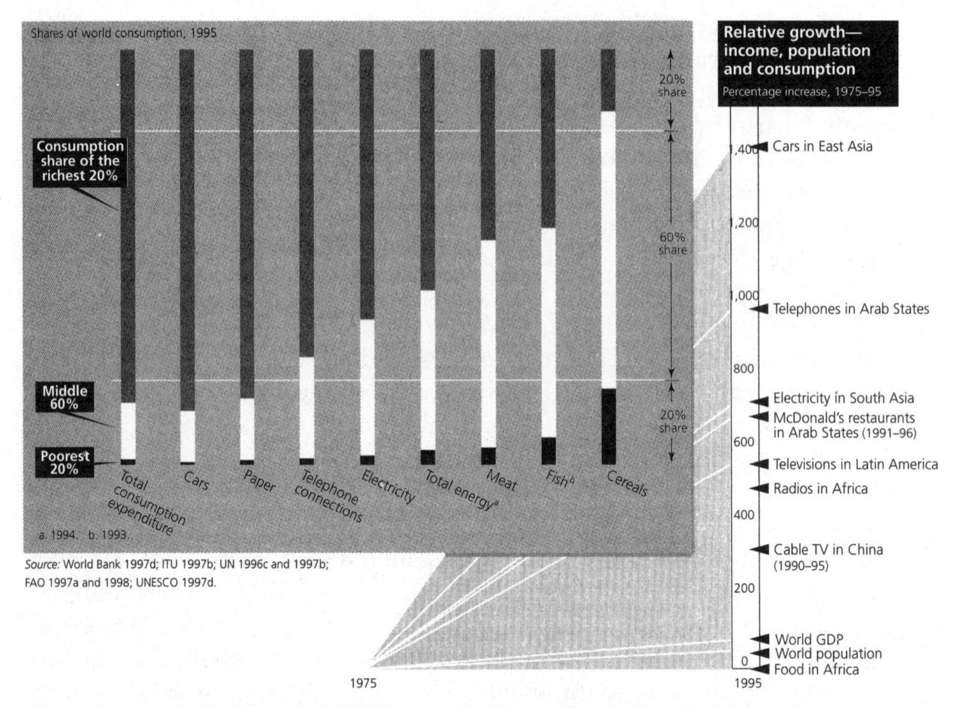

Figure III.3. Rapid Consumption Growth for Some, Stagnation for Others, Inequality for all—with Mounting Environmental Costs.
Source: World Bank 1997d; ITU; UN 1996c and 1997b; FAO 1997a and 1998; UNESCO 1997d.

ple. Among them is the well-noted tendency for population growth rates to decline with increases in economic security. Hammond (1998) notes that many of the raw materials needed to feed Northern appetites, such as metals, are disproportionately located in the South, and this is where the greatest environmental damage is often done.

Nathan Keyfitz (1998), like many writers on the subject, acknowledges the need to reduce population growth in the South and also to alter consumption patterns: reduce consumption by the North and reduce the aspirations for high-consumption lifestyles by the South. Alan Durning, one of the leading advocates of reductions in consumerism, points out both the hypocrisy and impracticality of demanding reductions in consumption in poor countries. "Limiting the consumer life-style to those who have already attained it is not politically possible, morally defensible, or ecologically sufficient." (Durning 1994, 46)

As Julka puts it, we face a clear choice between "marketism and sustainable development." "The market has been lauded as an engine of economic growth, the motivator of production. We know that production is meant to be a process of satisfying wants. But if in the process of production, a single satisfied want entails scores of unsatisfied wants, the process has a built-in accelerator. And, to the extent that production ultimately draws from nature, increasing production implies greater and greater demands on nature. The questions of resource exhaustion and ecological disruption would then be inescapable." (Julka 1997, 47)

Juliet Schor (1991) has pointed out that reductions in Northern consumption need not mean reductions in the quality of life. If productivity growth were channeled into shorter working hours instead of increased consumption, there could be a distinct improvement in both the quality of life and the impact of Northern lifestyles on the environment. (See the second volume in the Frontiers series, *The Consumer Society,* by Goodwin, et al., 1995, for a more complete discussion of the consumption issue.)

To be sure, there are many who still argue that there is no need to place limits on consumption in the North. Vincent and Panayotou, for example, state: "The root cause of environmental degradation is not the level of consumption, but rather market and policy failures that cause consumers and producers to ignore the full social costs of their decisions." (1997, 56) They urge continued economic and political liberalization combined with policy reform to reduce the environmental impacts of consumption. They argue that continued advances in technology will outpace resource depletion, particularly if externalities can be incorporated into pricing structures.

Two of the articles summarized in this chapter address the population-consumption issue as it relates to climate change. As Figure III.4 shows, there is little doubt that industrial countries, especially the United States, contribute a disproportionate share of the carbon emissions that contributes to global climate change.

A. Atiq Rahman directly challenges the "population myth," offering a detailed critique of the formulation first advanced by Paul Ehrlich in *The Population Bomb* that environmental impact is the product of population growth, consumption, and technology. He uses the case of climate change to show that such formulas oversimplify by treating environmental degradation as a simple result of population growth rather than the product of complex qualitative processes. **Anil Agarwal and Sunita Narain,** of the India-based Centre for Science and Environment offer their own critique of a much-publicized World Resources Institute (WRI) study that assigned greater blame for global warming to Third World countries. The authors assert that WRI misused available data in such a way as to overstate the contributions to global warming from Third World rice cultivation and livestock programs, which produce methane, and deforestation, which is a sink for carbon dioxide.

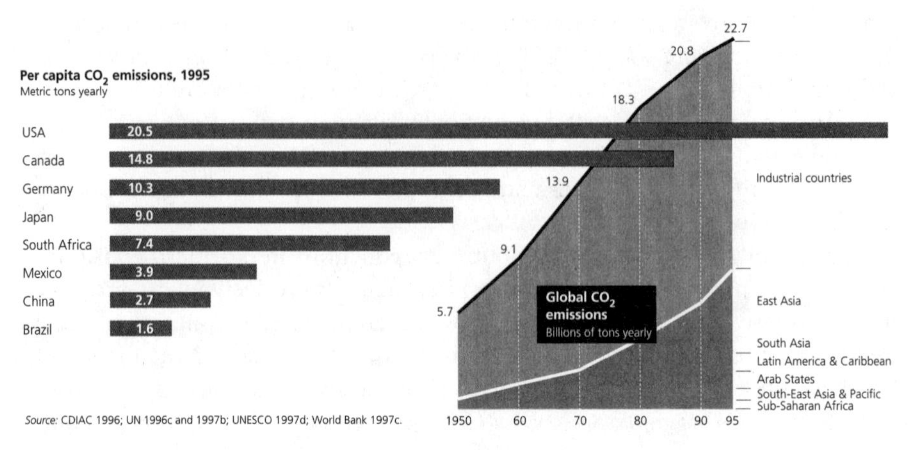

Figure III.4. Per Capita CO_2 Emissions, 1995.
Source: CDIAC 1996; UN 1996c and 1997b; UNESCO 1997d; World Bank 1997c.

Other North-South Fault Lines

> Few of the brave words spoken in Rio have been translated into action. It remains to be seen whether the negotiations in the future can take into account the interests of the poor—their rights and entitlements, their knowledge, the environmental space they need for future growth, the resources they provide for the world economy, and the issues that concern them here and now. (Anil Agarwal et al. 1999)[5]

Climate change, of course, is not the only area in which there have been discordant international negotiations over environmental issues. North-South disagreements have derailed or significantly hampered negotiations over hazardous waste, biodiversity, ozone depletion, and world trade. **Alain Lipietz,** in an article summarized here, tries to deconstruct some of those negotiations to identify ways in which the North-South divide might be bridged. He draws an interesting parallel, likening the present conflicts over the "global commons" to the European "enclosure movement" of the 14–16th centuries. Lipietz argues that just as peasants then were excluded from the land, current efforts at "global enclosure" threaten to exclude some nations and peoples from modernity. The biodiversity treaty negotiated in Rio de Janeiro, for example, represented a compromise acceptable to Northern and Southern elites, but effectively excluded indigenous peoples by failing to acknowledge their historic role as the "gardeners of biodiversity."

Lipietz argues for categories that go beyond North and South, pointing out that one needs to be clear who is most threatened by the environmental problem, who is most responsible, and who will be required to make the greatest sacrifices to resolve it. He notes that in early climate change negotiations there were signs of an alliance between what he calls the "Do Something North"—Japan and Northern European countries that are technologically advanced and comparatively moderate emitters of greenhouse gases—and the "Do Something South"—Bangladesh, India, and other countries that see themselves as the first victims of global warming and/or are such low emitters that any agreement would be unlikely to impinge on their development options. In the end, agreement was prevented by a "Do Nothing" bloc led by the United States, as the heaviest emitter, and some of the more rapidly industrializing developing countries, which wanted no constraints on their development. Despite the failure to achieve an implementable agreement, Lipietz sees signs of hope for breaking the traditional deadlock between Northern and Southern governments.

Interestingly, there is also reason for hope coming from some of the more critical Southern analysts. Many argue that negotiated compromises between governments violate fundamental principles of ecology and social justice. **Vandana Shiva** presents such a critique of international attempts to address issues of biodiversity. She argues that much of the planet's remaining biodiversity resides in the South, yet plans to preserve remaining genetic material rely not on the native populations that have nurtured such living wealth but on corporations, which demand the right to patent such life forms in order to save them. While it may seem ecological to declare the South's biodiversity "the common heritage of mankind," as some intellectual-property proposals assert, this effectively steals the South's accumulated biological wealth for the benefit of Northern corporations. She explains that this is not only unjust, but also that it ignores evolutionary biological processes and thus will fail in the long run to preserve biodiversity.

Wolfgang Sachs makes a similar point about global environmental negotiations in general, linking the issue to economic development strategies. Current attention to sustainable development, he argues, represents an attempt to better manage development, primarily through technology, rather than an acknowledgment of the urgent need to redefine goals and reapportion power. Within the existing paradigm, negotiations become battles over who gets to exploit nature and how, when instead they should be concerned with redefining the goals of development and humanity's relationship to nature. The United Nations Conference on Environment and Development (UNCED), Sachs states, was a "technical effort to keep development afloat against the drift of plunder and pollution" rather than a "cultural effort to shake off the hegemony of aging Western values and gradually retire from the development race." [245]

A similar warning comes from the India-based Centre for Science and Environment (Agarwal et al. 1999). They examine each of the major environmental negotiations in an attempt to demystify the politics of the environment. They argue the case for going beyond the positions of Northern and Southern governments, relying instead on the growing constituencies in Northern and Southern nations calling for sustainable practices. There is indeed the basis for a strong alliance between Northerners concerned about the environment and nongovernmental organizations in the South that articulate a comprehensive vision of sustainability—one that embraces both ecological protection and social development. For such an alliance to blossom, however, it is clear that Northerners will need to acknowledge both the responsibility for the ecological cost of their own lifestyles and the urgent needs of others.

Notes

1. From the "Statement of Shared Concern," issued in conjunction with the release of *Global Environmental Governance* by the Centre for Science and Environment. The statement is signed by seventy-three prominent researchers, policy-makers, and representatives of civil society organizations from the North and South. It is available on the CSE web site: http://www.oneworld.org/cse/html/eyou/gen1soc.htm.
2. Figures are from a May 1999 special issue of the *New Internationalist* magazine that focused on debt. Data come from the World Bank web site: http://worldbank.org or from the *World Development Report 1998/99* (Washington, D.C.: World Bank, 1999).
3. Ibid.
4. Ibid.
5. From "Statement of Shared Concern." See Note 1.

Summary of

Views from the Periphery: Encountering Development

by Philip W. Porter and Eric S. Sheppard**

[Published in *A World of Difference: Society, Nature, Development* (New York: The Guilford Press 1998), Ch. 6, 96–118.]

Development theory has distinctly Western, First World origins. It is based on the diffusionist notion that development is a common goal that can be reached by traveling paths that have already proven successful in the developed countries. Third World experiences of development have been quite different, gen-

**David Faust wrote part of the chapter summarized here. The authors acknowledge his contribution and thank him for it.

erating alternative schools of development theory. These have challenged not only prevailing theory but also ongoing practice, helping involve and give voice to those often marginalized in the development process. This book chapter surveys the evolution of these alternative development theories, which have contributed a great deal to our practical and intellectual understanding of the development process.

Encountering Development

The concept of development emerged from the developed world following World War II to pose a linear path to economic prosperity. Less-developed countries, often just emerging from colonialism, needed only to follow the same developmental road map as the colonial powers and they too would industrialize, modernize, and raise standards of living. Actual experience with development has been significantly different, often involving violence, impoverishment, and the destruction of indigenous cultures. As a result, significantly different views of development have emerged among Third World theorists. "As is so often the case with human ideas, a peripheral position has in fact facilitated a questioning or even deconstruction of what those in the core take for granted, resulting in this case in a revolution in development thinking." [97]

We can identify two distinct phases. First, national political independence from former colonial powers resulted not in a convergence of economic development but in growing disparities. This led theorists like Raul Prebisch, Andre Gunder Frank, Samir Amin, Dadabhai Naoroji, and others to argue that countries and peoples on the "periphery" of the world capitalist system were being actively "underdeveloped" by their links with global capitalism.

One can trace the origins of non-European alternatives to development theory to Vladimir Lenin's theory of imperialism, which represented a significant turning point in that it "argued that the penetration of capitalism into a non-capitalist society can lead to effects other than those already observed in other capitalist societies." [99] In the post-World War II period, the United Nations' Economic Commission for Latin America (ECLA), under the direction of Raul Prebisch, gained a reputation for its theories that the region's lack of development was the result of the unequal effects of trade. According to this theory, international trade exacerbates international inequalities, since developed countries capture a disproportionate share of the gains from technological innovation and trade. At a policy level, this approach advocated "import-substituting industrialization" with the state taking an active role in countering the effects of a country's peripheral position. When this strategy began to produce mixed economic results in the 1970s, "dependency theory" arose to assert a more radical interpretation.

Dependency theory found its most recognizable proponent in Andre Gunder

Frank. Based on his own empirical research in Chile and Brazil, Frank argued that capitalist penetration doesn't just hinder economic development, it causes underdevelopment. He combined Prebisch's theories of unequal exchange with Marxist theories of imperialism to build a new critique of modernization and contribute to radical movements for an immediate break with capitalist economic development processes.

Two other important contributions to dependency theory came from Theotônio dos Santos and Ruy Marini. Dos Santos tried to address the theory's tendency to explain all development problems in all countries by their peripheral status, ignoring local and national differences. He identified three forms of dependency: colonial; industrial-financial, in which core capitalists invest capital in peripheral countries to produce primary goods; and technological-industrial, with transnational corporations investing in production for local consumption in developing countries. Marini tried to articulate a more rigorous theory of dependent capitalism that could explain the different laws governing dependent development. He argued that Third World countries have little incentive to stimulate domestic consumption, which results in a self-sustaining cycle of dependency in which wages are kept low, productivity stagnates, and the underdevelopment of local markets inhibits investment in industrial production.

World System Theory

U.S. sociologist Immanuel Wallerstein coined the term "world system theory," but African economist Samir Amin simultaneously developed a similar framework. Both saw capitalism as the first global, interdependent economic system in history and attempted to identify the dynamics of the system as a whole. They recognized that both capitalist and noncapitalist production exist within this system, and they sought to understand the implications for development. Wallerstein sought to go beyond the concepts of core and peripheral countries to identify core and peripheral economic processes, allowing analyses that incorporated core processes in peripheral countries and vice versa. He also introduced the category of "semiperipheral country" for cases in which both processes coexist.

U.S. geographer James Blaut used a "world systems" approach to reexamine the origins of European capitalism itself. He called into question the assertion that European capitalism was the result of cultural, political, or economic superiority. Instead, he argued that medieval Europe was no more advanced than medieval Africa or Asia, and that geographic good fortune—in the form of access to the New World—stimulated the rapid shift from feudalism to capitalism, with the influx of capital accelerating development and giving European industrialists a decisive competitive edge. This contributed to the growing number of theorists questioning the assumption that Europe and North America represent a development model for everyone else.

Still other modern thinkers took the Third World as their starting point, noting that local experiences with development have seen the dissolution of indigenous cultures and institutions, growing gender, class, and ethnic inequality, and environmental destruction. This has provoked an attempt to apply postmodernist approaches in arguing that First World strategies are inappropriate to such radically different contexts. "In this view, the third world is an invaluable source of diversity in development thinking that is in danger of being overrun by development, much as the rain forest is a source of biodiversity that is being lost through the logging of tropical hardwoods." (98) Among the leaders in this school are Arturo Escobar from Latin America, V.Y. Mudimbe from Africa, and Edward Said and Vandana Shiva from Asia.

In this approach, the entire construction of development as a process of "catching up to the west" is disempowering to Third World people. First, it convinces them to devalue their own knowledge, values, institutions, and cultures. Second, the primary use of the development process was to intervene in the lives of those declared "underdeveloped" in the name of a higher evolutionary goal. Development became a technical matter for supposed experts, further disempowering those in the periphery. The assumption underlying development, which has been shared by neoclassical and Marxist economists, is that economic growth is the key to development and the agents of growth are corporations and the state. This implies an extension of markets and the state into areas of life previously governed by traditional relationships.

There are significant differences between the dependency theorists and the postmodernists. The dependency theorists focus on global interdependencies between development and underdevelopment, while the postmodernists address local contradictions. Still, both take as their starting point the thesis that development is linked with impoverishment, not growing prosperity.

"Dependency theory shows the intimate link that exists between development in the core and that in the periphery. From this we learn two things: (1) What is beneficial for the core can either inhibit, or completely alter the direction of, economic change in the periphery; and (2) the geographical and historical position of a nation (or region) within the evolution of global economic change can have a profound effect on the types of changes that are possible or desirable there. In short, there is no unified path to development, applicable to all nations." [118]

Summary of

Global Ecology and the Shadow of Development

by Wolfgang Sachs

[Published in *Global Ecology: A New Arena of Political Conflict*, ed. Wolfgang Sachs (London: Zed Books 1993), 3–20.]

Development is usually perceived as a process whereby all nations, Northern and Southern, move along the same path of increased economic production, although at different rates and from different starting points. In this article, the author argues that development has only widened the economic gap between the North and South and amplified Southern misery. Further, the term "sustainable development" has been co-opted to serve the interests of a Northern-dominated development process. The current view of international development agencies is that improving the management of development, rather than adopting different goals, is the cure for the environmental degradation and poverty that threaten the sustainability of the development process. This amounts to an extension of Northern global hegemony rather than true sustainability.

The Birth of "Development"

Since 1949, the objective of development policy has been to bring all nations into the global arena and get them to run in the race toward increased production. The worldview put forth by the North has been that catching up in the race is the only way to prosperity. Turning the South's societies into economic competitors meant not only capital injection and technology transfer, but also a complete deconstruction of their social fabric. Economic, political, and cultural institutions that were not compatible with market capitalism had to be revamped to achieve the textbook model of macroeconomic growth.

The result of this radical restructuring of Southern institutions has been the further widening of the gap between the North, where economic prosperity is concentrated, and the South, where the brunt of ever-increasing poverty and environmental destruction is felt. "During the 1980s, the contribution of developing countries to the world's GNP shrank to 15%, while the share of the industrial countries, with 20% of the world population, rose to 80%." [241] Why then do Southern nations stay in the race? Because "development" has created a strong global middle class composed of elites in the South and the majority in the North. "The internal rivalries of that class make a lot of noise in world politics, condemning to silence the overwhelming majority of the world's people." [241]

Ambiguous Claims for Justice

Ample evidence has demonstrated there is not enough room in the world for the environment to serve as the source for the inputs and the sink for the wastes of economic growth. However, the South fears that environmental concerns will limit its opportunities for economic growth, opportunities already exploited by the North, before the abuse of nature was a concern. Thus the South has used "justice" as a bargaining tool for concessions from the North in the form of development aid, clean technology, or access to bioindustrial patents. Unfortunately, in so doing the South accepts the notion of Northern cultural hegemony. All societies thus remain caught up in a race for ever-increased technical capacity and economic power. "Limits to road-building, to high-speed transport, to economic concentration . . . were not even considered at the 1992 Earth Summit in Rio." [242]

Earth's Finitude as a Management Problem

In 1962, Rachel Carson's *Silent Spring* introduced the idea that "development" and technological progress could be destructive to the environment, and gave rise to the environmental movement. But the 1980 "World Conservation Strategy" and the 1987 Bruntland Report promoted the idea that development itself is the only cure for developmental ills. The World Bank has defined sustainable development simply as "development that lasts" (1992). Thus the environment itself takes a back seat to anthropocentric concerns with increased consumption. The evolution of this worldview can be traced through historical stages:

- *The 1970s oil crisis*—Concern about the finiteness of natural resources and how it would affect growth overrode concern for the health of nature. Nature became a pawn to be manipulated to further long-term development.
- *The development of postindustrial technologies*—It became evident that growth could be pursued through less resource-intensive means, thus increasing the productivity of nature. "Limits to growth" was transformed into a technological challenge.
- *The discovery of environmental degradation as worldwide condition of poverty*—The poor who are dependent upon nature for survival have no choice but to destroy it. Humanity was branded the enemy of nature.

As a result, better managerial techniques became society's answer to environmental ills. Thus, the "sustainable development" promoted at the UN Conference on Environment and Development in 1992 amounts to a "technical effort to keep development afloat against the drift of plunder and pollution" rather than a "cultural effort to shake off the hegemony of aging Western values and gradually retire from the development race." [245]

Bargaining for the Rest of Nature

Although the 1980s saw the rise of a global environmental consciousness stemming from the universal threat to the global commons (the Antarctic, ocean beds, tropical forests, etc.), international diplomacy is inherently less cooperative. Instead of uniting to preserve the commons, nations bargain with each other, and not always fairly, for the largest share they can possibly secure for economic use. Thus, environmental concerns become bargaining chips in the struggle of interests.

Because so many nations are struggling for so few resources in international negotiations, "limits" are "identified at a level that permits the maximum use of nature as mine and container, right up to the critical threshold beyond which ecological decline would rapidly accelerate." [246]

The 1992 Earth Summit in Rio established an international "recognition of the scarcity of natural resources for development," rather than a commitment to a collective stewardship of nature. International environmental diplomacy thus encompasses four elements:

- *Rights to further exploitation of nature*—Who has access to dwindling genetic resources, tropical timber, ocean bed minerals, or wild animals?
- *Rights to pollution*—Who can pollute? Can pollution be "optimized" through the purchase and sale of pollution rights?
- *Rights to compensation*—Should the South receive compensation for the North's disproportionate resource use and cumulative environmental damage?
- *Overall conflict over responsibility*—Who carries the losses from restrained environmental exploitation? Who should foot the bill for transferring clean technology to the South?

Efficiency and Sufficiency

The net effect of the discussion on sustainability is that the idea of better management of development, or mastering nature's complexities, has replaced the notion of "limits to growth." The "efficiency revolution" aims to produce innovations that minimize the use of nature for each unit of output by "reducing the throughput of energy and materials in the economic systems by means of new technology and planning." [248] Although this revolution is meant to cure environmental ills, it only serves to further entrench current notions of economic development in the global system for the following reasons:

- It is difficult to use efficiency strategies in countries in the early stages of growth. To export the efficiency revolution to the South would require Northern capital as well as hegemony.

- The information and service society can only succeed on top of the industrial sector and in close proximity to it. "Gains in environmental efficiency often consist of substituting high technology for energy and materials, a process that presupposes the presence of a resource-intensive economy." [249]
- The technical knowledge required is concentrated in the North, and can be sold to the South, so the North again benefits to the detriment of the South.

The revision of means will not achieve environmental objectives without the revision of goals. In other words, the efficiency revolution is ineffective if it only serves to increase growth. For example, although today's cars are more fuel-efficient than ever, the growth in the number of cars used has eliminated these gains. Growth needs to slow down or the next round of growth will swallow the achievements of the efficiency revolution.

The Hegemony of Globalism

The claims of global management are in conflict with aspirations for cultural rights, democracy, and self-determination. The ambitious goals of global environmental management require the dominance of "global ecocrats" over local cultures and political systems. Technical data about resource flows and environmental impacts "provide a knowledge that is faceless and placeless, an abstraction that carries considerable cost: it consigns the realities of culture, power, and virtue to oblivion." [251]

Until recently, the North has been relatively unaffected by the negative consequences of the global development path, leaving its symptoms of sickness, exploitation, and ecological destruction for the South to absorb. However, today, for the first time, the North is feeling the unpleasant repercussions in the form of immigration, population pressure, tribalism with modern weapons, and the environmental consequences of global industrialization. Technocratic environmentalism is the Northern response—an attempt to manage the entire planet. "If there are no limits to growth, there are certainly no limits to hubris." [252]

Summary of

Biodiversity: A Third World Perspective

by Vandana Shiva

[Published in *Monocultures of the Mind: Perspectives on Biodiversity and Biotechnology* (London: Zed Books Ltd. 1993), Ch. 2, 65–93.]

The subject of biodiversity has become one of the focal points of North-South tension in world trade debates. Northern representatives have insisted on the incorporation of intellectual property rights into trade agreements. Such rights would include the patenting of life forms, such as plant germ plasm. Because much of the planet's remaining biodiversity resides in the global South, Southern critics have countered that such an approach both fails to preserve diverse ecosystems and amounts to a license to Northern corporations to profit exclusively from the South's biological resources. This chapter explains the Third World argument for a different approach to biodiversity.

The Crisis of Diversity

"Diversity is the characteristic of nature and the basis of ecological stability." [65] While societies have evolved over time in ways that both preserve and derive livelihoods from nature's diversity, today many of those societies and the ecosystems with which they have coexisted are under threat of extinction. Diversity is eroding dramatically with the loss of forest cover, where roughly half of the world's plant species reside. In marine ecosystems, biological diversity is being lost, with coral reef destruction comparable to deforestation rates. This has produced a decline in the fisheries base in many coastal regions.

The "Green Revolution" in agriculture has dramatically reduced the number of living crop varieties, supplanting diverse indigenous seed varieties with a small number of wheat and rice strains bred in Northern-dominated research institutes. Such monocultures increase susceptibility to pests. Livestock populations are being similarly homogenized. Jersey and Holstein cows are being systematically substituted for carefully evolved pure breeds in India, which had been locally bred for the specific eco-niches in which they had to survive. Flora and fauna have also gone extinct in agricultural areas, as chemical fertilizers and pesticides replace the diverse evolution of bacteria, fungi, pest predators, pollinators, and seed dispersers that have sustained agricultural production for centuries.

"The crisis of biodiversity is not just a crisis of the disappearance of species which have the potential of spinning dollars for corporate enterprises by serving as industrial raw material. It is, more basically, a crisis that threatens the life-support systems and livelihoods of millions of people in Third World countries." [68]

There are two primary causes of biodiversity destruction. The first is habitat destruction caused by internationally financed megaprojects, such as dams, highways, and mining operations in forest areas. Such projects have destroyed countless species and caused entire habitats to disappear. The second is the push to introduce homogeneity in forestry, agriculture, fisheries, and animal husbandry. "The irony of plant and animal breeding is that it destroys the very building blocks on which the technology depends." [70] The dominant paradigm of production calls for uniformity and monocultures, where plant improvement is based on the very biodiversity that it uses as raw material.

The dominant approaches tend to ignore these primary causes of biodiversity loss, preferring to focus instead on secondary causes such as population pressure. If societies are not displaced by dams, mines, factories, or commercial agriculture, populations will grow in harmony with their ecosystems. Thus, population pressure on biodiversity is a second-order effect of such displacement.

Biodiversity erosion creates both ecological and social vulnerability. Ecological vulnerability is well-known, as monocultures become a mechanism for fostering pests and weakening resistance to disease. Social vulnerability involves the disruption of the self-regulated and decentralized organization of diverse systems through the introduction of external inputs and external and centralized control. Where diversity ensures diverse livelihoods, homogenous production systems disrupt communities, displace people from diverse occupations, and create dependency on external inputs and markets. With a production base that is ecologically unstable and with unstable commodity markets, social vulnerability increases.

First World Bio-imperialism

European colonists enriched themselves by transferring biological resources from their colonies and introducing monocultures of raw material for European industry. Today, such dynamics have the same substance, even if they take a different form. The United States accuses Third World countries of engaging in "unfair trading practices" for refusing to adopt patent laws that grant corporations monopoly rights in life forms. Yet the United States has freely taken germ plasm from the Third World and turned it into millions of dollars in profits, none of which has been shared with Third World countries. In addition, corporations, governments, and aid agencies in the North are creating legal and political frameworks, under international agreements like the General Agreement on Tariffs and Trade (GATT), to ensure free access to Third World biological resources.

Unfortunately, most Northern approaches to biodiversity conservation are blind to the North's role in destroying Southern biodiversity. One example is "Conserving the World's Biological Diversity," a study by the World Bank, the World Resources Institute, the International Union for the Conservation of Nature Resources, and the Worldwide Fund for Nature (McNeely 1990). These groups neglect the primary causes of destruction. They fail to address the crisis of diversity in "production" spheres—forestry, livestock, and agriculture—problems caused by Northern development models and promoted by Northern governments and multilateral institutions. Instead, they focus on secondary causes: forest clearing and burning, overharvesting of plants and animals, and overuse of pesticides.

The Northern bias in this report is also clearly evident in how it chooses to value biodiversity. While recognizing that indigenous farmers and tribals are the original producers of the wealth of genetic diversity through generations of resource management, the authors fail to acknowledge that Northern corporations and scientists are primarily consumers, not producers, of this wealth. Instead, they divide biological resources into the following categories of economic value:

- *consumptive value*—products such as firewood, fodder, and game meat that are consumed without passing through a market.
- *productive use value*—products exploited commercially.
- *nonconsumptive use value*—indirect ecosystem functions, such as watershed protection, photosynthesis, and the like.

This framework defines those deriving their livelihoods directly from nature purely as consumers, while crediting commercial interests with being the producers. The logical conclusion is that Third World consumers are largely responsible for biological destruction, while the North alone has the capacity to conserve biodiversity. This obscures the true political economy underlying the destruction of biological diversity.

> Defining production as consumption and consumption as production also matches the demand for intellectual property rights of the North, and denies the intellectual contributions of those in the South who are the primary producers of value. [86]

This economistic bias reduces conservation efforts to financial values on the market, based on the biotechnology that transforms the planet's genetic resources into raw material for commercial enterprises. It justifies conservation only in "set-aside" areas where biodiversity is seen to serve those commercial interests. Conservation areas and *ex situ* preservation of germ plasm in high-tech gene banks may be an efficient way to preserve known, existing germ plasm, but it allows the continued destruction of the habitats within which such diverse life forms can further evolve and adapt.

From Bio-imperialism to Bio-democracy

The only sustainable and just approach to conserving biodiversity involves halting the primary threats to biodiversity, which are in the North, and strengthening those who produce based on biodiversity. Such an approach would involve the following:

- Stopping aid and incentives for habitat destruction by centralized and homogeneous systems of production in forestry, agriculture, fisheries, and animal husbandry.
- Recognizing community rights to biodiversity and valuing farmers' and tribals' contributions to its evolution and protection.
- Ceasing to finance biodiversity conservation by a small percentage of profits generated by biodiversity destruction.
- Recognizing the injustice of Northern demands that the South's biodiversity be treated as the "common heritage of mankind," as current GATT, World Bank, and other intellectual property proposals assert. This results in the South's biological wealth being patented, priced, and treated as the private property of Northern corporations.

The current regime based on bio-imperialism must be replaced with one based on bio-democracy, which "involves the recognition of the intrinsic value of all life forms . . . and the original contributions and rights of communities which have co-evolved with local biodiversity." [92]

Summary of

Global Warming in an Unequal World: A Case of Environmental Colonialism

by Anil Agarwal and Sunita Narain

[Published as *Global Warming in an Unequal World: A Case of Environmental Colonialism* (New Delhi: Centre for Science and Environment 1991).]

The North-South divide is nowhere more pronounced than on the issue of who should assume responsibility for controlling the emissions of greenhouse gases in order to address the problem of global climate change. Southern advocates, like the Centre for Science and the Environment in this selection, have argued effectively that the sacrifices must come first and foremost from those who caused the problem.

Global Warming in an Unequal World

"The idea that developing countries like India and China must share the blame for heating up the earth and destabilizing its climate, as espoused in a recent study published in the United States by the World Resources Institute (WRI)[1] in collaboration with the United Nations, is an excellent example of *environmental neocolonialism.*" [1] Based on politically motivated manipulation of available data, it is an attempt to shift some of the blame for global warming to the Third World while undercutting the argument that First World countries, which are responsible for the vast majority of greenhouse gas emissions, should take primary responsibility for reducing such emissions. Such an approach unfairly denies developing countries the right to develop by limiting both their energy production (much of which is from coal, which produces carbon dioxide) and their rice cultivation and livestock programs (which produce methane).

In fact, countries like India and China, which have over one-third of the world's people, consume far less than that proportion of the world's resources and contribute far less than that proportion of the greenhouse gases contributing to global warming. Those countries that consume more than their share and contribute more than their share of pollutants should be the ones to limit their development choices.

WRI's calculation are faulty in that they overemphasize the role and scope of deforestation and methane production in global warming and underemphasize the production of carbon dioxide from the use of fossil fuels such as oil and coal. Because developing countries are largely responsible for the former, this skews WRI's data and conclusions.

The India-based Centre for Science and Environment (CSE) has carried out its own analyses of existing data, which present a very different picture. On the issue of deforestation, WRI took the highest available estimates of deforestation for Brazil and India. In Brazil, WRI took the 1987 deforestation figure (8 million hectares) and used it as the annual average for the entire decade, even though existing data showed clearly that a variety of factors had contributed to exceptionally high deforestation rates that year. At such a rate, about one-fourth of Brazil's Amazon forests would have disappeared during the 1980s, a figure that clearly overstates the extent of deforestation by a factor of 3 to 5.

This is relevant to WRI's climate change calculations because deforestation contributes to global warming by eliminating natural sinks that absorb CO_2 emissions. So overstating deforestation in large Southern countries like Brazil and India exaggerates their contribution to the problem.

The same is true of methane emissions. Methane comes from a variety of sources, primarily fermentation in irrigated rice fields and stomach gas from livestock. But according to WRI, nearly 40 percent comes from other sources, such as leakages during hard-coal mining and natural gas exploration. The estimates of livestock and paddy-field emissions are quite unreliable compared with those for natural gas pipeline leakages. Yet WRI uses such estimates to attribute a high proportion of livestock emissions to developing countries, while failing to recognize that the way to reduce cattle herds, and methane emissions, is to reduce beef consumption, which mostly takes place in developed countries. Again, they have put the blame and responsibility in the wrong place.

Self-serving Approach

WRI was attempting to calculate each nation's contribution to global warming, but it took the wrong approach. The correct approach would involve calculating each nation's total greenhouse gas budget, taking into account both its sources of emissions and its terrestrial sinks—the forests, vegetation, and soils that absorb such emissions. One would then need to add to that budget each nation's share of oceanic and atmospheric sinks, which are a common heritage of humankind.

WRI in its report unfairly allocates the earth's ability to cleanse itself of greenhouse gases to different nations. If such sinks were allocated based on population, the figures would look quite different. India, with 16.2 percent of the world's population in 1990, was emitting just 6 percent of the world's carbon dioxide and 14.4 percent of the world's methane. Meanwhile, the United States, with just 4.7 percent of the world's population, emits 26 percent of the CO_2 and 20 percent of the methane. In addition, the United States and other developed countries have emitted large quantities of CFCs, which have no natural sink at all.

The WRI study instead makes the simple but misleading calculation that India is contributing to global warming in the same proportion as it is producing greenhouse gases, ignoring the allocation of sinks, or rather implicitly allocating those sinks based on a nation's production of greenhouse gases. The resulting figures are absurd. WRI effectively allocates sinks of 2,519 million tons of CO_2 and 35 million tons of methane to the United States, while India, with over triple the population, is given only 604 million tons of CO_2 and 26 million tons of methane sinks.

CSE did its own set of calculations, correcting for this misallocation of the global commons and correcting for WRI's overestimation of deforestation rates. The results present a very different picture of who is responsible for global warming. The United States's net contribution of greenhouse gases accumulating in the atmosphere goes up from 1,000 million tons of carbon equivalent to 1,532 million tons. Meanwhile, China's share goes down from 380 to 35 million tons, and India's falls from 230 to near zero. Together, India and China account for less than 0.5 percent of net emissions into the atmosphere, while WRI's study claims the two countries contribute about 10 percent. After adjusting Brazil's deforestation rate, its net contribution drops from WRI's estimate of 610 million tons to CSE's estimate of 197 million tons.

Tradeable Permits

These calculations are relevant to proposals for a system of tradeable permits as a way to reduce and control the emission of greenhouse gases. CSE believes such a system can work if countries are allocated permits in proportion to their population share and if the total quota for emissions equals the world's natural sinks. CSE's calculations are based on such assumptions. In addition to allowing the sale of excess permits by low-emitting countries to high-emitting countries, such a system should include a more substantial fine for emissions above available permits, which could go into a Global Climate Protection Fund. Based on CSE's calculations, developed countries would contribute the lion's share of capital for such a fund, which could contribute a great deal to global environmental protection. This would place responsibility where it belongs for cleaning up the global environment.

This does not mean that developing countries should not take steps to improve the environment. Deforestation should be controlled, for example, with afforestation rates eventually matching rates of wood use and burning. But "it is immoral for developed countries to preach environmental constraints and conditionalities to developing countries. They must first set their own house in order." [23]

Note

1. World Resources Institute (1990).

Summary of

Lifestyle Is the Problem

by A. Atiq Rahman

[Published in *Exploding the Population Myth* (Dhaka: University Press Limited 1998), Ch. 9, 103–107.]

Since the publication of Paul Ehrlich's *The Population Bomb* in 1968, great attention has been paid to the social, economic, and environmental problems associated with world population growth. The environmental impact of population growth was later summarized in the equation I=PAT, representing the theory that the negative impact on the environment (I) was the product of population growth (P), affluence or per-capita consumption (A) and the technology used to produce what is consumed (T). When it comes to international discussions of the environment, Northern researchers often place the emphasis on population growth, arguing that in the developing world it outweighs other factors. In this chapter, the author argues that on the issue of climate change such logic is flawed.

Taking a Comprehensive Approach

Simple formulae produce simple answers. The I=PAT equation and its many variants hide as much as they reveal about the causes of environmental decline. . . . There is an urgent need to broaden the base of analysis and try to address the more complex range of factors that bear down on the Earth. The challenge is twofold: first, to focus on the qualitative or systemic forces (poverty, gender, market mechanisms) that drive the quantitative factors involved in climate change (population, consumption, and technology); and second, to address the impact of population growth not only on the generation of greenhouse gases, but also on the ability of countries to adapt to the climate change made inevitable by past pollution. [103]

Population, like technology and consumption, is only an approximate cause of environmental change, and in many situations it is not the most important cause. The impact of population on the environment will depend on many variables: the number of people using a particular resource, the overall level of consumption, the manner in which a resource is extracted or protected, as well as social, institutional, and political factors. Where there has been a demographic transition to lower birth rates in developing countries, it has been achieved because social, cultural, and economic conditions favored the use of contraception.

There have been attempts to capture the wider complexities involved in environmental change. One World Bank analytical framework, for example, suggests that levels of indebtedness and ill-conceived economic policies are the real keys to understanding environmental degradation.[1] For example, it is often not population pressure but mismanagement of public lands that causes major environmental problems in developing countries. With such strong linkages between population, poverty, and environmental issues, the policies that make the most sense are ones that address unsustainable pressures simultaneously. Efforts to improve the rights and welfare of women are an example of such a win-win policy.

The other area in which we need to take a more comprehensive approach is in assessing the impact of population growth, consumption, and technology on the ability of countries to adapt to the climate change that is already inevitable, with global warming and the resulting rise in sea levels. In many countries with fragile coastal ecosystems, adaptation to climate change is more of a priority than reducing their own emissions. If population growth is slower, for example, some countries will find it much easier to respond to climatic changes. Early estimates suggest that the impact of climate change will be to make it much more difficult to assure food, clothing, and shelter for some ten billion people.[2]

Policy makers thus have a double reason to implement community-based development strategies, which are responsive to the ecological and human needs of each locality. Only by addressing the structural social, economic, and institutional problems that generate so much of today's impoverishment and environmental degradation can tomorrow's challenge of adaptation be met. [106]

Conclusions

The relationships between population growth and consumption are still poorly understood. This report has attempted to assess the arguments for enhanced population control as a means to halt climate change. We have found "a combination of muddled thinking and special pleading, which amount to the construction of a 'population myth.'" [106] Based on this assessment, we can draw the following six conclusions:

1. "Action to achieve a sustainable climate should be based on principles of equity (so that the polluter pays) and effectiveness (so that issues of least inertia are given priority). A clear distinction also needs to be drawn between the North's past and present unsustainable exploitation of shared global resources, such as the climate, and the South's potential for unsustainability (at much lower per capita intensity)." [106] Population growth is not the primary factor in climate change and is more difficult to change than consumption levels in the North.
2. It is important to control population growth in developing countries, and many are pursuing policies that are curtailing growth rates. But each nation and community needs to decide for itself how to best manage its demographic policies, recognizing that these are linked to a complex set of socioeconomic issues: poverty, social security, women's education and status, debt, unequal trade, and unproductive structural adjustment.
3. "The attempt by Ehrlich and others to assess the role of population growth in climate change has been used to obscure crucial qualitative elements that ultimately shape factors such as population and consumption growth. Furthermore, the assumption prevails that causality determines policy response: because of the complexity of the causes described above, even if population growth was the primary cause of climate change, increases in family planning programs first would not be the solution." [107]

4. We need to tackle the root causes of unsustainable trends in all areas. In the case of population growth in developing countries, this means addressing underlying socioeconomic issues. In the area of unsustainable consumption in the North, this "is rooted in a combination of market failures and a cultural premium placed on unsustainable growth." [107]
5. "Although developing countries bear little responsibility for climate change, some could be among the most affected by the impacts of global warming. The need to design effective strategies to respond and adapt to global warming should be used to reinforce the existing imperative to achieve community-based management of local resources, and stimulate the search for more subtle paths to fertility reduction, based on improving the status of women." [107]
6. "A new global commitment towards a convergence of equitable life style is urgently required. Wide support should be given to the initiative for a Global Poverty Convention, which should include measures to reduce over-consumption in the North as well as eliminate under-consumption in the South and in marginal sub-populations in the North." [107]

Notes

1. R. Paul Shaw (1992).
2. David Norse (1990).

Summary of

Enclosing the Global Commons: Global Environmental Negotiations in a North-South Conflictual Approach

by Alain Lipietz

[Published in *The North, the South, and the Environment,* ed. V. Bhaskar and Andrew Glyn (New York: St. Martin's Press 1995), Ch. 7, 118–145.]

With the growing importance of international negotiations to address global environmental issues, the North and South have come into conflict following predictable patterns. Overcoming the North-South divide is essential to reaching meaningful agreement on important issues. This article takes a deeper look at the way North-South conflicts have played out in environmental negotiations, highlighting the basis for potential alliances between developed and developing countries.

Enclosing the Global Commons

"With the ongoing negotiations around the international agreements on climate change and biodiversity, humankind is entering a new area. For the first time, we are involved in the collective management of *global* ecological crises." [118] These crises are characterized by diffuse causes and universal effects, quite distinct from local crises in which a local agent generally causes harm to local victims. Global ecological crises involve debates about national models and international justice, because the culprit may be an entire approach to development and the victims may live on distant continents.

The successful international negotiations on the depletion of the ozone layer set the pattern for future North-South dialogue. In this case, though, unlike future negotiations over biodiversity and global warming, the culprits (high emitters of CFCs and other gases) were in the North, and the first obvious victim (Australia) was a Northern high-consumption country. Tension came when the South, specifically India and China, complained that they had not yet gained the benefits—refrigeration—now limited by the Northern agreement. This initiated a new round of negotiations in which large Southern countries exercised considerable influence by their threat to refuse to sign the agreement.

"The 'North-South' aspect of the problem then arises from the fact that . . . the economic consequences may differ in the extreme according to the initial positions of different States, and more precisely according to their historical level of development." [121]

The "enclosure movement" that resulted from the European crises of the 14th–16th centuries was a response to the economic, social, demographic, and ecological crises of the time. The present crises of the global commons may require some sort of "global enclosure" as well, which may have as one of its features the exclusion of "less efficient nations" from some aspects of modernity, much as peasants were "proletarianized" by being excluded from the land. This "political economy of the global commons" is the root of North-South conflict in these negotiations.

The Biodiversity Negotiations

"The biodiversity battle expressed in a caricatured way the North-South character of the global environmental negotiations." [121] Here again the "value of biodiversity" is unknown at present, and it is asserted as part of the global commons. Yet, in contrast to the ozone agreements, surviving biodiversity is largely found in the global South, where standardized and industrialized production has yet to take hold. Thus the North-South tension is heightened, with biodiversity as a raw material in the South and the industries that use it in the North. Hence the U.S. position in the Rio negotiations and in GATT that any molecule in the world can be patented as intellectual property by anyone who "discovers" it, and the South's position that biodiversity is a local resource like oilfields that belongs to the host country and that its use-value should become a common good of humanity.

The negotiations, of course, were more complex. Because a global regulation to protect biodiversity appears to restrict the right to modernize by setting aside protected areas, the opponents included the "productivist elites" in newly industrializing countries like Malaysia and Brazil. The final agreement, which the United States refused to sign in Rio (it was later signed by the Clinton administration) was a compromise acceptable to Northern and Southern elites, because Northern firms won the right to patent living stock while Southern elites won the right to new royalties on their territories in exchange for their agreement not to exploit certain areas. Indigenous peoples were the main losers, as the agreement failed to recognize their role as "gardeners of biodiversity."

The Climate Conflict

While the climate change negotiations had many of the features outlined earlier in the discussion of the ozone agreements, there were two important differences. First, the victims are mainly in the South, where wetter weather will disproportionately affect peasant agriculture and where projected sea-level increases will be a disaster for India, China, Bangladesh, and other countries with large seashore populations. Second, the burden of the policies needed to address the problem would fall mainly on the North, where the majority of greenhouse gas (GHG) emissions originate. But these policies could also limit the South's development options.

We can thus anticipate the factors likely to lead countries to adopt a "Do Nothing" position in global negotiations:

- The anticipated direct impact of global warming on the country, with those least affected being the most likely to "Do Nothing."
- The extent of current GHG emissions, with the largest polluters most resistant to taking action.
- The cost of achieving GHG-efficiency, with developed countries that have the technology to reduce emissions relatively inexpensively more likely to be willing to take action.

This topology of factors produces a useful picture of some of the complexities in the climate change negotiations and explains the range of positions taken. Countries generally group into four categories:

1. Northern "Do Nothing"—The list is headed by the United States, where the dangers are limited and the costs of reducing emissions are marginally low but absolutely high because of the extent of current emissions. The interest in negotiations here is to block agreement. This group also includes other fossil-energy wasters, like South Africa and the ex-socialist countries.
2. Southern "Do Nothing"—Where impacts are relatively limited, current GHG emissions are relatively low, and the cost of GHG-efficiency is high, countries will oppose action, accusing the North of causing the problem and limiting the South's development options. Malaysia articulated this position in negotiations, but this group also includes Brazil, Mexico, and China—countries that are "too poor to be GHG-dangerous" or "GHG-efficient." [130] While it seems opposed to the U.S. position, in negotiations this position serves the interests of blocking agreement.
3. Northern "Do Something"—These are the rich and GHG-efficient countries of the sociological North, including most of Europe and Japan. "These countries are ready to implement the precaution principle: they have or can get the technologies, they are already at a relatively low (yet unsustainable) level of emissions." [130] Their position is: "We have the capacities to fight the greenhouse effect and it is in our interests to offer it to the world." [132]
4. Southern "Do Something"—These are countries that have reason to believe they would be the first victims of global warming: Bangladesh, India, Maldives, and much of Africa and South America. Particularly where GHG emissions and GHG-efficiency are low, they may believe that under any agreement they will retain a wide margin for GHG-sustainable development. In effect, they are saying "We need to fight the greenhouse effect, and we could afford it with some help from the North." [132]

In the actual negotiations leading up to the Rio conference, these differences produced a new basis for compromise, with the "Do Something North" (Japan and Northern Europe) and the "Do Something South" (Southern Asia) presenting a proposal that broke through some of the North-South divide. Presented as the "European Commission Communication to the Council" one year before Rio, the proposal included a carbon-tax in developed countries while allowing Southern countries to increase emissions if they increased GHG-efficiency. The carbon tax was defeated at the Maastricht conference, and that, combined with the failure of the biodiversity treaty, doomed the emergence of a strong "Do Something" alliance at Rio. Instead, it had the appearance of other North-South conflicts, with China, India, and Malaysia in their usual "accusing" position.

Still, the Rio Conference and the climate negotiations were not a total failure. The final agreement captured many Southern demands and European proposals, in particular, leaving out methane emissions (associated with Southern rice and cattle production) and focusing on carbon, as well as recognizing that developed countries had to take the first step. The negotiations also showed the potential to overcome traditional North-South divisions in global environmental negotiations.

BUILDING BLOCKS OF SUSTAINABILITY

PART IV
Population and Urbanization

Overview Essay

by Jonathan M. Harris

Population in the Twenty-First Century

For the last two hundred years, population growth has accompanied economic growth. The relationship between the two has often been in dispute. Some have argued that population growth promotes economic growth, others that economic growth promotes population growth. Thomas Malthus and his followers famously maintained that population growth must ultimately act as a check on economic growth and living standards, while many economists and proponents of technological progress have dismissed this view as unwarranted pessimism. A related debate has been over whether there is an identifiable planetary "carrying capacity" for the human population and, if so, whether we are approaching or have surpassed this maximum level (Marquette and Bilsborrow 1999; Cohen 1995).

Theorists of sustainable development have generally rejected the concept of unlimited growth, whether of population or of economic production. Even if a specific carrying capacity for humans is difficult to identify, resource and environmental constraints will eventually be reached, if they have not been already. A sustainable society, it is widely thought, must ultimately imply a stable level of population. But what does this mean in more specific terms for local, national, and global development policy? Where are we now in terms of population growth or stabilization, and where should we seek to be?

According to the theory of demographic transition, falling birth rates should eventually stabilize population levels. This process has been completed in Europe, which now has a stable population with only slight rates of growth or decline in individual countries. The global demographic transition is different both in quantity and quality. Much larger total numbers, and the still-rapid growth rates in many areas, mean that the global demographic transition is far from accomplished, and the future course of global population growth is still uncertain (Figure IV.1).

Net annual additions to global population peaked around 1990 but are projected to decline only slowly over the next several decades (Figure IV.2). This

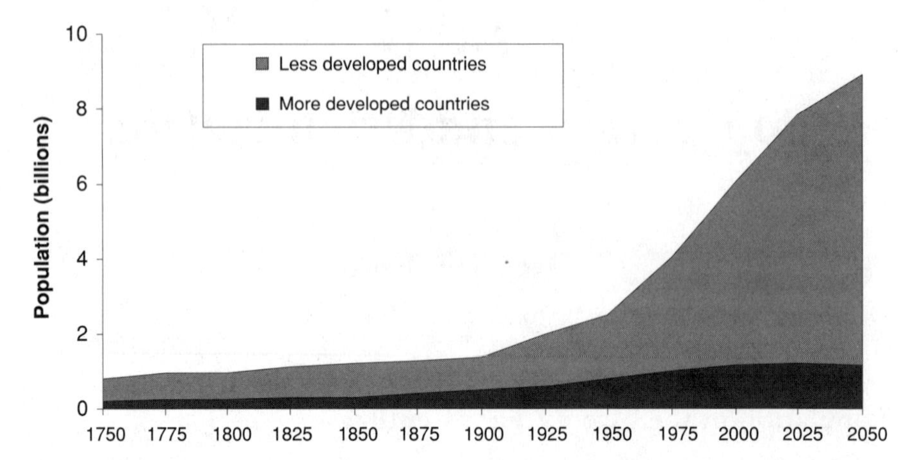

Figure IV.1. Population Growth in Developed and Developing Countries, 1750 to 2050.
Source: United Nations Economic and Social Council, World Population Prospects: The 1998 Revision.

means that while world population is probably headed toward stabilization by the middle of the twenty-first century, there will be very large increases in absolute numbers throughout the developing world before stabilization is reached. For example, in India, where population reached 1 billion in 2000, an additional 400 million people are expected to be added by 2025. Africa, with a population of 790 million in 2000, is projected to reach 1 billion by 2012, then will add nearly 300 million more by 2025.

Continuing global population growth has clear social, resource, and environ-

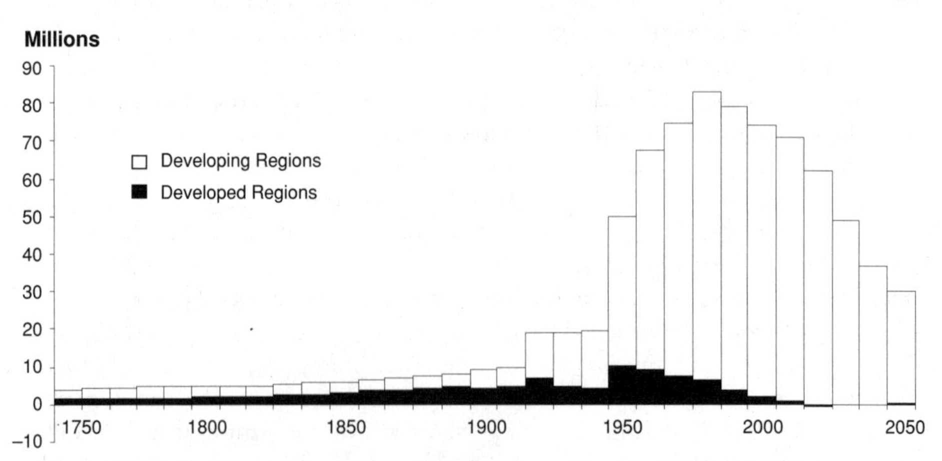

Figure IV.2. Net Annual Increase in Population per Decade, 1750 to 2050.
Source: United Nations Economic and Social Council; World Population Prospects: The 1998 Revision, and Repetto (1991).

mental implications for developing nations and for the world as a whole. From one point of view, stabilizing population is a natural concomitant of economic development. Indeed, the record shows that in almost every country growth in per capita economic output has been accompanied by slowing rates of growth in population. But for countries still facing substantial added numbers, resource and environmental limits are a significant problem. Globally, many resource systems are under significant stress from a combination of population pressures and poor management. Examples are water supplies in South Asia, China, and arid parts of North America, fisheries and grazing lands in many areas of the world, as well as the pollution absorption capacity of ocean and atmospheric systems (Brown 2000).

Even more significant than absolute limits, however, are the interactions of social, economic, and environmental factors. Economists have often pointed out that a resource base that is adequate to support a given population can easily be squandered by inefficient and wasteful patterns of use (Panayotou 1998). Unfortunately, such inefficient use is more often the rule than the exception.

Economic inequality, both international and intranational, also contributes to environmental damage in two ways. First, the consumption of the rich imposes excessive resource demands, meaning that the ecological impact of consumption by affluent nations often extends well beyond their physical boundaries (Wackernagel and Rees 1996). Second, the poor are often forced by their circumstances to adopt destructive resource use patterns. Demand pressures on higher-quality lands force poorer farmers to move onto marginal lands, including hillsides, forests, and arid areas. Lack of credit and market access makes it difficult for small farmers to invest in land- and water-conserving techniques. At the same time, subsidized inputs often encourage more affluent farmers to waste resources. Problems of inequity and inefficiency are not a result of population growth (though population growth can sometimes exacerbate them), but they combine with expanded human numbers to cause rising environmental pressures.

For this reason, standard economic approaches that neglect population must be altered to take specific account of the interrelationships between population, economic activity, and the environment. An extensive, interdisciplinary literature has developed to address this question. In general, simple models have proved unsatisfactory. An adequate understanding of the role of population in sustainable development requires insights from both ecology and economics, as well as from social and political theory. The articles summarized and reviewed in this section draw insights from a variety of disciplines, and from empirical observation, to develop a richer analysis of population and development issues.

Differing Perspectives on Population Growth

Economists and ecologists have different perspectives on population growth, economic growth, and the environment. Economists emphasize the role of institutions and incentives, while ecologists emphasize human/environmental interactions and the far-reaching consequences of ecosystem damage. A third perspective focuses on the importance of social and cultural factors both in determining the course of population growth and in responding to its impacts. The three views are not incompatible. In fact, it is essential to combine insights from all three perspectives to understand the issue and devise appropriate responses.

In an article summarized here, **Nancy Birdsall** suggests that a careful analysis of the relationship between economic growth and population growth can identify policies for promoting sustainability. Birdsall's earlier work includes a thorough overview of economic perspectives on population (1989). Based on this overview, she suggests that the effects of population growth may be neutral, beneficial, or detrimental depending on specific circumstances and existing institutions.

Birdsall focuses on the link between high fertility and poverty, which creates a "vicious circle" of negative social and environmental outcomes. She identifies a significant range of policies that can bring benefits both in slowing population growth and improving economic efficiency and output. Prominent among these are the promotion of education and other social programs, improvement in the status of women, and improved health care, including contraceptive availability.

Birdsall also recommends policies that promote broad-based economic growth as an effective means to reduce both poverty and high birth rates. This is consistent with the evidence showing that people who are more economically secure favor smaller family size. However, Birdsall's analysis clearly differs from the more simplistic assertion that economic growth will solve population problems. If economic growth leads to highly inequitable social systems, the vicious circle of population, poverty, and environmental degradation will worsen. Specific social and economic interventions are essential to avoid this outcome, but these interventions can often be good both for the economy and for stabilizing the population.

Ecologist **C. S. Holling** offers a less optimistic perspective. He maintains that ecologists have good reason to be "gloomy Malthusians." Unlike economists, whose models provide no upper bound on economic growth, physical scientists and ecologists are accustomed to the idea of limits. Natural systems must exist subject to the unyielding laws of thermodynamics, and the science of population ecology has explored the implications of these laws for living organisms. From an ecological perspective, sustainability must involve limits on population and consumption levels. These limits apply to all biological systems. While humans may appear to evade them for a time, they must ultimately accept the boundaries of a finite planet.

However, this simple assertion of limits does not fully capture the contribution of ecologists to the discussion of sustainability. What Holling identifies as a third axiom of ecology has even more significant implications. The third axiom "concerns processes that generate variability and novelty"—the generation of genetic diversity and the resultant processes of evolution and change in species and ecosystems.

Genetic diversity gives rise to *resilience* in ecosystems. Resilience is a "bounce-back" capacity that enables a system to respond to disturbances or damage. For example, a forest ecosystem may recover from a pest infestation through an increase in the population of predators that control the pest, an expansion of species unaffected by the pest, and possibly a development of pest resistance in affected species. The patterns of response will be widely variable, but the resilient ecosystem will maintain its effective functions and its capacity to respond to further environmental changes. The key to resilience is the existence of a wide variety of species interacting with each other and providing a reservoir of genetic forms that provide the potential to adapt to changing conditions.[1]

The importance of the ecological perspective is increasingly evident, as more of the critical problems facing humanity arise from failures of ecological resilience resulting from human impact. The resurgence of diseases due to the development of antibiotic resistance, the disruption of ecosystems by introduced species, the formation of "dead zones" in coastal waters, and the multiple ecological threats related to climate change and increased climate volatility: all testify to the impacts of expanding human economic activity.

The horrifying impact of AIDS, most especially on the African continent, may be a case of feedback effects from the increased size and mobility of human populations. AIDS probably originated in rainforest primates and spread to humans through human intrusion into the forest. Rather than remaining isolated in small communities, it then spread worldwide through global commerce and travel, like many other destructive viruses and pests. Population checks through such drastic ecological backlash are, of course, familiar to ecologists. But they have generally been far from the thoughts of the economists and policy-makers who up until now have shaped our conceptions of development.

How can we balance the more optimistic perspective set forth by Birdsall with Holling's compelling ecological pessimism? Clearly there needs to be some combination of economic and ecological analyses. In Part I of this volume, Giuseppe Munda and others argued that an ecological economics approach requires setting aside a linear view of economic development in favor of a concept of *codevelopment* of economic, ecological, and social systems. At the macro level, we need a more accurate estimate of the impact of economic expansion on the environment—something that is generally completely lacking in macroeconomic theory, as Herman Daly has emphasized (1991). At the micro level, we should consider the interactions of social, economic, and environmental factors in specific situations. Here we will consider some ef-

Box IV.1. The Ecological Footprint

How can we measure the impact of human activity on global ecosystems? One approach is to recognize that all life is dependent on photosynthesis. Green plants convert solar energy into a usable form, which may then be used by animals. An increasing portion of this fundamental energy supply, the net primary product (NPP) of photosynthesis, is used by humans. Human activity also preempts some photosynthetic potential by converting land to urban, residential, and transportation uses.

Ecologist Paul Ehrlich and colleagues have estimated that humans are now directly or indirectly appropriating about 40 percent of the energy supply available from photosynthesis. (Vitousek 1986; Ehrlich 1994.) Clearly, a doubling of this demand, as might well be implied by a 33 percent increase in population (to 8 billion) and a 50 percent increase in per capita consumption by 2050, would leave little room for any other species on the planet. This perspective can be used as a kind of macro framework for the analysis of ecological sustainability (Haberl 1997; Wackernagel et al. 1999).

A related approach, the ecological footprint analysis, has been proposed by Mathis Wackernagel and William Rees (1996). Rather than focusing on NPP, they focus on the land use required to support human consumption. By computing the land use directly or indirectly associated with economic activity, they compute a "footprint" expressed in acres per capita, or total acres for a country, city, or region. By comparing the size of this "footprint" to the available land area within the region under consideration, they compute an ecological surplus or deficit associated with that region. For example, the United States is estimated to have a deficit of 10 acres per capita. Supporting the consumption of the residents of the United States requires 9.8 million square miles of land, whereas the actual land area of the United States is 5.7 million square miles. Thus the total ecological deficit is 4.1 million square miles of land.

The ecological footprint approach has the significant advantage of focusing on consumption levels. This gives essential insight into the relationship of population and resource use. The resource demands of a typical citizen of the United States or Canada, for example, are ten times those of an average Indian (Table IV.1).

Authors from developing nations have often pointed out that Northern consumption, rather than Southern population levels, impose the main burden on the earth's ecosystems (see Rahman in Part III of this volume). At the same time, the ecological footprint approach indicates the extent of added stress that will be imposed on ecosystems as the developing nations increase their consumption requirements toward Northern levels. China and India, despite much smaller per capita footprints, already have overall ecological deficits (Wackernagel et al. 1997).

Critics of the ecological footprint analysis have pointed out that it is a very crude measure of ecological impact, failing to differentiate between destructive and relatively benign uses of land (e.g., an acre used for organic agriculture and an acre used for a toxic waste dump would be rated the same). In addition, it measures hypothetical rather than actual land use, for example calculating the number of forest acres that *would* be necessary to store carbon from fossil fuel emissions (van den Bergh and Verbruggen 1999). This limits its usefulness as an analytical measure and also its policy relevance. However, the ecological footprint retains considerable value as a "tool for communicating human dependence on life-support ecosystems." (Deutch et al. 2000)

Table IV.1. Consumption and Ecological Footprints

Consumption per capita, 1991	Canada	United States	India	World
CO_2 emissions (tons/year)	15.2	19.5	0.81	4.2
Paper Consumption (kg/yr)	247.0	317.0	2.0	44.0
Fossil Energy Use (gigajoules/yr)	250.0 (237 excluding exports)	287.0	5.0	56.0
Fresh water withdrawal (m^3/yr)	1,688.0	1,868.0	612.0	644.0
Ecological Footprint (ha/person)	4.3	5.1	0.4	1.8

Source: Wackernagel and Rees 1996, p. 85.

forts to develop analyses both at the macro and micro levels, looking first at the overall impacts of population and consumption and then at specific development cases.

Rural Population Growth, Agriculture, and Resource Degradation

The relationship between population growth, agriculture, and the environment is complex. Responses to population growth include extensive and intensive agricultural expansion, innovation, migration, and changing fertility patterns. It is the balance between these that determines social outcomes and environmental impacts. Population growth can stimulate innovation and increased agricultural productivity (Boserup 1965, 1981), but there are also many ways in which population growth can contribute to environmental degradation.

Michael Lipton examines the interaction between Malthusian pressures leading to agricultural resource degradation and mitigating factors such as incentives, innovation, and migration. He notes that migration typically increases cultivation on marginal lands, creating both environmental damage and social conflict between residents and immigrants. The situation can be much improved by appropriate policies, including land redistribution and incentives for conservation techniques. Agricultural intensification also can have differing results depending on the policy environment, technology, and price incentives.

In Lipton's view, economic growth is not a panacea, as the forces it unleashes can result either in improved technology and land management, or in increased resource degradation. Institutional issues such as effective management of common property resources, or a transition to secure and equitable private property rights, play a central role. Population growth places stress on existing institutions, but this stress is not necessarily unmanageable. While the issues that Lipton discusses are more specifically related to agriculture, his conclusions are consistent with the approach set forth by Birdsall. Both assert that policy responses are crucial in determining whether population growth has benign or destructive impacts.

Sara J. Scherr also finds that economic and institutional factors play the central role in shaping the outcomes of population growth. Population has grown significantly faster than agricultural land area, resulting in a steady decline in arable land per capita. Increased yields have made possible rising food production per capita, but this has been accompanied by widespread land degradation. Patterns of land use vary between irrigated, high-potential, and marginal lands. In all cases significant environmental problems exist, but in some areas these are more easily remediable through sustainable management techniques. The

Box IV.2. Population Growth and Migration

With an annual growth in world population of over 80 million per year, it might be expected that international migration would play a significant role in redistributing population among countries. In fact, migration flows are relatively small on a global scale (Zlotnik 1998). Among developed nations, the country receiving the largest amount of immigration from the developing world is the United States (see Table IV.2). Total immigration from developing nations to the five developed nations experiencing the largest immigrant flows is only about 1 million annually.

In the United States, about half of current annual population growth of 1.6 million is from natural increase and half is from immigration. But by 2020 almost all net population growth in the United States will be from post-2000 immigrants and their descendants. Without immigration, the population of the United States (274 million in 2000) would peak at about 290 million by 2025; with immigration continuing at current rates, it will grow to 335 million by 2025 (Bouvier and Grant 1995; Population Reference Bureau 1999). Immigration, primarily from the developing world, is thus a significant contributor to population growth in the United States as well as to population growth in Canada and Australia. In Europe, the proportion of immigrants from the developing world is much smaller—only about 0.1 percent—and population growth is also lower—also about 0.1 percent.

Table IV.2. International Migration to Selected Countries

Country	Annual Immigration 1995–96 (Canada 1990–1994)	Immigration from Developing Countries
United States	813,730	653,518
Canada	235,509	184,547
Australia	102,030	74,190
Germany	397,935	70,099
United Kingdom	53,900	37,100

Source: Zlotnik 1998.

greatest problems exist in marginal areas where degradation tends to be more severe and the institutions and incentives for better management techniques are weakest.

Like Birdsall and Lipton, Scherr emphasizes the importance of economic incentives. But she also sees a direct role for population growth rates in determining outcomes. As people struggle to respond to higher demands on the land, slower population growth allows crucial breathing space—time to innovate and adapt. Higher population growth rates can push rural communities over the edge into neo-Malthusian collapse—not because of an absolute limit on carrying capacity but because the means and incentives to adopt new techniques were not forthcoming in time. Scherr's article strikes a fine balance, indicating both the urgency of the situation in marginal rural areas and the potential for effective policy responses.[2]

Urban Growth: Poverty, Infrastructure, and Environmental Crises

Population growth is most rapid in urban areas. The world's urban population grew from 1.54 billion in 1975 to 2.58 billion in 1995 and is projected to reach 5 billion by 2025 (World Resources Institute 1996, 150). While rural growth rates worldwide are about 0.8 percent per annum, urban growth rates, driven by natural increase and by in-migration, are 2.5 percent. In Africa, urban growth rates are 4.4 percent per annum, and in Asia 3.3 percent. These growth rates imply a doubling in population size within a generation. They usually reflect a combination of natural increase and of rapid in-migration, and pose a huge challenge for twenty-first century development.

Rapid urban growth has led to major social and infrastructure problems in rapidly growing cities in developing nations. Inadequate housing and sanitation, congestion, air and water pollution, disruption of water cycles, deforestation, solid waste problems, and soil contamination are typical of large cities in the developing world.

Sai Felicia Krishna-Hensel discusses the role of urbanization in Indian development. A staggering combination of health, environmental, and social problems affect India's megacities. Water supply is one of the most critical issues, with limited availability exacerbated by serious pollution of existing supplies. Meanwhile, population growth is relentless. The population of Calcutta doubled from 1950 to 1980, and will have doubled again by about 2010. For India as a whole, urban population is projected to rise from 250 million in 1995 to 630 million in 2025 (World Resources Institute 1996,151). How can urban authorities cope with existing problems, let alone manage this continued massive growth?

Krishna-Hensel sees some positive trends: projects by governmental and nongovernmental agencies, as well as by private firms, to improve health, nutrition, and sanitation, as well as self-help movements among the urban poor. Strategies of decentralization, service privatization, and community-based development have been successful. However, the point made by Sara Scherr regarding rural growth and adaptation surely applies with even greater force to the urban situation. The struggle to respond to massive social and environmental problems is clearly made far more difficult by continuing rapid and unplanned growth. Moderation of overall population growth, and possibly also of in-migration, will have to be an essential component of efforts to achieve urban sustainability.

Priscilla Connolly examines the case of Mexico City, where urban population tripled between 1950 and 1970 from 3.1 to 9.3 million and then grew by another 60 percent to reach 15 million by 1995. Growth rates have now slowed, however, with net in-migration becoming negative. The city will continue to grow as a result of natural increase, but at a slower rate. Major problems include overdraft of water supplies and serious air pollution. Unfortunately, the transportation system of the city has been heavily oriented toward automobile transport, with the number of cars rising much more rapidly than the population. This trend has been supported by direct or indirect government subsidies for road-building and fuel. Similarly, water management policy has been oriented toward augmenting supply rather than toward limiting demand or increasing conservation.

While the picture Connolly presents of the current situation in Mexico City is not encouraging, she does not see current population growth as the most crucial issue in urban problems. The central problem is poor policies that have contributed to worsened housing, health, sanitation, and environmental conditions. In theory, then, policy reversals could lead to significant improvements in well-being. The main barriers that Connolly sees to the development of a better policy environment include concentrations of political power, corruption, and income inequality.

The studies by Krishna-Hensel and Connolly frame the issue of urban sustainability, showing once again the interaction between underlying population growth and policy responses. The relative weighting differs depending on the

Table IV.3. The World's Twenty Largest Cities and Their Average Annual Growth Rates*

	1995 Population (millions)	Average Annual Growth Rate, 1990–95		Population (millions)	Average Annual Growth Rate, 1990–95
Tokyo, Japan	26.8	1.41	Jakarta, Indonesia	11.5	4.35
São Paulo, Brazil	16.4	2.01	Buenos Aires, Argentina	11.0	0.68
New York, USA	16.3	0.34	Tianjin, China	10.7	2.88
Mexico City, Mexico	15.6	0.73	Osaka, Japan	10.6	0.23
Bombay, India	15.1	4.22	Lagos, Nigeria	10.3	5.68
Shanghai, China	15.1	2.29	Rio de Janeiro, Brazil	9.9	0.77
Los Angeles, USA	12.4	1.60	Delhi, India	9.9	3.80
Beijing, China	12.4	2.57	Karachi, Pakistan	9.9	4.27
Calcutta, India	11.7	1.67	Cairo, Egypt	9.7	2.24
Seoul, S. Korea	11.6	1.95	Paris, France	9.5	0.29

*Dhaka, Bangladesh, is twenty-third in size but is growing at 5.74 percent per annum.

Source: World Resources, 1996; O'Meara 1999.

particular situation. Cities such as Mexico City, Buenos Aires, and Rio de Janeiro, as well as major cities in the developed world, are now experiencing relatively low growth rates, which may provide the opportunity for policy reform in the direction of sustainability. Other major cities such as Bombay, Shanghai, Beijing, Calcutta, Jakarta, Karachi, Dhaka, and Lagos are still experiencing rapid population growth (Table IV.1). In almost all cases, the problems are urgent, but the situation in the still rapidly growing cities may be more critical.

Fortunately, there is extensive experience with effective urban policy reform, albeit rarely at a scale commensurate to the size of the problems. **Molly O'Meara Sheehan** reviews efforts to improve urban policies and institutions

Box IV.3. Sustainable Urban Management in Curitiba, Brazil

"Curitiba, Brazil, has received international acclaim as a city that works—a good example of sustainability and exemplary urban planning. In 1950, however, all trends indicated that Curitiba was likely to become yet another city overwhelmed by rapid population growth and urban environmental problems. From 1950 to 1990, Curitiba mushroomed from a town of 300,000 to a metropolis of about 2.3 million. Migrants, pushed from the land as a result of agricultural mechanization, flocked to the city and settled in squatter housing at the urban periphery. . . . How did Curitiba manage to turn itself into a positive example for cities in both developed and developing countries? . . . The most important feature of Curitiba's success is its emphasis on integrating transportation and land use planning." (World Resources, 1996)

Curitiba used a radial pattern of public transit to channel development outward, keeping the central city as a pedestrian zone. Automobile use has been kept to a minimum. Effective water and sewer networks have been developed. Green space and parks have been protected by legislation, and an effective recycling program handles two-thirds of the city's trash. Publicly subsidized loan programs encourage the development of low-cost housing.

Key to the success of Curitiba has been a responsive, democratic government oriented toward public participation. Incentives have been developed for the urban poor to participate in housing, streetcleaning, and recycling programs. Street vendor activity has been licensed. City programs for children offer food, shelter, and work opportunities to streetchildren, as well as a network of daycare centers for working parents and sports centers for after-school activity. An "Open University of the Environment" provides education and opportunity for involvement in city planning, and is establishing an economic and environmental database for the city. Despite continuing poverty, the city has low crime rates and has been able to maintain a high quality of life and sense of civic involvement.

Sources: World Resources 1996,120–121; Rabinovitch and Leitmann 1993; McKibben 1995.

governing patterns of water, waste, food, energy, transportation, and land use. These include

- Investing in transportation infrastructure, water, and sewer systems
- Decentralized, community-based systems for water supply, sanitation, and recycling
- Urban food production
- Use of decentralized and renewable energy technologies
- Effective public transportation systems and road pricing
- Tax and zoning systems that reward urban land improvement and reduce sprawl
- Effective use of service fees and municipal bonds

O'Meara notes the emergence of new networks providing databases and information exchange on effective urban policies, including the Urban Management Program, a joint effort of multinational development organizations, and the nongovernmental International Council on Local Environmental Initiative (ICLEI). The importance of these and similar institutions will certainly increase as the urban portion of the world's population grows from 45 percent to over 60 percent during the next quarter century.

Toward Consensus on Policy Solutions?

It is apparent that population growth has been a major factor in shaping development patterns during the second half of the twentieth century and will continue to play a central role during the first half of the twenty-first. Its role has generally been neglected in economic theory. As we have seen, however, when the perspectives of economists, ecologists, and other social theorists are brought to bear on issues related to population, new insights emerge. This interdisciplinary analysis will be crucial in shaping policies for sustainable development for the foreseeable future.

Despite the wide array of differing analytical perspectives, some consensus has emerged regarding responses to population growth. Extreme pro-natalist perspectives have generally been discredited: there is broad agreement that moderating population growth is an important goal. Top-down population control policies have also been discredited both on human rights grounds and as failing to alter basic incentives regarding fertility. Sen (1999) points out that the voluntary reduction of birthrates in Kerala, India, associated with higher levels of basic education and health care, has actually been more effective than China's draconian "one-child family" policy.

Improving nutrition, health care, and education, especially women's education, are seen as key factors in lowering fertility rates. Improved employment possibilities (for women and in general), improved pension systems, access to contraception, and better information on methods and benefits of family planning, are all "win-win" policies. Sound macroeconomic policies, improved credit markets, and improved terms of trade for agriculture are important in promoting broad-based growth and poverty reduction, which in turn is essential to population/environment balance.

Significant differences among the various perspectives remain, and we will make no attempt to resolve them here. However, it is evident that a serious focus on population issues is essential and can serve as the basis for a better understanding of the relationship between population, resources, economic growth, ecosystem health, and human well-being.

Notes

1. In *The Diversity of Life,* E.O. Wilson asserts that "biological diversity is the key to the maintenance of the world as we know it" (Wilson 1992).
2. A similar argument concerning the urgency of implementing sound population policies before ecological thresholds are crossed has been made by Robert Engelman (1999).

Summary of

Government, Population, and Poverty: A Win-Win Tale

by Nancy Birdsall

[Published in *Population, Economic Development, and the Environment,* ed. Kerstin Lindahl- Kiessling and Hans Landberg (New York and Oxford: Oxford University Press, 1994), 173–198.]***

Concern that rapid population growth may slow development has been reinforced by awareness of environmental stresses in developing nations that could be exacerbated by population growth. However, there is no consensus on whether government should intervene aggressively to lower population growth rates. Some question whether population growth is a fundamental cause either of slower economic growth or of environmental degradation. Others are con-

***Also published Cassen and Robert. *Population and the Environment: Old Debates, New Conclusions.* Ch. 14. (New Brunswick and Oxford: Transaction Publishers, 1994).

cerned that government intervention in family decision-making will have a high cost to human rights and well-being.

This article reviews the main concerns about the impacts of population growth in developing countries: slower economic development, greater environmental damage, and greater poverty and income inequality. It then describes the kinds of interventions that are justified by welfare theory. These interventions turn out to be "win-win" in the sense that they can be justified as having beneficial effects independently of the population issue. The main reason for this is the well-established link between high fertility and poverty. Breaking the vicious circle of poverty and high fertility is beneficial in terms of increasing the welfare of the poor and is also effective in lowering population growth rates.

Concerns About Rapid Population Growth

The negative effects often attributed to rapid population growth arise from *externalities* associated with childbearing decisions. The decision to have a child has external effects, including reducing the per capita allocation of publicly owned and common-property resources such as fisheries and aquifers. Where such resources are already mismanaged or overused, population growth makes matters worse. This is true for natural resources. It also may be true for social programs such as education. With high fertility and absolute increases in the size of the school-age population, it is more difficult for governments to maintain adequate educational spending. If the increase in social spending is insufficient (to maintain quality, for example), society as a whole may lose out, because higher levels of education augment growth.

In addition, population growth may depress wages, especially in rural areas, creating a *pecuniary externality* that reduces the incomes of the poor and exacerbates problems of poverty and income inequality. Also, to the extent that people seek larger families in order to increase their relative status or security, the effort becomes self-defeating if all families do likewise, and average family size rises without any individual family gaining advantage.

What Kind of Interventions?

The main criterion for effective population policy should be to seek interventions that simultaneously induce lower fertility and improve the welfare of the poor. Four of the following five policies satisfy this criterion:

- *Taxes on children and fertility-reduction incentives*—These are problematical because taxes will damage the welfare of the poor and would penalize children born into large families. Incentives are less directly damaging but could constitute a form of coercion on poor families desperate for additional income and are therefore considered unacceptable in many countries (they have been used in China, India, and Bangladesh).

- *Education and other social programs*—Expanded educational opportunities for girls raise the opportunity cost of childbearing and reduce fertility while improving the income prospects of women. The availability of local education encourages families to invest in educated children (with greater income-earning potential), rather than in large numbers of children. Pension programs also increase the welfare of the poor while reducing the need for children as providers of old-age security.
- *Availability of contraception*—There is significant unmet demand for fertility reduction among women in developing countries. Poor information about contraception, fear of social sanctions, and high prices for contraception all contribute to this gap, which affects the poor more than the rich. Subsidized family planning services for the poor can close the gap and improve the welfare of the poor by increasing their choices. In some cases, the issue is not so much the need for subsidy as for the removal of government-sponsored restrictions on contraception. Examples from Latin America shows how, despite such government restrictions, the poor seek out modern contraception wherever possible.
- *Supply of information*—Higher returns to education, better health care and declines in child mortality, and improved employment opportunities tend to reduce desired fertility. To the extent that people are unaware of these factors, government educational policies can be effective in reducing fertility and increasing welfare.
- *Good economic management*—Poorly functioning capital markets, inflation, and economic insecurity increase the desire for children as one of the few dependable assets for the poor. Broad-based growth has led to dramatic reductions in poverty in Indonesia, Thailand, Botswana, Chile, and Costa Rica. Policies to promote broad-based growth include universal access to basic education, lower taxes on the agricultural sector, and an emphasis on nontraditional exports, which encourage innovation and create employment opportunities.

The latter four of these five are all win-win types of intervention, which reduce the externalities associated with childbearing while providing high social and economic returns. Public policies and programs to raise human welfare and reduce population growth can thus be fully consistent.

Summary of

An Ecologist View of the Malthusian Conflict

by C.S. Holling

[Published in *Population, Economic Development and the Environment*, ed. Kerstin Lindahl-Kiessling and Hans Landberg (New York and Oxford: Oxford University Press, 1994), 79–103.]

Ecologists and economists often have different points of view on issues of population growth and the carrying capacity of the planet. In part this arises from the different theoretical paradigms they base their views on, but it may also reflect misunderstandings that arise on both sides. This article attempts both to establish some areas of agreement and to explain the basic principles and some of the subtleties of the ecological approach.

Economic and Ecological Perspectives

At a recent meeting of economists and ecologists sponsored by the Beijer Institute of the Royal Swedish Academy of Sciences, participants were able to overcome stereotypes regarding the limitations of each others' fields and discover elements of convergence between the disciplines. In particular, both economists and ecologists agreed that current problems often had a similar profile in the following respects:

- Human influences on air, land, and oceans trigger threshold effects that can threaten health and the environment.
- There is an increasing globalization of biophysical phenomena accompanying the globalization of trade and large-scale movements of people.
- Problems emerge suddenly in several different places rather than in one location at a speed that allows for a considered response.
- Both the ecological and the social effects of these problems are novel and unfamiliar, and thus are inherently unpredictable.

The perspective of ecologists on these developments can be understood by examining three questions:

- Why are ecologists so gloomily Malthusian?
- Why has the world not collapsed long ago?
- Why worry about the negative impacts of growth in human populations and activities?

Why Ecologists Are Gloomy Malthusians

Ecologists generally accept two axioms: populations have the inherent propensity to grow exponentially, and the environment sets ultimate limits to growth. Species of organisms can be loosely divided into two groups based on their survival strategies: those that specialize in growth, often colonizing recently disturbed habitats, and those that can outcompete the growth specialists and persist for long periods. Populations of both types encounter limits, giving rise to the theory of ecological succession according to which a particular set of "climax" species form a stable assemblage that remains until some disturbance allows new "pioneer" species to establish themselves and restart the succession process.

Mathematical ecology, initiated by A.J. Lotka (1925), uses differential equations to establish the relationship between exponential growth and environmental limits, often generating population oscillations around a stable point. If time lags are introduced into a system of predator-prey relationships, the oscillations can become destabilized, leading to the extinction of species. This highly deterministic formulation, however, must be modified by a third axiom: the continual propagation of variability and novelty, which is the basis for the theory of evolution. The interaction of this third axiom with the first two gives a more subtle picture of continual experimentation in a changing environment.

Why Has the World not Collapsed Long Ago?

Given the possible instability of predator-prey interactions as well as external physical variability, the key to system persistence lies in spatial heterogeneity and biotic diversity. These characteristics can make an ecological system *resilient*—able to withstand internal imbalances or external disturbances.

Ecological models show a very wide range of complex behaviors, with multiple stable states, boom-and-bust cycles, and even chaotic behavior. Plant- and animal-specie fluctuations on a local scale interact with geophysical variables on a much larger scale to generate robust and resilient ecosystems. Human population growth and economic activity affects the local-scale relationships in ways that can profoundly change overall ecosystems.

The resource management concepts of maximum sustained yields (e.g., of fish populations) and fixed carrying capacities (e.g., of terrestrial herbivores) have been discredited by these more sophisticated views of broad ecosystem function. The very success of achieving management yield goals tends to reduce variability and damage ecosystem resilience.

Part of the answer to the question "why has the world not collapsed?" lies in the resilience of ecosystems. The other part lies in human creativity and adaptive behavior. Human adaptability is the key to economists' optimism about our ability to substitute for scarce materials and develop successful responses to environmental problems. However, the resilience of natural systems is not unlimited, and human adaptability is limited by specific environmental contexts.

Why Worry about Negative Impacts of Growth in Human Population and Activities?

Studies of ecological management and policy have shown that human exploitation of ecosystems decreases resilience, making the system more vulnerable to surprise and crisis. Resilience decreases due to loss of species diversity and an increase in spatial homogeneity. Management institutions that are economically efficient are often rigid and unresponsive to resource dynamics. For example, short-term success of spraying programs to control spruce budworm has led to more extensive outbreaks over larger areas. Use of fish hatcheries for salmon spawning has resulted in a precarious dependence on a few artificially enhanced stocks. Both ecosystems and local timber and fishing industries are thus endangered.

Such problems are now widespread. "Increasing human populations in the South and the planetary expansion of their influence combined with exploitative management in both North and South, reduces functional diversity and increases spatial homogeneity not only in regions but on the whole planet." [93] Critical attributes of ecological resilience are now being compromised at the planetary level.

Negative impacts of human activities on a global scale include increases in climatic variability, possible climate warming, increasing soil aridity in large areas, and increased ultraviolet radiation that is damaging plankton productivity in the oceans. Less well-known examples include

- *Increasing costs of natural disasters*—This results from more people moving to more vulnerable places; loss of resilience as a result of human construction such as dikes, dams, and drainage canals; and increased climate variability resulting from deforestation and fossil fuel emissions.
- *Decline in migratory bird populations*—Habitat fragmentation in temperate regions and landscape transformations caused by human pressures in neo-tropical regions have led to a decline of almost 50 percent since 1965 in migratory bird flights between North and South America. This in turn increases insect outbreaks that affect forest ecology.
- *Emergence of new human diseases*—There is growing evidence that new infectious diseases are initiated by environmental change. Viruses such as AIDS, Ebola, Marburg, Rift Valley fever, and the Hantaan virus, all of which have recently appeared in human populations, have existed in other hosts or in benign forms for centuries. Recent human landscape disturbances have created opportunities for these viruses to cross over to hu-

> mans, and extensive human movements have then spread them on a planetary scale. HIV, for example, may have existed in isolated populations for centuries, causing little illness. But rapid urbanization and the global movements of people created the conditions for the emergence of a much more lethal virus and a resulting global pandemic.

Population growth and greater land use contribute to the intensification and expansion of these and other destructive trends. The resulting destabilization of global ecosystems will have inherently unpredictable consequences on a global scale.

Summary of

Accelerated Resource Degradation by Agriculture in Developing Countries?: The Role of Population Change and Responses to It

by Michael Lipton

[Published in *Sustainability, Growth, and Poverty Alleviation: A Policy and Agroecological Perspective,* ed. Stephen A. Vosti and Thomas Reardon (Johns Hopkins University Press, 1997), Ch. 6, 79–89.]

What is the relationship between rural development patterns and resource degradation? People are usually rational, and do not seek to destroy the resources on which they and their communities depend. However, there are national and international pressures that may lead them to do so. At the national level, these pressures include population increases and declines in common property resources. At the international level, forces such as interest rate changes and technology transfers can affect resource use. The international forces are dealt with elsewhere (by Lipton in Chapter 11 of *Sustainability, Growth, and Poverty Alleviation,* and in Part VII of this volume). This article addresses the relationship between population growth and resource degradation, including a consideration of how population growth interacts with management of common property resources.

Population Growth and Environment

There is a broad academic consensus regarding causes and effects of population growth in developing countries. It is generally accepted that most couples act rationally in setting family size norms, subject to societal pressures. However, individually rational decisions may not lead to socially optimal consequences. As

Sen has pointed out, poor couples who have more children in the hope of eventual higher incomes may find that overall labor supply increase drives down wages, leaving them worse off. Lack of information, low levels of female education, and poor employment prospects for women can all contribute to fertility levels that are higher than is socially desirable.

The discussion of the population/resource relationship starts with Malthus. While Malthus did not deal explicitly with issues of resource degradation, he saw both diminishing returns in agriculture and wage-lowering labor surpluses as leading to immiseration. He acknowledged extensive and intensive agricultural expansion as possible responses, but believed that both would have limits. In his later work, he placed greater hope in willingness to restrain fertility in response to economic growth and educational opportunity. However, both he and his modern successors give insufficient attention to incentives and to changing patterns of behavior, including migration and innovation, in response to population change.

Extensification and Migration

In many countries low-potential marginal areas are showing faster population growth than areas of high agricultural potential. The Rajasthan Desert in India, for example, has experienced a population growth rate significantly higher than India as a whole, with cultivation being extended onto fragile arid areas where water scarcity limits yields. This is a seemingly paradoxical trend, since one might expect more outmigration to more fertile areas. But in practice rural-to-rural migration to Green Revolution areas, such as the Punjab in 1967–1973, is often not sufficient to prevent big rises in real wage-rates. This induces more use of labor-displacing methods (tractors, threshers, reaper-binders, etc.) and a shift of land from small labor-intensive farms into bigger, mechanized, labor-displacing farms—both by amalgamation and by landlords resuming sharecrop tenancies. So, demand for labor migrants in Green Revolution areas often falls off later. In the longer term, rural-to-rural migration involves net movement toward fragile areas, as displaced rural workers and smallholders search for new land. Other examples of labor movement away from Green Revolution areas and toward extensive margins include migration from southern to northwest Brazil, and in Bangladesh from Comilla to the Chittagong Hill Tracts.

Incoming migrants may come into conflict with local populations who use land more sustainably but who lack clear property rights systems. Intensification in the receiving areas often results in lower elasticity of employment with respect to output so that poverty reduction from intensification is lower than might be expected. Extending cultivation can thus lead to a socially and environmentally unsustainable situation.

These negative impacts can be moderated by government support for appropriate technologies and price incentives for conservation techniques that "substitute employment for environment"—for example, by planting and maintaining tree cover, by terracing, or by erecting vegetative or contour-plowing barriers against erosion. Investment in communication, innovation, and research on sustainable techniques are also important. While these are not strictly public goods, they are unlikely to be adequately provided by private enterprise.

Land redistribution in the more fertile areas can also be a major remedy for excessive pressure on marginal lands. This has been true in many parts of Asia (including China, Taiwan, Kerala, and West Bengal) as well as in Africa (Kenya, and currently possibly Zimbabwe and South Africa). Rural-to-urban migration, in contrast, has a rather small impact on net resource and farmland availability; any increase in per-capita farmland supply is balanced by the tendency of urban growth to pull land and water into nonfarm uses.

Intensification and Technical Progress

Intensification can raise farm output per person, both in currently farmed areas and on the extensive frontier. However, the production increases resulting from intensification may or may not be sustainable. Population growth may engender Boserup-type responses, in which innovation and investment raise per person output, or Hayami-Ruttan-Binswanger (HRB) responses, in which increased labor availability serves as an incentive to labor-using technical change. But empirical evidence indicates that HRB responses to rural population growth in parts of South Asia are weak, and Boserup responses are very weak in sub-Saharan Africa. These problems are related to the need for appropriate early rural investment, for example in water management. Such investment may or may not be forthcoming, and the role of price and technology incentives is crucial.

"Intensification can lead to eroded dust bowls—or to the use of fertilizers and composts to regenerate depleted soils. Extra labor can repair bunds and plow along contours—or harvest more and more high-yielding cassava until the soil is destroyed." [86] If population growth creates heavy pressure to raise agricultural yields in the short term, longer-term conservation goals are likely to be neglected.

Fertility Responses to Population Growth and Economic Growth

The changes in income, incentives, and information linked to rural modernization eventually lower fertility rates, but evidence from India suggests that this can be a long process. HRB-type technical progress raises the returns to child and adolescent labor well before better opportunities for women, which create an incentive for lower fertility, become available. Only at a later stage in the development process do the benefits of education for women and their children predominate over the income-generating and income-security effects of larger families.

It is not necessarily clear that eventual slower population growth will lead to less resource degradation. The process of earning extra income may degrade local resources faster than simple population growth. Forest products, for example, have an income elasticity greater than one, meaning that demand for forest resources will increase disproportionately with income growth. It is important to understand the relationship of income growth and resource depletion rates. This relationship will be determined by relative prices and available technologies, which will tend to be more exogenous ("imported") rather than endogenous (locally created) as communities become more integrated into the national and global economies.

Common Property Resources

Common property resources (CPRs) are important in developing countries, especially in arid and semi-arid areas. In India, CPRs have declined in area and productivity since the 1950s, to the detriment of the rural poor. Population growth is positively correlated with CPR decline for two reasons. First, an increased number of claimants to the CPR decreases benefits per person and makes rule enforcement more complex. Second, more people living near a CPR increases the likelihood of poaching or illegal use. While it is definitely untrue that CPRs necessarily lead to the "tragedy of the commons," some resources may be better protected if there is a shift to other forms of ownership.

The structure of property rights, however, is not the key issue. Rather, the crucial variables are prices, technology, and incentives, which can lead to the use of resource-degrading or of resource-conserving techniques. It is the combination of higher population and incentives for short-term resource management that poses a threat to sustainability.

Summary of

People and Environment: What Is the Relationship Between Exploitation of Natural Resources and Population Growth in the South?

by Sara J. Scherr

[Published in *Forum for Development Studies,* No. 1 (1997).]

Over the past century technological and institutional innovations have dramatically raised global food production capacity, causing the specter of Malthusian famine to recede, although the challenge of equitable food distribution remains. However, a neo-Malthusian pessimism suggests that while food production potential may keep pace with population, our capacity to maintain environmental integrity may not.

This article examines the connections between population, agriculture, and natural resource management. A broad overview of land-quality issues is presented. The example of land management under population pressure in tropical hillsides is then used to analyze the sustainability of agricultural production and the maintenance of environmental services. Somewhat surprisingly, the evidence indicates that the effect of population on land quality is indeterminate, with other economic and institutional factors being more significant in determining outcomes.

The Nature and Scale of the Problem

Since the early 1960s, cultivated area has increased by 18 million hectares (ha.) in Asia, 28 million ha. in South America, and 31 million ha. in Africa. Area in permanent pasture expanded even more, while forest and woodland area declined. While deforestation is likely to continue, area in crop production is likely to expand only another 12 percent over 1997 levels by 2010. Rising total population means that arable land per capita has actually declined from just under 0.5 ha./cap in 1950 to under 0.3 ha./cap in 1990, and it is expected to continue to decline to between 0.1 ha./cap and 0.2 ha./cap by 2050.

Average crop yields have grown rapidly since the 1970s, and yield gains are projected to continue, though at lower rates. Per capita yields rose much more slowly due to the effect of rapid population growth. In addition to food requirements, rural land and resources were increasingly needed to provide other livelihood needs. Although rural growth rates have declined from 2.2 percent per annum in 1960–1965 to 1 percent per annum in 1990–1995, the absolute number of rural dwellers will continue to increase at least through 2015. In 2015, the developing world will have 3 billion rural dwellers, 94 percent of the world's rural population.

The Global Land Assessment of Degradation (GLASOD) estimates that nearly 2 billion hectares of agricultural land, pasture, forest, and woodland (22.5 percent of the total) have been degraded since the mid-1900s. About 3.5 percent is severely (perhaps irreversibly) degraded, and just over 10 percent is moderately degraded, requiring significant on-farm investment to restore. Nine percent is lightly degraded and could be restored through good land husbandry.

Damage to forests and agricultural lands is most widespread in Asia, and to pasture, in Africa. Poor agricultural practices, overexploitation of vegetation, overgrazing, and deforestation all contribute to land degradation. However, GLASOD figures may overstate the scale and impact of the problem, since damages may be compensated by nonland inputs or by land-improving measures such as terracing, nutrient enrichment, contour hedges, or agroforestry. While there is considerable local evidence documenting land degradation, there is less data on productivity or income effects.

Dynamics of Change

Four different patterns of land use change can be identified for different types of rural land:

- *Irrigated lands*—Irrigated area grew by 100 million ha. between 1961 and 1990, mainly in developing countries. Irrigated areas are now 17 percent of cropland but produce a third of world food output. Green Revolution gains are concentrated on irrigated lands, but so also are problems of salinization, groundwater decline, waterborne diseases, and water pollution from fertilizer, pesticide, and animal waste runoff.
- *High-potential rainfed lands*—Intensification in areas with fertile soils and reliable rainfall has made permanent cropping (sometimes export cropping) possible, with increased fertilizer application and improved seeds. This pattern is characteristic of parts of India, Zimbabwe, the uplands of Java, and the Kenyan highlands. Environmental concerns include mechanization damage to soils, acidification, pesticide pollution, and deforestation. The resources in these areas are generally resilient, and with proper incentives technologies for sustainable management can be adopted.
- *Long-settled marginal lands*—In areas with less-favorable environments such as drylands, rain forests, shallow or acid soils, or steep slopes, population and demand growth have forced a transition from long-fallow to short-fallow or permanent cropping systems. As a result, these lands are threatened by soil erosion, fertility depletion, and devegetation. Further ecological impacts include loss of biodiversity and degraded watersheds. Frequent crop failures can lead to depletion of forests and other resources to meet consumption needs. Poor infrastructure and an inability to undertake land-improving investments compound problems in regions where poverty and subsistence production is widespread. Some promising technical approaches for such areas have been developed, but they have received only limited research attention.
- *Frontier marginal lands*—Migration of landless and displaced people has led to new settlements and land-clearing in marginal environments. Long-fallow cropping and livestock production have replaced traditional shifting cultivation systems, with resulting deforestation, biodiversity loss, and watershed degradation. Problems of lack of infrastructure and inputs are even

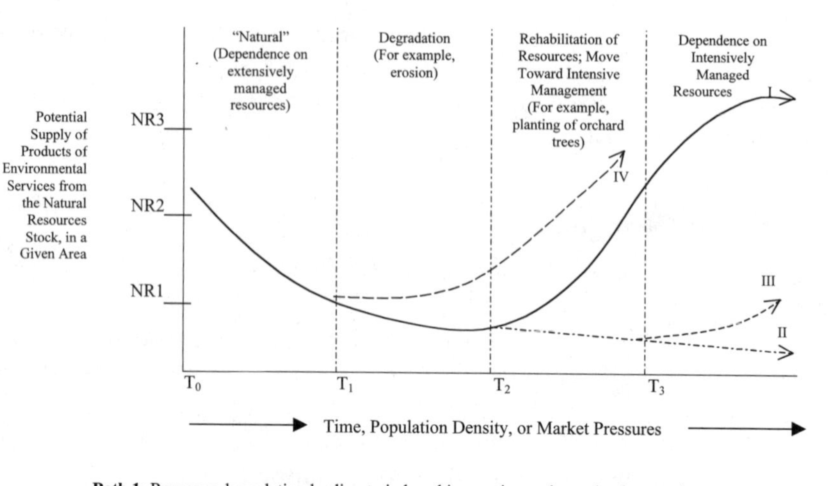

Path 1: Resource degradation leading to induced innovation and sustained output increase
Path 2: Overexploitation and continued resource decline
Path 3: Delayed resource conservation with limited potential
Path 4: Accelerated innovation leading to successful intensive management

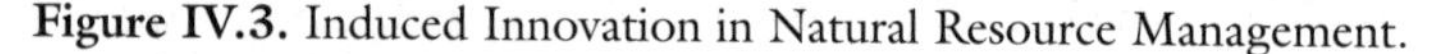
Figure IV.3. Induced Innovation in Natural Resource Management.

more acute in these regions. Examples include the Amazon and the Atlantic zone of Central America, West African drylands, and the Indochinese hill country.

While total population is highest in the two higher-potential areas, as of the mid-1980s half of the poorest people in developing countries were living in marginal lands. Pathways of development vary widely under all four land patterns, with intensification sometimes leading to land degradation and sometimes to land improvement. Improvement of productivity, human welfare, and protection of the natural resource base are dependent on interactive relationships between population and economic growth, technology, institutions, investment, and infrastructure. In some circumstances, the end result can be a neo-Malthusian collapse of the resource base, while in others resource-conserving intensification can lead to a sustainable outcome (Figure IV.3).

Land Quality and Population Growth in Tropical Hillsides

Around 500 million people now live in tropical hillsides and mountains. Extensive empirical studies of such areas find that increases in population density are associated with higher rates of deforestation, but also with increases in planted tree density. Similarly, a denser population leads to shorter fallow and more frequent cropping, but also to more land-conserving investments. Overgrazing tends to occur at moderate population densities, but at higher human and livestock densities there is a shift from land-intensive to labor-intensive feeding methods.

"Most of the environmental impacts of production increases in the hills and mountains thus depend on whether sufficient microeconomic incentives exist for people to choose production systems . . . that enhance land characteristics or, at least, retard their degradation." [42] Successful intensification has been documented in the Machakos district of Kenya and in the mixed crop-livestock systems of western Kenya. Increased market opportunities and better product prices are conducive to farmer investment in tree-planting and soil conservation. Secure land tenure and cooperative land improvement associations also help to sustain production at higher population levels without excessive resource degradation.

"Higher rural population is not directly associated with degradation. However, a slower rate of demographic growth or decline allows people more time to innovate or adapt products, technologies, property rules, and collective management. Evidence suggests that endogenous processes of induced innovation are widespread, but they are too slow, relative to high current rates of population growth, to achieve sustainable outcomes (and avoid irreversible degradation) in the short to medium term." [44] Policy initiatives are therefore essential both to slow population growth and to promote innovation and investment.

Research and extension programs must integrate agricultural and ecological analysis. Land improvement strategies need to be designed with farmer input. Improved or subsidized credit may sometimes spur improvements, but price incentives that make innovation intrinsically profitable, together with appropriate extension and farmer organization, are more effective and reach more producers. Secure property rights, women's land rights, and involvement of local organizations and nongovernmental organizations in land management are essential. Improvement of transport infrastructure and market access, removal of price distortions that lower land values, and proper pricing of irrigation water, logging concessions, and farm inputs can create markets for higher-value products that are environmentally suitable.

"A neo-Malthusian scenario may indeed threaten large parts of the rural South, but only if public policies fail to provide a supportive environment and long-term commitment for land-improving investment. National and international attention to this challenge is long overdue." [47]

Summary of

Population and Urbanization in the Twenty-First Century: India's Megacities

by Sai Felicia Krishna-Hensel

[Published in *People and their Planet: Searching for Balance,* ed. Barbara Baudot and William R. Moomaw (London: Macmillan; New York: St. Martin's Press, 1999), Ch. 10, 157–173.]

Megacity growth is a central feature of Indian development. Cities that were already large at the time of independence, such as Calcutta, Bombay, and Madras, have continued to increase in size until they have reached the category of megacities—urban areas with populations exceeding eight million.[1] The same locational and economic forces that drove the original growth of cities at favorable geographic sites have continued to attract migrants in search of opportunities. While urban agglomeration is often seen as evidence of successful modernization, there are increasing problems associated with growth.

The difficulties facing megacities are potentially overwhelming. They appear to be too large to be efficiently administered through one central agency, requiring the development of effective local institutions to provide education, health, hygiene, and other services. But in addition to being a huge challenge for administrators, megacities are also laboratories for ingenious local experimentation.

Patterns of Growth

The shift from an agrarian to an industrial economy has encouraged population movement toward urban areas engaged in industrial activity and trade. Small-scale manufacturing enterprises provide many job opportunities. People also come to the city seeking skills and education. While the growth of megacities is partly a function of natural population increase, large-scale migration is a more important cause. The partitioning of the subcontinent in 1947 and the partition of Pakistan in 1971 also led to large refugee movements to cities such as Calcutta.

Civic authorities in Calcutta and Bombay were poorly prepared to handle large influxes of migrants. In Calcutta, "pavement residents" create major health and safety problems. Bombay has nearly half its population residing in squatter or slum housing. In Madras and Delhi, the record is somewhat better. Madras, with growth more evenly divided between in-migration and natural increase, has a relatively small proportion of slum housing. In Delhi, regional planning authorities have consistently encouraged dispersal of growth into surrounding townships.

Urban Problems

Serious problems surround the expansion of shantytowns in urban areas. Residents often pirate electricity from power lines and illegally tap into public hydrants. Flimsy construction poses fire and health hazards, and untreated sewage often enters the general water supply. Education and health-care facilities are generally lacking. Crowded conditions are a breeding ground for disease, environmental degradation, and crime. Between 40 and 70 percent of urban residents reside in slums or poor-quality tenement housing.

Megacities are characterized by "high land prices, air and water pollution, and a growing climate of corruption and extortion." [162] Organized crime is active in the real estate market, extorting a 10 percent levy on building projects. The absence of housing codes means that many renters live in derelict and unsafe housing. Legal restrictions often prevent effective property development while creating opportunities for bribery and illegal profiteering.

Urban air pollution has become a major health hazard, leading to a rising rate of respiratory disease. Vehicular emissions are steadily increasing, with 700 new vehicles added daily in Delhi alone. World Health Organization pollutant limits are regularly exceeded. The use of biomass fuels and wood contributes both to air pollution and to regional deforestation, which in turn lessens vegetal absorption capacity and thus worsens pollution. Data on air quality are often unreliable or lacking entirely.

Water supply is a serious problem, with demand far exceeding supply. About half the sewage in Delhi is completely untreated; only about 20 percent is fully treated. Solid waste is dumped outside the city in open pits, posing a further health hazard. Local rivers have high levels of untreated sewage pollution. These conditions are conducive to the spread of diseases, including cholera, diarrhea, gastroenteritis, and malaria. In 1993, an outbreak of plague in northern India created widespread alarm and spurred some improvement in sanitation measures.

The enormous problems of the megacities have led to the emergence of locally based initiatives attempting to respond to social, health, and environmental issues. Some positive examples include

- The UNICEF project for improving the nutritional quality of infant feeding in Delhi, emphasizing the importance of early intervention in preventing malnutrition.
- The Exnora International organization, which collects about 20 percent of Madras's garbage for a low fee.

- Assistance from the World Bank, which has allowed cities such as Bombay and Calcutta to almost double their investment in sewage management projects.
- Support from international agencies for community projects, which has encouraged the creation of creative grassroots movements by residents of severely stressed urban communities.

This awakening of self-reliance among the urban poor is a global phenomenon, and is a source of hope for the future of megacities.

The Planning Challenge

Rapidly growing populations provide a major challenge for urban planners. Census data only reflects changes long after their impact has been felt in terms of exacerbated urban problems. For example, the population of Calcutta doubled from 4.4 million in 1950 to 9 million in 1980, and during the next 10 years it added another 2.8 million inhabitants. The population was estimated to reach 15.7 million in 2000. Improved medical care and "green revolution" food supplies have reduced death rates, but these very successes magnify the problems of urban growth.

Planners are finding that decentralization, privatization of services, and the development of community-based support groups are the best strategies for providing basic services and improving the quality of urban life. A concerted effort by the public and private sectors is needed. Development permits should be linked to the provision of necessary services in low-income and unplanned neighborhoods.

Authorities in Bombay and Calcutta have attempted to decentralize urban activities, developing small townships to relieve central city congestion. Delhi planning includes a National Capital region, including outlying districts and counter-magnet cities to draw growth away from the center.

There is a need to harness technology to improve urban transportation systems. A more extensive underground transportation network in Calcutta would relieve surface congestion and pollution. Urban food supply systems are also inefficient. Decentralized markets and kitchen gardens are helping to meet needs of outlying developments and to shorten supply chains. However, many of these efforts are temporary or lack legal property rights.

Accurate assessment of growth trends is crucial for effective megacity planning. More reliable data collection is needed, especially in the area of pollution. Cooperation between nongovernmental organizations and municipal agencies will be needed to promote effective decentralized development. Finally, the importance of education in mobilizing popular support is an essential component of urban sustainability strategies.

Note

1. Some authorities use a figure of 10 million to define a megacity. See, for example, O'Meara, in this volume.

Summary of

Mexico City: Our Common Future?

by Priscilla Connolly

[Published in *Environment and Urbanization,* 11, 1 (April 1999), 53–78.]

Although Mexico City is neither the largest nor the most populated metropolitan area in the world, there are serious environmental threats to its survival as a viable city, as well as regional and even global spillover effects. While some problems are being addressed, others are deteriorating. This article puts the problems of Mexico City in perspective with respect to the following:

- Population growth and distribution, and their effect on poverty-related environmental problems.
- The water cycle, deforestation, and subsoil contamination.
- Atmospheric pollution by industry and transport.

Some of the issues discussed are common to all urbanization processes, while some are exacerbated by the scale of growth in Mexico City or by its specific historical and geographic conditions. It is important to distinguish between the common and the localized problems. It is also important to distinguish between poverty-related environmental deprivation and ecological degradation by economic progress.

Too Many People?

A focus on simple demography can be misleading, since population growth per se is not the problem. Neither is controlling city growth necessarily a solution. A review of trends over time in population and economic growth reveals a more complex picture (Table IV.4).

Rapid population growth and in-migration rates peaked after 1970. The demise of import-substituting industrialization, sharp reductions in fertility rates, and decentralization of educational opportunities all contributed to declining trends in Mexico City's growth rate. The financial crisis of 1982, the subsequent recession, and the ruthless opening-up of the country's economy led to a de-industrialization of Mexico City, decentralization of employment,

Table IV.4. Selected Basic Statistics for Mexico City

Mexico City Metropolitan Zone	*1950*	*1970*	*1990*	*1995*
Population (millions)	3.1	9.3	14.7	15.6
Mean annual growth rate (%)	6.7	5.6	2.3	1.9
Mexico City as percentage of total national population (%)	12.0	19.0	18.0	17.0
Total Mexico State and Federal District	*1950*	*1970*	*1988*	*1993*
Total GDP as percentage of national GDP (%)	34.0	37.0	33.0	35.0
Manufacturing GDP as percentage of national GDP (%)	Nd.	50	42	43
	1977	*1990*	*1993*	*1997*
Total private sector formal employees as percentage of national total (%)	40.0	33.0	31.0	28.0
Private sector formal employees in manufacturing as percentage of national total (%)	47	33	32	25
Total National	*1950*	*1970*	*1990*	*1997*
Population (millions)	25.8	48.2	81.2	94.7
	1960–70	*1970–80*	*1980–90*	*1990–97*
Average annual percentage population growth over decade (%)	3.3	3.3	1.9	1.6
GDP per capita average annual percentage growth (%)	3.5	4.1	0.1	0.6

and a reduction in the city's socioeconomic advantages over other areas. These trends coincide with a rising awareness of environmental deterioration, and especially traffic congestion and air pollution.

Between 1987 and 1992, out-migration from the city exceeded in-migration. Mobility is still high, but among the millions of people flowing in and out of the city, those leaving are predominantly the more skilled, while the newcomers tend to be poorer and less educated. Overall, population projections for Mexico City have proved to be drastic overestimates; nonetheless, environmental conditions have worsened. The demographic transition has led to smaller families and higher housing demand. Inflated land prices, combined with falling incomes have driven more low-income families to unserviced sites on the expanding city edge. Central areas, despite existing services and infrastructure, are losing population faster than before. The problems are accentuated by a politi-

Table IV.5. The Two Mexico Cities

	Federal District	Mexico State
Population 1997	8.5 million	8.7 million
Mean annual growth rate 1990–1995	0.5%	3.3%
Percentage of houses without inside running water	28%	45%
1992 local government budget	c. US$5,500 million	c. US$1,300 million
Annual investment in public transport 1992	c. US$1,000 million	US$ 80 million*

*State plus municipalities

cal and administrative division that places many outlying municipalities outside the better-financed central Federal District in poorer Mexico State (Table IV.5).

Water: Too Much and Too Little

Mexico City is situated in a closed basin with 1,000 millimeters of rainfall every year between May and September—under natural conditions the area would be a lake. A massive drainage system takes sewage water combined with storm drainage out of the basin, making it difficult to store or recycle water, and necessitating pumping in drinking water via aqueducts from 60 to 150 kilometers away. A growing expanse of concrete and tarmac prevents absorption of water into subsoil. Sixty percent of the city's water is provided by local wells, and depletion of the aquifer has caused drastic soil subsidence.

Soil subsidence in turn damages water pipes, leading to drinking water contamination and leakage. Desiccation of soils has also led to loss of vegetation, erosion, and wind-borne particle pollution. Poor sanitation conditions mean that the dust includes fecal matter, contributing up to 30 percent to the toxicity of the city's atmospheric pollution.

Policies needed to respond to this worsening problem of water waste and overdraft include

- Retention and recycling of water within the valley, to reduce the need to pump fresh water in and waste water out.
- Repair of the leaks in the water pipe system.
- Water conservation measures, such as the substitution of 6-liter for 16-liter toilets.

Thus far there has been little effort to increase the efficiency of the water system to limit consumption, perhaps because building aqueducts at public expense is more profitable and politically more popular.

Air Pollution and the Energy-Matter Cycle

Atmospheric pollution is strongly related to energy consumption. As an oil-producing nation, Mexico has placed little emphasis on energy efficiency or on the development of alternative energy sources. Electricity is subsidized, and liquid petroleum gas (LPG) is made available to households at a price 55 percent below cost. Fifty-six percent of the fuel energy consumption in Mexico City is gasoline, diesel, and LPG used in transportation, while 23 percent is industrial use of natural gas and LPG.

Rising awareness of the seriousness of air pollution, bolstered since 1988 by a network of air quality monitoring stations, has led to progressively stricter emissions control policies. This has resulted in significant declines in maximum levels of industrial and transportation pollutant emissions. However, average pollution levels are still unacceptable by Mexico's norms, which have been adjusted to meet international standards. Mexico City's location in a high, enclosed valley worsens pollution: thermal inversions trap pollution and low oxygen content means that internal combustion engines are 23 percent less efficient than at sea level. While transport is the largest contributor to atmospheric pollution (75 percent by weight), industrial emissions rise in significance if toxicity is taken into account.

Transportation and Automobile Use

Private cars and taxis are responsible for 67 percent of transportation emissions. However, more than 70 percent of total passenger trips are by public transport, dominated by privately franchised minibuses. Although minibuses are neither energy-efficient nor clean, high use rates mean that their per-passenger pollution emissions are far lower than private vehicles. The least-polluting vehicles are high-capacity buses, trolleys, and the subway. However, the political power of subsidized minibus and taxi cartels has impeded rational planning of public transit. Federal district support for buses has decreased as a result of budgetary restrictions and as a way of suppressing militant labor unions. Use of the subway has also declined as residential patterns have become decentralized and because of competition from the minibuses.

Public transport policy in Mexico City is effectively decoupled from planning and building legislation, including the provision of parking. The use of pricing policy to modify transport behavior and reduce pollution has not been attempted. Thus, individual transport in cars and taxis has been allowed to dominate, and in effect has been supported by government investment policies.

The number of cars in Mexico City has grown much faster than the population, almost doubling from 1.3 million in 1986 to 2.4 million a decade later. Thirty-seven percent of households in Mexico City have at least one car. There have been virtually no efforts to curb car ownership and use. An emergency one-day-a-week automobile ban backfired as people acquired second (or third) cars to evade the ban on the first car.

New and wider roads and the construction of overpasses encourage more car traffic. Parking is widely and cheaply available, and politicians have actively promoted the development of new car parks. Shopping malls with restaurants and cinemas—inaccessible to pedestrians but convenient for cars—are multiplying all over the city. A strong political dynamic is set in motion: "the population that is supposed to abandon their cars in favor of some potentially improved public transit system is always someone else. No one who has a car at present seriously envisages a future without one." [78]

Conclusion

The environmental problems facing Mexico City are both complex and dynamic. Solutions will need to be found in measures not directly associated with environmental policy, especially housing and land use, but also health, education, and retailing. Further political reform is also essential to counter cartel power and clientilistic systems. Reducing poverty is also a key factor, since the economic struggle for survival will take precedence over environmental issues. These interrelated social, political, and environmental issues will be more important than sheer population size in determining the future of Mexico City.

Summary of

Reinventing Cities for People and the Planet

by Molly O'Meara Sheehan

[Published as Worldwatch Paper No. 147 (Washington, D.C.: Worldwatch Institute, 1999).]

In 1900, only 160 million people, or 10 percent of the world's population, lived in cities. By 2006 about 3.2 billion people, half of the estimated total population of 6.4 billion at that date, are projected to be urban dwellers. Cities occupy only 2 percent of the world's surface but consume over 75 percent of key resources and account for 60 percent of human use of water.

Many of the world's social and environmental problems are concentrated in cities. At least 220 million people in cities in the developing world lack clean drinking water, 420 million lack basic sanitation, 600 million lack adequate housing, and over 1 billion suffer severe air pollution. China alone reported 3 million deaths from air pollution between 1994 and 1996; children in Mexico City, Beijing, Shanghai, Tehran, and Calcutta inhale the equivalent of two packs

Table IV.6. Metropolitan Areas (population in millions)

1900		2000	
London	6.5	Tokyo	28
New York	4.2	Mexico City	18.1
Paris	3.3	Bombay	18
Berlin	2.7	São Paolo	17.7
Chicago	1.7	New York	16.6
Vienna	1.7	Shanghai	14.2
Tokyo	1.5	Lagos	13.5
St. Petersburg	1.4	Los Angeles	13.1
Manchester	1.4	Seoul	12.9
Philadelphia	1.4	Beijing	12.4

of cigarette per day. Patterns of water, waste, food, energy, transportation, and land use in urban areas are critical determinants of the balance between human activities and the environment.

Urbanization Trends

At the beginning of the twentieth century, all of the world's ten largest cities were in Europe, the United States, or Japan. By 2000, seven of the world's ten largest cities will be in developing nations (Table IV.4). Rapid urban growth continues throughout the developing world. In just five years, between 1990 and 1995, the cities of the developing world grew by 263 million people—the equivalent of another Los Angeles or Shanghai forming every three months. By

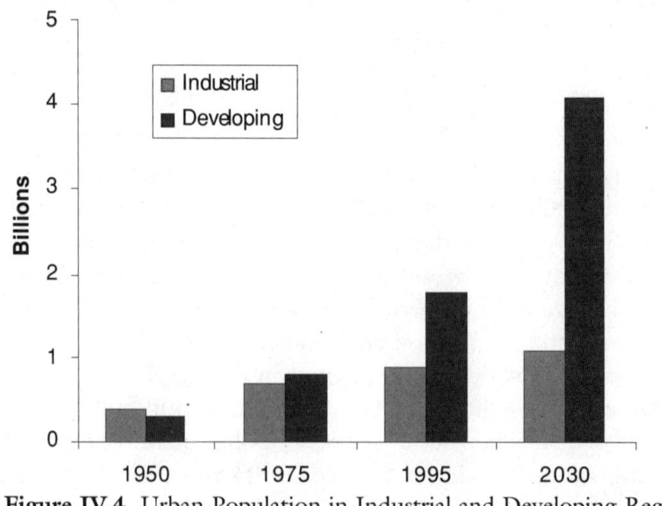

Figure IV.4. Urban Population in Industrial and Developing Regions.

2030, the urban population in developing regions is projected to more than double over 1995 levels (Figure IV.4).

This explosive urban growth includes "megacities" of over 10 million (fourteen in 1995), cities of 5 to 10 million (twenty-three in 1995), and numerous large cities of over 1 million and "megavillages" of several hundred thousand. Over half of the largest cities (populations over 5 million) are in Asia. The most urbanized region within the developing world is Latin America, where 73 percent of the population lives in cities (a proportion comparable to Europe and North America). The fastest rates of urban growth are in Africa, which is characterized by one megacity, Lagos, and many rapidly swelling megavillages. Many of these urban areas present management problems that are outrunning the capabilities of local authorities.

Water and Waste

While agriculture is the world's largest user of water, urban and industrial demands for water are growing rapidly. Urban growth in water-scarce areas such as the western United States and northern China, together with existing agricultural demand, has led to water overdraft and declining water tables. Overpumping of aquifers often leads to subsidence; parts of Mexico City have sunk more than 9 meters. Comprehensive watershed protection and conservation strategies are needed for urban areas, including adequate water pricing, collecting rainwater, and recycling wastewater.

Urban waste streams can lead to "trash mountains," such as New York's Fresh Kills landfill, which rises higher than the Statue of Liberty. Organic waste, which accounts for 36 percent of the waste stream in industrial countries, pollutes waterways with excess nitrogen but can be converted into a valuable resource through composting. Some European cities now recover more than 85 percent of these wastes. Urban solid waste recycling rates are often low in the United States but have been boosted to the 45–60 percent range in some cities through curbside recycling and "pay-as-you-throw" systems for unsorted garbage. In developing nations, decentralized and community-based innovation in water supply, sanitation, waste collection, and recycling have been successful in cities such as Karachi, Pakistan, and Cairo, Egypt.

Food and Energy

Most urban areas rely on long food and energy supply lines, with significant processing and transportation costs (for food) and transmission and distribution costs (for energy). Centralized power plants and industrialized agriculture have many negative environmental impacts, including nitrogen and sulfur emissions, fertilizer and pesticide runoff, and genetic uniformity of crops. But alternative techniques involving shorter supply chains and less environmental damage are available. "Homegrown food and clean, locally produced energy can not only green a city but also increase income and security for its inhabitants." [33]

Urban agriculture has a long history in Asia, where cities such as Hong Kong and Singapore produce large proportions of their own vegetables, poultry, and meat. As recently as the 1980s, China's largest cities produced 90 percent of their vegetables and over half of their meat and poultry, but these numbers have declined due to intensive building. Urban agriculture is widespread in Africa, where it is an essential survival strategy for many residents, and is on the rebound in many European cities.

Decentralized energy technologies such as solar panels, geothermal heat pumps, and small cogenerating gas turbines make it possible for city buildings to be self-sufficient in energy or to sell power back to the grid. Improved energy codes for buildings can cut heating and cooling requirements by up to 70 percent. Landscaping, including trees and roof gardens, can reduce energy use and provide environmental and aesthetic benefits.

Transportation and Land Use

Automobile use in U.S. cities is the highest in the world and continues to rise, increasing by 2,000 kilometers per person between 1980 and 1990. While most of the rest of the world is far below U.S. levels, urban automobile use continues to increase almost everywhere. Sprawling automobile-based conurbations diminish the vitality of street life, promote congestion and pollution, and contribute to a toll of fatalities estimated at 885,000 worldwide per year the equivalent of ten fatal jumbo jet crashes per day. They also require more water and sewer pipes, power lines, roads, and building materials. Wasted fuel and lost productivity from traffic jams cost $74 billion annually in the United States.

Integrating transportation and land-use planning can achieve more livable cities with an emphasis on public transit, bicycles, and high-density development. Curitiba, Brazil, and many European cities have successfully reduced automobile traffic while improving the flows of people and goods. Portland, Oregon, has taken steps to limit outward growth of the city and to reduce the need for cars within city boundaries. Tolls and pricing policies can be used effectively to discourage automobile congestion; the city-state of Singapore leads the world in regulating traffic through fees that rise at rush hour. Proper pricing for the extensive parking spaces that cars require, and the abolition of parking subsidies, also help to make drivers aware of the true costs of automobile use in the city.

Shifting property taxes from buildings to land can help promote compact development and avoid urban blight. This tax system rewards owners for productive improvement of urban land while reducing inefficient land uses. Protection of surrounding fields and forests can be combined with policies to encourage "infill" development. Provision of good public transit systems can be complemented by public-private partnerships to provide fleets of bicycles for public use, or car-sharing networks, both of which have been successful in European cities.

Financing the Sustainable City

Lack of local control over financing is a significant problem for city management. Most cities are dependent to some degree on transfers from state or national governments, but these authorities often do not favor sustainable urban priorities. National policies supporting road-building rather than mass transit can thwart sustainable urban policies. Subsidies for water and energy can undermine city building codes aimed at promoting energy and water efficiency. Cities may be able to close the gap by using fees—on water supplies and sewer linkage, and on garbage collection and parking, for example.

Municipal bonds can provide an additional source of funds. Revenue bonds can be linked to fees for specific projects, while general obligation bonds depend on future tax revenues, of which land value taxation is the best suited to cities. Unfortunately, most local authorities in developing countries do not have access to a reliable municipal bond market. Attacking corruption and establishing effective tax and fee collection systems is the essential prerequisite to city creditworthiness. Ahmedabad, India, has recently been able to float India's first municipal bond after reducing corruption and improving revenue collection.

Public-private partnerships to provide urban services can raise additional funds while also providing badly needed employment opportunities. Small-scale lending programs and community reinvestment can help promote entrepreneurship and small business development

Helping existing local businesses to expand is usually more successful than trying to lure outside businesses to impoverished areas.

Information Needs and Political Strategies

Demographic and environmental data are important in addressing urban problems. The Urban Management Program, a joint effort of UNDP, The U.N. Centre for Human Settlements (UNCHS, or Habitat), and the World Bank, has developed a "rapid urban environmental assessment" survey to identify key environmental indicators in cities of the developing world. UNCHS has created a database of such indicators for 237 cities in 110 countries. Map data provided by geographic information systems (GIS) can aid urban planners and provide a valuable tool for citizen activist groups. Direct information exchange between cities can leapfrog national government bureaucracy and divisions; the

Toronto-based International Council on Local Environmental Initiative (ICLEI) links over 2,000 cities in 64 countries working on local Agenda 21 programs for environmentally sound development. Greater awareness of interrelated problems can help to promote metropolitanism—cooperation between cities and suburbs to deal with mutual problems of energy, transportation, water, waste, and economic development.

PART V

Agriculture and Renewable Resources

Overview Essay

by Jonathan M. Harris

The earth's renewable resource systems have suffered major damage during the last century, with the severity and scale of ecosystem impacts increasing significantly between about 1950 and the present. Examples of civilizations undermining their resource base are common in history (Box V.1), but the present global scale of the damage to soils, water, forests, and atmospheric and ocean ecosystems is unprecedented. The United Nations Environment Programme report *Global Environmental Outlook 2000* finds that "environmental gains from new technology and policies are being overtaken by the pace and scale of population growth and economic development" (UNEP 1999, xx–xxii). Among the major environmental problems cited by UNEP are climate change caused by fossil fuel emissions and other greenhouse gases; pollution and soil damage from intensive agriculture; degradation and loss of forests, woodlands, and grasslands; biodiversity loss; water contamination and overdraft; degradation of coastal areas by agricultural and industrial runoff; and overexploitation of major ocean fisheries.

The World Resources Institute documents these and similar impacts, but also notes that a shift to sustainable management can significantly alter these trends: "Already, the transition to more environmentally benign ways of growing food, producing goods and services, managing watersheds, and accommodating urban growth has begun in many far-sighted communities and companies. How fast this transition to more 'sustainable' forms of production and environmental management will proceed, and whether it can effectively mitigate the effects of large-scale environmental change, is the real question." (WRI 1998, 140)

In this section we explore some of the issues involved in a large-scale shift to sustainable management of natural resource systems. To use the terminology introduced in Part I of this volume, we will discuss the conditions for the conservation of natural capital rather than its depletion. In the areas of agriculture, biodiversity, fisheries, forests, and water, we review the nature and scale of current problems. The questions that concern us are: What sustainable management techniques are available? What economic incentives and political institu-

Box V.1. Exceeding the Limits: The Collapse of a Civilization

The first literate civilization in the world collapsed because of its failure to recognize ecological limits. Around 3000 B.C. the Sumerians of southern Mesopotamia, between the Tigris and Euphrates rivers, built a complex society based on irrigated agriculture, and invented wheeled vehicles, yokes, plows, and sailboats, as well as accounting and legal systems.

But their growing population placed too heavy of a demand on the natural resources of the region. Deforestation and overgrazing led to heavy soil erosion. Irrigation caused the underground water table to rise, depositing salts that poisoned cropland. Eroded soils loaded the rivers with silt, leading to catastrophic flooding.

"The limited amount of land that could be irrigated, rising population, the need to feed more bureaucrats and soldiers, and the mounting competition between the city states all increased the pressure to intensify the agricultural system. The overwhelming requirement to grow more food meant that it was impossible to leave land fallow for long periods.

"Short-term demands outweighed any considerations of the need for long-term stability and the maintenance of a sustainable agricultural system. . . . Until about 2400 B.C. crop yields remained high, in some areas as high as in medieval Europe and possibly even higher. Then, as the limit of cultivable land was reached and salinization took an increasing toll, the food surplus began to fall rapidly . . . by 1800 B.C., when yields were only about a third of the level obtained during the Early Dynastic period, the agricultural base of Sumer had effectively collapsed." (Ponting 1993)

The process of irrigation, salinization of soils, and agricultural collapse was repeated twice more as later societies attempted to rebuild in the same region. Finally, the land was exhausted. "Once a thriving land of lush fields, it is now largely desolate, its great cities now barren mounds of clay rising out of the desert in mute testimony to the bygone glory of a spent civilization" (Hillel 1991).

tions are needed to promote sustainable techniques? And what are the economic and political barriers to their implementation?

Agriculture: Consumption, Production, and Environment

The discussion about world agriculture has shifted from a debate between optimists and pessimists about overall adequacy of food supplies to a focus on distributional and environmental issues. Technological productivity and yield increase outran population growth in the second half of the twentieth century, but at the cost of significant environmental degradation and without solving the problem of hunger. The number of seriously malnourished people remains stubbornly high—over 800 million worldwide (FAO 1996)—while soils, water supplies, and ecosystems have suffered widespread damage.

To project a successful future for world agriculture in 2000–2050 and beyond, several major issues must be addressed:

- Is it possible to expand food production to accommodate the needs of 8–9 billion people at higher per capita consumption levels without worsening environmental damage?
- Can food consumption patterns become more equitable or will the needs of the affluent for meat and luxury foods squeeze out the world's poor?
- Can local and regional food production in the developing world come close to self-sufficiency or will import dependence continue to grow?

A comprehensive study by the International Food Policy Research Institute (IFPRI 1995) projected that in the medium term (through 2010) "world food supply would probably meet global demand, but that regional problems could occur" with South Asia and sub-Saharan Africa most likely to suffer shortfalls. While contributors to the IFPRI study generally leaned toward the optimistic side in projecting food availability (Mitchell and Ingco 1995; Agcaoli and Rosegrant 1995), they also acknowledged that environmental problems posed important future constraints.

According to Pierre Crosson, "maximum realization of potential land and water supplies at acceptable economic and environmental cost in the developing countries still would leave them well short of the production increases needed to meet the demand scenario over the next 20 years." (Crosson 1995) This picture is borne out in a more recent overview of global population-supporting capacity by researchers at the U.S. Department of Agriculture Office of World Soil Resources: "From a global land-productivity point of view, the specter of Malthusian scenarios seems unwarranted. Sadly, however, local and regional food shortages are likely to continue to occur unless mechanisms for adequate food distribution, effective technical assistance, and infusions of capital for infrastructure development are implemented in some developing countries." (Eswaran et al. 1999)

Area in cereal crops, a fundamental mainstay of global nutrition, is no longer growing (Harris 1996), and grain area per person has been declining since the 1950s as a concomitant of population growth (Brown 1999). Maintaining yield growth is therefore critical to feeding a world population projected to grow by about 2 billion people in the next generation. But there is some evidence that yield increases, the key to maintaining and increasing per capita consumption in developing countries, are slowing (Harris and Kennedy 1999). While a Malthusian scenario is not likely, it is apparent that maintaining an adequate food supply while preventing further environmental damage presents a major challenge.

Per Pinstrup-Andersen and Rajul Pandya-Lorch argue that food security can be achieved but that this will require significant reforms in the structures of production and distribution. Since major cropland expansion is not possible, yield increase is key, but the current high-input techniques of the Green Revolution that have worked in prime agricultural lands are not suited for many of the world's more marginal areas. In addition, they have caused unacceptable environmental damage. Thus a different paradigm of agricultural development is needed that is more environmentally friendly and oriented to the needs of small farmers and the rural poor.

The dimensions of this new paradigm for sustainable agriculture are discussed by **Gordon Conway.** The essential issue is the perception of agroecosystems as modified versions of natural systems rather than as rural production units converting inputs to marketable outputs. Within the agroecological perspective, there are many possibilities for multicropping, agroforestry, biological pest controls, and crop/livestock systems. Few of them fit well into the economic formula of monocrop output, because they have multiple and not always easily measurable products, including biodiversity conservation. All involve significant reduction of external inputs such as fertilizers and pesticides.

Issues of production techniques are interrelated with issues of equity and social relations, as local management is often key to successful agroecosystems, while large-scale production of monocrops is conducive to centralized and concentrated agribusiness. Local food systems can often provide both greater food security and greater equity (Campbell 1997; Anderson and Cook 2000; Barkin 2000).

There is an extensive literature on the principles and economics of sustainable agriculture. The 1989 National Research Council (NRC) report *Alternative Agriculture* first gave authoritative support to the proposition that "Alternative farming methods are practical and economical ways to maintain yields, conserve soil, maintain water quality, and lower operating costs." (NRC 1989). A study by Lockeretz et al. (1981) indicated that yields in organic agriculture in the United States and Canada were comparable to those of farms using mainstream high-input techniques. The *American Journal of Alternative Agriculture* provides continuing detailed analysis of such issues as providing alternatives to pesticide use (den Hond et al. 1999) and the productivity of organic agriculture (Hanson et al. 1997). Comprehensive assessments of the economics and agronomics of sustainable agriculture in both developed and developing nations are provided by Edwards et al. (1990) and Lampkin and Padel (1994). Pretty et al. (1995) focus on the agroecology of low-input and community-based agriculture in developing nations, and case studies of the implementation of sustainable practices are reviewed in Thrupp, ed. (1996).

A major economic barrier to the adoption of sustainable agricultural techniques is the existence of an extensive network of subsidies for agricultural inputs and energy, as well as for crop and livestock production (Panayotou 1998, 68–71). Both in developed and developing countries, these subsides typically provide the greatest benefits to affluent farmers and agribusiness, who constitute a powerful political lobby for their continuation. At the same time, subsidy removal may create hardships for small farmers and low-income consumers. Thus it is important to integrate the process of subsidy removal, which is well justified on both economic and ecological grounds, with policies aimed at supporting small farmers and promoting equity. These could include agricultural extension and informational outreach on low-input techniques to enable farmers to maintain or improve productivity while reducing fertilizer, pesticide, and water use. Targeted food subsidies and social investment strategies may help low-income consumers while reducing the budget costs of broad-based subsidies (see Donaldson 1991; Hopkins 1991; and summarized article by Heredia in Part X of this volume).

Biotechnology in Agriculture

Among advocates of sustainable agriculture, there is a division of opinion on the acceptability of biotechnologies. Some, like Pinstrup-Andersen and Conway, see biotechnology as an essential complement to agroecological systems. Others warn of the potential of biotechnology to disrupt and destroy natural ecosystems (Rifkin 1998). Modern genetic technologies differ in nature and scope from traditional crossbreeding techniques, and raise novel issues concerning the social governance of technology (Krimsky 1991; Krimsky and Wrubel 1996).

In agriculture, genetic technologies involve a whole spectrum of issues such as pest, disease, and herbicide resistance, nitrogen fixation, and animal growth hormones. While the advocates of biotechnology see great promise for directly or indirectly increasing yields, **Jane Rissler and Margaret Mellen** argue that the potential for unintended negative consequences from genetically modified organisms is great. Once introduced into the environment, traits such as pest-resistance may "jump" from crops to weedy relatives, thereby creating superweeds.[1] Pesticides genetically engineered into crops may hasten the development of resistant pests, already a major problem in "modernized" agriculture, or may kill beneficial species (already reported in the case of genetically modified corn and monarch butterflies).

Mae-Wan Ho (1998) emphasizes the importance of the interconnections of genetic patterns in natural ecosystems. While genetic engineers concentrate on the development of a single trait, they ignore the more complex effects on the organism as a whole and on the ecosystem. New and more powerful viruses may be unintentionally created, some of which may affect humans as well as plants.

Advocates of the rapid adoption of biotechnology argue that it will help solve urgent food supply problems in developing nations (Paarlberg 2000). However, the social impacts of genetic engineering imply an expropriation of genetic resource from the South and increased marginalization of small farmers (see article by Vandana Shiva in Part III of this volume). In Europe, resistance to corporate control of genetic material has grown, and together with health and environmental concerns has led to strong consumer resistance to the introduction of genetically modified foods (Lappé and Bailey 1998).

Perhaps the most far-reaching threat associated with genetic engineering in agriculture is the destruction of natural biodiversity. Rissler and Mellen point out that the introduction of nontransgenic crossbred crops has already led to numerous cases of extinction of wild relatives. The greater power of transgenic techniques implies greater potential ecosystem impacts. While such impacts are unlikely to enter into the profit-oriented calculations of agribusiness corporations, they can create irreversible changes that would undermine the long-term sustainability of agroecosystems. To get a sense of the importance of this issue, we need to examine the concept of biodiversity and its implications for the future of human systems and of the natural world.

Biodiversity and Natural Genetic Resources

Biodiversity is one of the most important elements of natural capital, as well as one of the most difficult to measure and value. Biodiversity refers both to the total number of species and to the rich complexity of species' interrelationships in natural settings. It can thus be degraded by simplification of the species patterns in ecosystems, as well as permanently diminished by species extinction.[2]

One way of understanding the value of biodiversity is to consider the tremendous complexity of genetic codes. The genetic code of each species is a store of information built up over millennia, adapted specifically to planetary ecosystems, and irreplaceable once lost through extinction. Biodiversity is one of the most valuable resources on the planet, "the key to the maintenance of the world as we know it" (Wilson 1992). Using the economic terminology introduced in Part I, an excellent case can be made for defining biodiversity as "critical natural capital."

To some extent biodiversity can be conserved by preserving genetic specimens, for example, in seed banks. However, in addition to the genetic code itself, the structure of complex ecosystems involving many types of symbiosis is an essential part of biodiversity. Further, most complex ecosystems have a minimum size below which species loss occurs due to ecosystem fragmentation.

Thus in many cases the only effective method for preserving biodiversity is to maintain natural ecosystems in an unchanged or minimally modified form.

Achieving this goal presents major problems in a production-oriented economic system that values resources only insofar as they contribute directly to meeting human consumption demands. Many traditional agricultural and production systems have achieved coexistence with minimally disturbed ecosystems, but are threatened by competition with more "productive" but biologically simplified techniques (see article by Shiva summarized in Part III).

Estimates of species loss indicate a continued rapid decline in biodiversity, despite publicity and some efforts to address the problem by NGOs (nongovernmental organizations) and multinational institutions.

Paul Ehrlich and Gretchen Daily show that the scope of the problem is greater than implied by species extinction figures. The loss of local populations, even when members of the species survive elsewhere, has left huge areas of the world, including Europe and much of Asia, "biologically depauperate." Impoverished ecosystems possess diminished life-support capability and adaptability. The loss of ecosystem services such as water retention and detoxification, nutrient recycling, and carbon fixation have significant economic costs (Daily 1997). The esthetic and spiritual losses associated with a diminished natural world are more difficult to measure, but nonetheless real. More extreme consequences, such as the spread of plant blights and animal and human diseases, can lead to the kind of devastating ecological backlash discussed by C.S. Holling in Part IV of this volume.

Kamaljit Bawa and Madhav Gadgil point out that the people most exposed to declines in ecosystem services are the rural poor. At the same time, local communities can be the most effective stewards of ecosystem diversity, given the right incentives (Bawa and Gadgil 1997). Like Barkin (2000), Bawa and Gadgil point out the need to devise systems that provide economic incentives for the maintenance of biodiversity in agriculture, forestry, and fisheries as well as through ecotourism or other sources of income from conservation.

In the area of biodiversity, the limitations of standard economic theory for resource conservation are brought into sharp relief. Theories and techniques that work well for mineral deposits are ill adapted to the much-greater complexity of biological systems. This is especially true on an international scale; the globalization of markets threatens existing ecosystems with degradation, species extinction, and the introduction of invasive species, whether natural or engineered. Attempts to value ecosystem services are relatively crude, but indicate that even in standard economic terms their value is very large (see Costanza and Folke 1997, Costanza et al. 1998, and discussion in Part I of this volume). Policies based on the "internalization of externalities" constitute a relatively weak response compared to the broader goal of ecosystem preservation.

Water, Forests, and Fisheries

Economic theory is a better guide to the overexploitation of common property resources. The well-known economic principle of the "tragedy of the commons" is misnamed, however—it should be "the tragedy of open-access." While common property resources have often been well-managed through traditional rules and local government oversight, broadening markets and advancing technology has brought more powerful extractive techniques to bear on water, forests, and fisheries, with predictable results of overuse and depletion.

Where property rights are well-defined, exploitation is more rational and sometimes more restrained, but here too ecological values and future interests are poorly represented. The private owner of a forest has some incentive to replant after an economically optimal rotation period, but little incentive to conserve biodiverse forest systems as opposed to monocrop plantations. Proper pricing of water will help reduce wasteful use, but it may actually increase incentives to deplete rivers and aquifers to meet consumption demands. For all these systems, principles of economic rationality must be matched with principles of ecological conservation.

World fisheries are a glaring example of the perverse incentives and institutional failures of the current economic system, as outlined by **Anne Platt McGinn.** Both national policies and international agreements have so far been ineffective in preventing the depletion of the world's major fisheries. Aquaculture can fill some of the gap between growing demand and limited supply, but many forms of aquaculture are also associated with major environmental problems. In part the problem of overfishing arises from growing population and per capita demand, in part from failures of national policy, and in part from globalization of fishery access without comparable global regulation. The principles for sustainable management of fisheries are not hard to define, but the institutional barriers to achieving effective conservation are great.

The problems of forest management are somewhat different. Wooded land is often privately owned, which should in theory promote efficient resource use. However, private owners usually "see" only a limited portion of the true values provided by forests. They may manage efficiently for timber and perhaps for some nonwood products, but they generally ignore the many environmental services provided by forests. Given the imperative to produce commercial returns, private owners will also have a bias toward plantation forestry—monocultures of fast-growing species. An old-growth forest will be "mined" for one-time profit, then abandoned or turned to ecologically degraded uses.

Norman Myers provides a broader perspective on forest ecology and forest decline. He sees an inadequate public presence in forestry management both at the national and international levels. Too often, government agencies serve essentially to promote the interests of timber producers, making forest tracts available for commercial exploitation at low cost, thereby squandering potential revenues as well as encouraging deforestation, siltation, and floods (Panayotou 1998, 69). Myers outlines the scope of the challenge confronting new global forestry institutions, as well as existing government agencies. Most important, these authorities must accept a definition of sustainable forestry that goes well beyond maintaining timber revenues to include broad ecological and social functions. Improved scientific inputs to forestry decision-making are essential to complement the economic perspective that has characterized management in the past. Perhaps the single most pressing issue is the removal of widespread subsidies that continue to promote rapid forest destruction.

Water systems are another area in which economic analysis can be a useful guide to policy only if it is combined with an understanding of the ecological function of water cycles. Postel (1999) offers an alarming overview of the worldwide decline in the capacity and function of water systems. Overuse and overdraft of water have led to escalating problems of soil salinization, declining water tables, and ecological damage. Dams, intended to increase water supply, have contributed to drying up rivers and in many cases have rapidly lost capacity to siltation. Worldwide, irrigation is the largest source of water demand, and this mainstay of world agriculture is threatened by its own excesses, combined with growing urban and industrial demand.[3]

Postel suggests a range of solutions to water problems that draw both on traditional systems and modern technology. Water harvesting and small-scale irrigation are low-tech solutions; efficient modern micro-irrigation is more expensive but effective in water-short developed areas like Israel. Here also economics plays a crucial role: only with proper water pricing can incentives for efficient use be created. A range of vested interests militate against effective water policies—large users insist on maintaining subsidies, government bureaucracies respond to public pressure for big new projects and cheap, or free, water. Local communities may be well placed to implement effective water management but need both technical and institutional support. Much of the current institutional and economic infrastructure is oriented in the wrong direction, toward centralized control and supply augmentation. Technological and economic solutions are at hand but are held back by political constraints.

Conclusion

While the scope of the problems associated with damage to natural resource systems is staggering, the possibilities for sustainability-oriented policies is also great. The first principle must be the recognition of ecological limits. As we saw in Part I, a shift is needed from viewing natural resources as inputs to economic production toward adapting economic production to ecological realities. In the analysis of the management of natural systems, a key issue is broadening the scope of analysis from demand, supply, and production to the management of ocean and forest ecosystems, watersheds, and other biomes.

Many useful technological applications and economic incentive systems can be brought to bear within such a framework. Without appropriate institutional control and direction, however, the logic of the market tends to select destructive technologies and create incentives for more rapid exploitation. In all cases, the scope of the problems has grown steadily during the last fifty years. This implies that significant institutional and attitudinal changes will be required to reverse course in the twenty-first century toward reclamation of damaged resource systems.

Notes

1. A case of triply resistant weeds has already been reported in Canada as a result of the planting of genetically engineered weedkiller-resistant canola.
2. For an in-depth treatment of biodiversity and an assessment of current losses and future threats to biodiversity, see E.O. Wilson 1988, 1992.
3. For a comprehensive overview of the state of the world's water resources, see Gleick 1998.

Summary of

Food Security and Sustainable Use of Natural Resources: A 2020 Vision

by Per Pinstrup-Andersen and Rajul Pandya-Lorch

[Published in *Ecological Economics,* 26, (1998), 1–10.]

At the threshold of the twenty-first century, widespread poverty, food insecurity, and environmental degradation cause severe human suffering and threaten to destabilize the world's economies and ecosystems. More than 800 million people—20 percent of the world's population—are food insecure, lacking economic and physical access to the food required to lead healthy and productive lives. The article summarized below presents the 2020 Vision, developed by the International Food Policy Research Institute, which looks forward to the elimination of malnutrition through efficient and low-cost food systems that are compatible with sustainable management of natural resources. The authors argue that this vision can be realized if individuals, communities, businesses, governments, and the international community as a whole change their behavior, priorities, and policies and work together to take action. Priority actions include further investment in poor people, acceleration of agricultural productivity, sound management of natural resources, strengthened capacity of developing-country governments to perform appropriate functions, and increasing and realigning international development aid.[1]

Poverty and Human Resource Development

Thirty percent of the world's population currently lives in absolute poverty, on less than $1 per day. The poor generally do not have the means to grow or purchase an adequate food supply. Thus, the extent to which food needs are converted into effective market demand will depend on the purchasing power of the poor. Agricultural production can meet the food needs of the poor by acting as both a source of food and as an employer, providing incomes that allow the poor to buy food.

Sustained action is necessary to improve the productivity, health, and nutrition of poor people and to increase their access to employment and productive assets. Primary education, health care, clean water and sanitation, skill development, and incentives for gender equality are essential goals for government and nongovernmental organizations. Further, improved access to productive resources by the rural poor, especially by women, can be facilitated through land reform and sound property rights legislation, strengthened credit and savings institutions, more effective rural labor markets, and infrastructure for small-scale enterprises.

Population Growth and Movements

In the next twenty-five years, an estimated 70–80 million people will be added annually to the world's population, 98 percent of them in developing countries. Rapid population growth and urbanization could more than double the urban population in developing countries by 2020. Developing countries are projected to increase their demand for cereals by 80 percent between 1990 and 2020, and for livestock products by 160 percent.[2] Providing access to reproductive health services and eliminating high-fertility risk factors such as high rates of infant mortality are essential to curbing population growth. Female education and measures to improve income security for women are among the most important investments for assuring food security and sustainable resource use.

Food Supply

Food production increases did not keep pace with population growth in more than fifty developing countries in the 1980s and early 1990s. Growth rates in yields of rice and wheat have begun to stagnate in Asia, and production from marine fisheries appears to have peaked. Significant expansion of cultivated land area is not economically or environmentally feasible in most of the world. To increase food production, more efficient use must be made of land already under cultivation. In many areas where the poor are concentrated, food yields are low and variable. Low-income developing countries are grossly underinvesting in agricultural research compared with industrial countries, even though agriculture accounts for a much larger share of their employment and incomes.

To overcome this problem, agricultural research and extension systems, the growth of which has declined in recent decades, need to be strengthened in and for developing countries to increase the productivity of land and agricultural workers and thus lower the costs of food production, processing, and distribution. Biotechnology research should be expanded to support sustainable intensification of small-scale agriculture.

Natural Resources and Agricultural Inputs

Degradation of natural resources such as soils, forests, fisheries, and water systems undermines food production capacity. Since 1945, approximately 2 billion of the 8.7 billion acres of agricultural land, permanent pastures, and forest and woodlands have been degraded by overgrazing, deforestation, and poor agricultural practices. The causes of this degradation include inadequate property rights, poverty, population pressure, inappropriate government policies, lack of access to markets and credit, and inappropriate technology. Crop productivity losses from degradation are significant and widespread. Poverty, population growth, and continued food insecurity will continue to promote deforestation and soil degradation, especially in Africa, unless more effective ways are found to meet food needs.

Overcoming water-related problems is central to achieving the 2020 Vision. New water sources are increasingly expensive due to increasing construction costs of dams and reservoirs and environmental concerns, and current efficiency of water use in agriculture, industry, and urban areas is low. Pollution of water resources by industrial effluents and runoff of agricultural chemicals and sewage is a growing problem. In most areas water for irrigation is essentially unpriced. Policies that effectively manage water as a scarce resource are essential.

The depletion of soil nutrients, a significant cause of soil degradation, is a critical constraint to food production in sub-Saharan Africa. While there are negative environmental effects from overuse of fertilizers, in most developing countries the task is to promote a balanced and efficient use of plant nutrients from both organic and inorganic sources.

Overuse of pesticides creates a threat to human health and the environment and leads to the evolution of resistant and secondary pests, eventually causing decreased food production. More research and farm-level experimentation is needed to promote integrated pest management with minimal pesticide use.

To achieve these goals, investments need to be made by the private and public sectors in infrastructure, market development, natural resource conservation, soil improvements, and primary education and health care, as well as in expanded agricultural research in areas with large agricultural potential, fragile soils, and high poverty concentrations.

Natural resources also must be priced according to their value to ensure their efficient use. Local farmers and communities should be provided with incentives to protect natural resources and restore degraded lands. Local control over resources must be strengthened, and local capacity for organization and management improved. Integrated soil fertility, water management, and pest control programs can best be implemented at the local level.

Markets and Infrastructure

Governments of developing countries need to provide support for efficient agricultural input and output markets. However, the recent transition from controlled to market economies generated confusion about the appropriate role of government and weakened its capacity to perform needed functions. Each government must decide which agricultural policy functions should be strengthened and which are best relinquished to the private sector.

Additionally, governments need to facilitate a social and economic environment that provides all citizens the opportunity to assure their food security by implementing long-term national strategies for food security and nutrition, agricultural development, and natural resource management. Specifically, governments should

- Maintain adequate exchange rates and monetary and fiscal policies for accelerated economic growth.
- Gain access to international markets.
- Lower food marketing costs through investment in improved transportation, infrastructure, and marketing facilities.
- Phase out inefficient, state-run firms in agricultural input and output markets.
- Create an environment conducive to effective competition among private agents.
- Remove policies and institutions favoring large-scale, capital-intensive enterprises over small-scale ones.
- Develop and maintain public-goods infrastructures such as roads and electrical facilities, or effectively manage private-sector investment in these areas.
- Develop and enforce standards, weights and measures, and other regulatory instruments essential to market development.

- Facilitate development of small-scale credit and savings institutions.
- Provide technical assistance and training to create and strengthen small-scale, rural enterprises.

Domestic Resource Mobilization and International Assistance

Domestic resource mobilization through savings and investment is imperative to achieving the elements of the 2020 Vision. Currently, however, low income in poor countries leads to low savings, low investment, low growth, continued poverty, and continued low savings. Private capital flows to developing countries benefit a handful of medium-income countries, bypassing the poorest countries, while official development assistance has declined. This downward trend in international development assistance must be reversed, with all members of the United Nations moving toward spending a target of 0.7 percent of their GNP on foreign aid.

Development assistance should complement national and local efforts and should be made widely available to those countries whose governments have demonstrated a commitment to and a strategy for reducing food insecurity. In response, each recipient country should develop a coherent strategy for achieving its goals related to overcoming food insecurity and should identify the most appropriate uses of international assistance.

Notes

1. International Food Policy Research Institute (1995).
2. See Rosegrant et al. (1995) and Harris (1996).

Summary of

Sustainable Agriculture

by Gordon Conway

[Published in *The Doubly Green Revolution: Food for All in the 21st Century* (Ithaca, New York: Cornell University Press, 1998), Ch. 9.]

> Agriculture is a science which teaches us what crops are to be planted in each kind of soil, and what operations are to be carried on, in order that the land may produce the highest yields in perpetuity.
>
> —Marcus Terentius Varro, *Rerum Rusticarum*

Thus did Marcus Terentius Varro, a Roman landowner of the first century B.C., define sustainability in agriculture. In our time, Varro's clarity of meaning has been lost as "sustainability" has become a highly politicized term. The diversity of opinions as to its meaning put forth by agronomists, economists, and environmentalists may be useful in gaining consensus for radical change, but is often too abstract for farmers. This article discusses the specific meaning of sustainability in terms of agricultural practice.

Ecology and Sustainability

The emergence of ecology as a sophisticated discipline gives a basis for a definition of agricultural sustainability that is scientific, open to hypothesis-testing and experimentation, and also practicable. Population, community, and ecosystem ecology provide a better understanding of the complex dynamics that arise in agriculture, for example, in crop populations, multiple-cropping systems, agroforestry, and range management. Research centers such as the Centro Internacional de Agricultura Tropical (CIAT) in Colombia, and the Multiple Cropping Center at Chiang Mai University in Thailand have begun to develop a body of research on complex agricultural systems such as the rice fields of northern Thailand and the savanna ecosystems of Zimbabwe.

Agricultural systems such as the Thai rice fields can easily be recognized as modified ecological systems. Each field is formed from the natural environment, with a ridge of earth serving as its boundary. Inside, the great diversity of the original wildlife is reduced to a limited set of crops, pests, and weeds but still retains some of the natural elements, such as fish and predatory birds. Natural

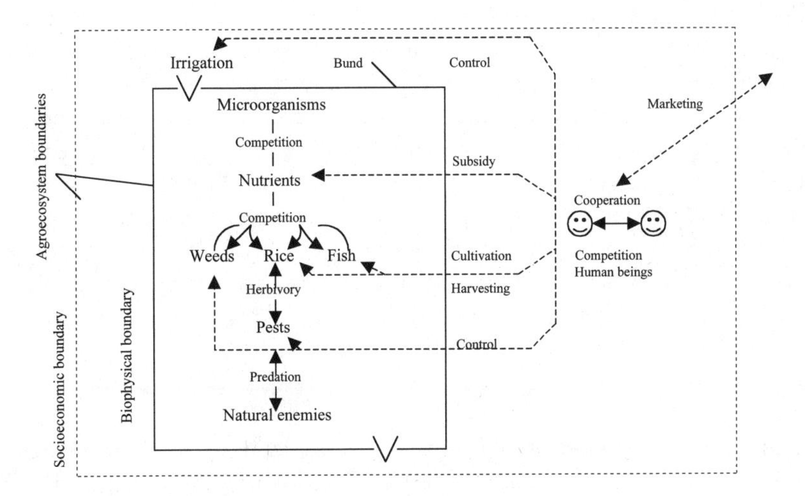

Figure V.1. The Rice Field as an Agroecosystem.

ecological processes such as competition between rice and weeds, herbivory of the rice by pests, and predation of the pests by fish and birds are overlaid with the agricultural processes of fertilization, control of water, pests, disease, and harvesting. These agricultural processes are, in turn, regulated by economic and social decisions. The boundaries of the socioeconomic system are not as easy to define as the biophysical ridge boundary of the field, but together they form an agroecosystem.

More formally, an agroecosystem is "an ecological and socio-economic system, comprising domesticated plants and/or animals and the people who husband them, intended for the purpose of producing food, fiber, or other agricultural products" (Conway 1987). An agroecosystem can be evaluated in terms of *productivity, stability,* and *sustainability.* Productivity measures output per hectare, stability measures variability of production from year to year, and sustainability can be assessed as the ability of ecosystems to maintain productivity over time and to rebound from stress or shock.

One way of improving sustainability is to protect the ecosystem from stress, for example, by using ridges to protect fields from flooding. For pest control, the development of inherent genetic resistance and the use of biological or integrated control methods tend to be more sustainable than the use of pesticides. The stress caused by removal of nutrients during harvesting can be countered by applying fertilizers or by boosting natural fertility through nitrogen-fixing legumes or composts. Shifting cultivation and the use of fallow periods also restore soil fertility. Economic sustainability may be promoted by producing a mix of crop and livestock products, using labor-saving cultivation techniques in response to emigration, or switching to higher-value, lower-volume products to lower transportation costs.

An important issue is the choice between external resources, such as fertilizers and pesticides, and internal resources, such as natural pest predators, algae, bacteria, green manures, agroforestry, multiple cropping, and indigenous tree, fish, and crop species. Internal resources, which are often free to the farmer, have an economic advantage over purchased external resources. Dependence on external resources may be both costly and risky, since it puts the farmer at the mercy of sudden changes in price and availability. External resources such as the Green Revolution package of hybrid seeds, fertilizers, and pesticides may also lead to changes in farming systems that make them more vulnerable to the vagaries of the local environment.

Equitability and Trade-Offs

In addition to productivity, stability, and sustainability, the performance of an agroecosystem can be evaluated in terms of *equitability.* "An African village that has a high, stable yield of sorghum and is using practices and varieties that are broadly resistant to pests and diseases might be regarded as more successful than another village having lower, less stable and sustainable yields. However, it is not only the pattern of production that is important, but also the pattern of consumption. Who benefits from the high, stable, and sustainable production? How is the harvested sorghum, or the income from the sorghum, distributed among the villagers? Is it evenly shared or do some villagers benefit more than others?" [17] In commercialized agroecosystems, benefits are divided between producers and nonproducers. Trade-offs between productivity, sustainability, and equitability, and between producer and consumer interests, are common.

The Green Revolution has generally favored high productivity at the expense of the other indicators. We are now entering a new phase of development in which much greater attention will have to be paid to stability, sustainability, and equitability in addition to productivity. The object of the Doubly Green Revolution is to minimize the trade-offs between objectives.

An example of effective minimization of trade-offs is found in the home gardens of Indonesia. The most prominent characteristic of home gardens is their great diversity relative to their size. In a Javanese home garden of little more than half a hectare, fifty-six different species of useful plants were found, with uses for food, condiments and spices, medicine, and livestock feed. The plants are grown in an intricate relationship with one another, so that the garden seems like a miniature forest. The food produced goes primarily for home consumption, but some is bartered or sold. These home gardens display high productivity, stability, sustainability, and equitability.

Farm Households and Livelihood Goals

In addition to diversity, home gardens epitomize the importance of farm household decision-making. Complex multigenerational and extended family structures are common in farm households. The role of women is particularly important both in production and in access to food and nutrition security. The displacement of women by agricultural mechanization has had severe effects on poor households, lowering household income and worsening children's access to food. Time stress on women can also adversely affect breastfeeding and children's nutrition.

The balance of livelihood goals involves a complex decision-making process, which takes place within the structure of traditional customs, rights, and obligations. Livelihoods often involve a combination of land husbandry, natural resource harvesting, off-farm employment, and handicrafts. Amerindian groups in central Brazil, for example, divide their time between gardening, hunting, and fishing, with variable patterns depending on local habitats. The sustainability of these livelihoods depends on their diversity, and potential innovations such as new crops should not conflict with existing effective patterns of activity.

Livelihood analyses should form an essential part of any development program. Unfortunately, few such analyses have been undertaken. While there is great potential in agricultural innovation, including biotechnology, this potential must he harnessed wisely in the interests of poor communities. Development planners, farmers, and field and laboratory scientists must collaborate in responding to the socioeconomic needs of poor households. "We need a shared vision based, above all, on partnership, among scientists and between scientists and the rural poor." [182]

Summary of

Environmental Risks Posed by Transgenic Crops

by Jane Rissler and Margaret Mellen

[Published in *The Ecological Risks of Engineered Crops* (Cambridge, Mass.: MIT Press, 1996), Chs. 3, 5, and 6, 27–70 and 111–128.]

The commercialization of transgenic crops may pose a spectrum of risks—from ill effects on humans and animals that consume engineered crops to the disruption of ecosystems. In these chapters, the authors discuss two major categories of environmental risks—those posed by the transgenic crops themselves and those associated with the movement of transferred genes (transgenes) into other plants. An assessment of these risks leads to specific recommendations for stronger regulatory controls on genetically engineered crops.

Transgenic Crops May Become Weeds

The definition of the term "weed" depends on context and human values. It is used to refer to plants that interfere with agricultural and other human activities, as well as to invasive species that disrupt wildlife habitat. Nearly all food and fiber crops have close relatives that are regarded as weeds in some areas. Many plants purposefully introduced as food or forage crops or as ornamentals have later become weeds, including crabgrass, kudzu, purple loosestrife, tamarisk, and water hyacinth. These have all caused major ecosystem disruption and displacement of native species. Invasive weeds reduce agricultural output, clog waterways, and have adverse health impacts on humans and animals. In the United States, billions of dollars are spent annually to control weeds, and hundreds of millions of pounds of herbicides are applied.

Weeds are characterized by *persistence* and/or *invasiveness.* Once introduced, weeds are difficult to extirpate and may spread rapidly to other sites. Research has suggested that transgenes may confer or enhance these "weedy" qualities in some crops. Alteration of traits such as seed dormancy and germination, stress tolerance, and growth patterns could increase a plant's ability to outcompete other species. While some crops such as corn are unable to survive without human cultivation and are therefore unlikely to become weeds through gene modification, other crops already possess some characteristics of weeds. Crop plants that are "on the edge of weediness" could be pushed over the edge by the addition of one or two genes. Crops such as sunflowers, strawberries, rye-grass, and broccoli have already established themselves as weeds in some areas. Others such as rice, barley, lettuce, oats, potatoes, sorghum, wheat, and alfalfa have close relatives that are weeds. Small genetic changes such as the insertion of disease- or insect-resistant genes might dramatically increase persistence or invasiveness in such plants.

Ecosystem Impacts of Transgenics

The introduction of genetically modified plants into an ecosystem may set off *cumulative* or *cascading effects* in an ecosystem. For example, salt-tolerant rice planted near coastal wetlands might invade nearby salt-water ecosystems, displacing native salt-tolerant species and setting off cascading effects on other organisms such as algae, microorganisms, insects, arthropods, amphibians, and birds. While complex ecosystems often possess resilience that allows them to absorb perturbations, species with transgenic traits may stress natural systems beyond their ability to react and recover. If gene-transfer technology becomes widespread, hundreds of transgenic organisms will be released into the world's ecosystems, with unpredictable outcomes.

Some transgenics are of special concern due to their ecotoxicity. When plants are engineered to produce pesticides and other drugs, it is predictable that non-target species will be affected. Beneficial insects and soil fungi may be harmed by the *Bacillus thuringiensis* (Bt) toxin and fungicidal transgenes now being engineered into a wide array of crops. The spread of Bt-engineered crops could also accelerate the development of pest resistance to Bt, thereby destroying the benefits of this widely used organic insecticide. (When Bt is engineered into a plant, it creates continuous pest exposure rather than the carefully limited applications characteristic of organic farming.) Similar pest-resistance problems are likely to develop from the insertion of other pest-control genes into plants.

Movement of Transgenes into Other Plants

Once transgenic crops are planted in large numbers near wild or weedy relatives, transgenes will almost certainly flow via pollen to these other plants. If the hybrid plants thus created produce viable seed, the transgenes will enter the gene pool of the wild population. The transgenes most likely to be retained are those that confer a competitive advantage, creating new or more persistent weeds in farm and nonfarm habitats. These new organisms may alter habitats, community structure, and food-chain composition, ultimately affecting genetic and biological diversity. For example, the transfer of disease-resistance into a minor weed whose population has been controlled by plant pathogens could convert it to a major threat to crops or ecosystems. Herbicide-tolerant genes engineered into crops might transfer to weedy relatives, making the weed even more difficult to control and canceling out the economic advantage of herbicide-tolerant crops.

There are a number of precedents for gene transfer to weedy relatives: in some cases new weeds are created that mimic the original crops and are therefore difficult to identify and control. Gene flow from rice, sorghum, millet, corn, and sugar beets has led to mimetic weeds in India, North and Central America, Africa, and Europe. Transgene characteristics are likely to create similar mimetic weeds possessing the same competitive advantages that the genetic engineers sought to create in the crop. The cumulative and cascading effects of transgenes that enter wild populations may be worse that those created by the original crop plants, since the weedy relatives are likely to be hardier and more invasive.

Threats to Rare Species and Crop Diversity Centers

Gene flow from crops can cause species extinction by overwhelming small wild populations. Hybridization with nontransgenic crops has already led to the extinction of wild crop relatives of hemp, corn, pepper, and sweet pea. Transgenics that convey tolerance to cold, heat, drought, or salt may lead to the extension of cropping to areas previously beyond pollination distance, endangering fragile ecosystems in these areas.

The threat of genetic degradation of ecosystems is particularly severe in the *centers of crop diversity,* which serve as natural gene banks for future agricultural use. In Canada and the United States, such centers of diversity exist for berries, sunflower, Jerusalem artichoke, pecan, black walnut, and muscadine grape. Most other centers of diversity are in the developing world, and not all have yet been identified. These centers contain genetic resources that will be essential in responding to future environmental change or disease outbreaks. They are already losing genetic variability at an alarming rate. If they cannot be protected from invasive transgenics, the impact on future agricultural resilience could be profound.

Viral Resistance and New Viral Diseases

Many agricultural biotechnology labs are working to create virus-resistant plants. This may have an unintended effect of producing new strains of viruses or exacerbating existing viral diseases. This could occur through recombination—the exchange of nucleic acids between transgenic plants and viruses. Recombination has already been observed in transgenic plants, for example, between the cauliflower mosaic virus (CaMV) and a CaMV gene in transgenic turnips. Other genetic transfer processes may make it possible for viruses to broaden their host range or to increase their rates of crop infection. Transgenic viral products may also interact with other viruses to cause more severe diseases.

Some scientists contend that the risks of transgenic effects on viruses are no greater than those already associated with natural interaction among viruses. However, the difficulty of predicting the consequences of a technology in advance of its implementation should argue in favor of caution with respect to this and other potential risks of transgenics. Our understanding of physiology, genetics, and evolution is limited. Unexpected effects of new genes in a gene pool cannot be ruled out. Direct transfusion of new functional DNA into plants may be utterly new from an evolutionary standpoint, making risk assessment difficult and implying that as-yet unknown risks may exist.

International Implications and Policy Recommendations

Seed, pesticide, and biotechnology companies are now introducing transgene technology on a global basis. Engineered crops are potentially harmful to the environment. The dangers may vary with location; for example, a transgenic insect-resistant soybean might pose minimal risk in the United States but serious risks in China, where numerous wild relatives of soybean are endemic. Similarly, pest-resistant corn might not be a problem in the United States but if used in Mexico could pose a serious danger to the genetic diversity of teosinte, the wild relative of corn whose genetic reservoir is an irreplaceable agricultural resource. Unfortunately, most countries are not prepared to control the ecological risks of transgenic crops.

To protect against the ecological risks of transgenic plants, the following recommendations are made for United States and United Nations policies:

- Federal regulatory programs should be strengthened.
- All transgenic crops should be fully evaluated for ecological risk and ecotoxicity.
- The National Academy of Sciences should prepare an assessment of the risks posed by engineered crops to the international centers of crop diversity.
- All transgenic seeds approved for the United States use should bear a label stating that U.S. approval carries no implication of safe use in other countries.
- The United Nations should develop international biosafety protocols to protect against the risks of genetically engineered crops.

Summary of

Population Extinction and Saving Biodiversity

by Paul R. Ehrlich and Gretchen C. Daily

[Published in *Ambio,* 22, 2–3 (May 1993), 64–68.]

Biodiversity is the array of populations and species on earth, and the communities, ecosystems, and landscapes of which they are component parts. Extinction may threaten local populations or entire species. While attention has focused on the problem of species loss, in many parts of the world the loss of local populations may be the most important aspect of the degradation of biodiversity. This paper discusses the importance of population extinction relative to species extinction, offers a preliminary assessment of its importance, and examines policy implications.

Defining Species and Populations

Nature at any given moment presents a "snapshot" of the process of diversification of populations. It is not always possible to have fixed guidelines for defining species and subspecies. A species is usually defined by breeding isolation—members of the species do not breed with members of other species. But local populations may also be defined by geographic isolation. The millions of known species may represent billions of genetically distinct populations, contributing to a pattern of biodiversity that varies across tropical, temperate, subarctic, and arctic regions. While there is less species diversity in nontropical zones, species in temperate, subarctic, and arctic regions have larger ranges and therefore almost certainly more local populations per species.

Rates of Population Loss

Considering only species extinctions may greatly underestimate the rate of loss of organic diversity. Species extinctions, especially in tropical areas, have gained attention, but population extinctions, which predominate in the temperate zone, are often ignored. In Britain, for example, changes in land-management practices since 1940 have led to a drastic decline in butterfly populations. Paving over of habitat, drainage of marshes, replacement of deciduous woodlands with conifer plantations, and use of fertilizers, herbicides, and insecticides, as well as climate change and acid rain, have contributed to a general decline of the butterfly fauna. Six percent of species have disappeared and an additional 29 percent have suffered massive population extinctions.

Britain, as well as much of Europe, is now "biologically depauperate." Most of temperate Asia, especially China, is in even worse condition, and North America is traveling the same course. Physical habitat destruction and modification, as well as importation of exotic species, are major causes of the imperiled status of 150 North American bird species and subspecies, as well as the population declines of other species not yet recognized as threatened.

The Significance of Population Extinctions

Why does it matter if populations become extinct? There are moral considerations surrounding the spreading destruction of the natural world. Loss of aesthetic value affects both those who are aware of the loss, such as naturalists and birdwatchers, and also those who suffer the opportunity cost of never experiencing or developing an appreciation for natural beauty. The direct economic value of fish and other harvested species is lost. In addition, population loss reduces genetic diversity, leaving species and ecosystems more vulnerable to environmental change. Reduction of genetic diversity also means that medicinal and agricultural resources are lost. Genetic variability is the "raw material" for selective breeding. Interpopulation variability is a key component of this diversity, one that cannot be duplicated in zoos or botanical gardens.

Perhaps the most important reason to care about population extinctions, however, is the ecosystem services they provide. Carbon fixation, water retention, flood prevention, nutrient recycling, pest control, detoxification of wastes, and many other services depend on natural populations of fauna and flora. An example is provided by the salinization of Australian wheatlands. In southwestern Australia, transpiration by native trees and shrubs maintained groundwater at relatively low levels. When the native species were cleared for wheat cultivation, the groundwater level rose, bringing with it salt concentrations that killed the wheat.

Dependence of ecosystem productivity on local populations is often subtle, yet crucial. The productivity of lakes, for example, is determined by the precise species balance of small crustacea and other organisms. Ecologists are just beginning to understand the degree to which extirpation of a population of one species can lead to a cascade of extinctions. In one case, removal of a predatory starfish allowed one mussel species to outcompete and exterminate numerous other species. Species interactions are often complex. Swallow species in Colorado depend on the co-occurrence of aspen trees, willow trees, red-naped sapsuckers, and a fungus that attacks aspens. The sapsuckers excavate nest cavities in fungus-affected aspens, which then become available as nest sites for the swallows within reach of food sources provided by the willows. Loss of species diversity in an area disrupts the web of species connections, causing a loss of "ecosystem plasticity" or adaptability.

The Economics of Preservation

The literature on the economics of species preservation is well-intentioned but unsatisfactory, since the uncertainties associated with valuation of existing species are overwhelming (see Bishop 1978). Further, in a context of rapid global change it is extremely difficult to determine the minimum viable population levels of species. Economic evaluation of the benefits and costs of conserving any single species is an exercise in "crackpot rigor"—detailed mathematical analysis without ecological foundation. Analysis of population losses brings in additional uncertainties regarding the uniqueness of the population, the reversibility of the loss, and the impact on the viability of the species as a whole.

The solution to this dilemma is for both ecologists and economists to focus on the overall values of ecosystems and to eschew analysis of the costs and benefits of extinctions. Populations of species in an ecosystem are analogous to rivets in an airplane wing. Some are more critical than others in maintaining system function, but the continued deletion of populations, like continual popping of rivets, will eventually lead to collapse. "Both the great uncertainties in how much biodiversity is required to maintain humanity's life support systems and the irreversibility of any mistakes call for an extremely conservative approach. The burden of proof should be shifted to those who promote the loss of biodiversity for short-term gains. In addition, economists should focus on strategies to monetize the known values of ecosystems so that ways can be found to internalize them." [67]

Policy Implications

Habitat preservation is crucial both to saving biodiversity and to preserving ecosystem function. No destruction or fragmentation of habitat should be taken lightly. The population extinctions resulting from habitat loss and fragmentation should be considered at least as important as species losses. Legislation should provide for the maintenance and restoration of habitat. Emphasis should also be placed on restoration ecology—restoring degraded lands and increasing population diversity. Urban and regional planning should focus on ways to reduce the human "footprint" on the planet and to maximize the areas that can be devoted to the preservation of population, species, and ecosystem diversity. "The health of the human ecosystem must be assumed to depend as much or more on the maintenance of population diversity as it does on the maintenance of species diversity. Any other assumption amounts to taking a gigantic gamble with the future of civilization." [67]

Summary of

Rocking the Boat: Conserving Fisheries and Protecting Jobs

by Anne Platt McGinn

[Published as Worldwatch Paper No. 142
(Washington, DC.: Worldwatch Institute, 1998).]

The fishing industry across the globe is affected by severe resource depletion. Fishing disputes between nations are common, and many traditional fisheries have been disrupted by exposure to global market forces. A majority of the world's marine fish stocks have reached peak production and many are in decline. The fishing industry is heavily overcapitalized, with huge factory ships capable of decimating stocks. Government subsidies create further incentive for overfishing, while important breeding areas such as coral reefs, tidal estuaries, and ocean-floor environments are being ravaged by indiscriminate fishing methods and continental runoff pollution. This article explores the dimensions of the problem and suggests policy remedies to conserve fish stocks and create a sustainable industry.

The State of the World's Fisheries

Fishery declines due to overfishing are nothing new. The U.S. government established its first conservation agency, the Commission of Fish and Fisheries, in 1871 in response to overfishing off the coast of New England and in inland lakes. The whaling industry had severely reduced whale species by the mid-1800s. But the global nature of the problem today is unprecedented. Eleven of the world's fifteen most important fishing areas and 60 percent of the major fish species are in decline. Whereas no fish stocks were in urgent need of management in 1950, today a majority of the world's fisheries require urgent action to rehabilitate damaged resources.

The dimensions of the problem are not apparent to consumers, since fish supplies appear ample. This is a result of a rapid increase in fish catch since the 1950s, and a more recent boom in aquaculture. However, there is strong evidence that the wild fish catch has reached a plateau, and expanding aquaculture is bringing significant environmental problems in its wake. Still-growing total fish production masks a pattern in which high-tech fishing fleets rapidly drive individual fish species to critically low levels and then move on to other species and different parts of the world. In addition to devastating individual species, this process disrupts the oceanic food chain, making it difficult for species to recover, and promotes adverse ecological changes.

In addition to stock reduction through harvesting, significant damage is being done to ocean ecosystems through pollution. Fish nurseries in coastal areas are threatened by pollution from cities, farms, and industries and habitat degradation. Excessive nutrient runoff causes algal blooms that drain oxygen from the water and often release toxins. Fishing gear also causes major ecological damage to ocean floors, while indiscriminate fishing practices destroy millions of tons of "bycatch"—unwanted species that are discarded, including fish, marine mammals, seabirds, and turtles. Small-mesh nets and the use of cyanide poison, common in tropical areas, cause widespread mortality among nontarget species.

Overcapacity and Economic Decline

The open-access nature of fishing encourages overcapitalization and excessive harvesting. Once the maximum sustainable yield is exceeded, both total harvest and profitability of the fishery decline. Technological increases in productivity then only worsen the problem of overfishing. This economic principle has been played out on a global scale since 1950, as electronic navigation systems, surveillance technologies, and sonar have enabled huge factory trawlers to harvest entire fish populations. The establishment of 200-nautical-mile exclusive national fishing zones under the U.N. Convention on the Law of the Sea placed some limits on international exploitation of local resources, but it also encouraged the development of major new fishing fleets in developing countries and did nothing to curb overfishing by nations in their own waters.

By the late 1980s, the world's large-scale fishing fleet had a fishing capacity that exceeded the maximum sustainable yield of all commercial fish stocks by 30 percent (see Figure V.2). Since then, the world's fleet has continued to increase in numbers and capacity, while catch rates have steadily declined.

As a result of the perverse economic logic of open-access, fishing fleets have taken billion of dollars in losses, but these losses have been balanced by government subsidies estimated at $14 to $20 billion. At least 20–25 percent of global fishing industry revenues come from government subsidies.[1] In addition to political pressure for subsidies, the fishing industry has successfully lobbied governments to maintain excessive fishing quotas, thereby perpetuating a vicious circle of overcapacity and strain on the resource base.

Impacts on Local Fisheries

In many developing countries, fish is an essential food for meeting minimal nutritional needs. People in developing countries consume an average of 9.2 kilograms of fish per person per year, as compared to 27.9 kg/person/year in industrial countries. However, this represents a much larger portion of their

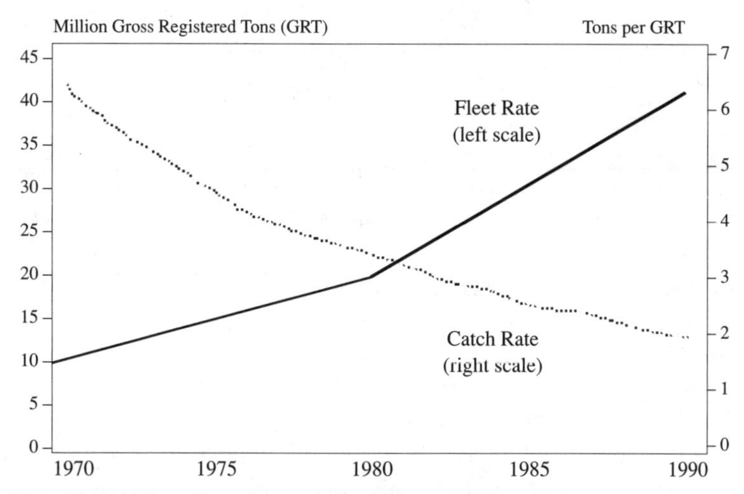

Figure V.2. Global Fleet Capacity and Catch Rate, 1970–1989.

animal protein. In many areas, traditional fisheries have come under pressure from mechanized trawlers and export-oriented aquaculture. While fishery exports can bring in needed revenues and raise incomes for some, they also threaten food security for those previously dependent on local fisheries and for other local consumers who cannot afford the higher prices resulting from export demand.

More than 200 million people worldwide depend on fishing for their income. But traditional fishers often find themselves squeezed between offshore factory trawlers and migrants from the interior attempting to enter the fishery. Depletion of fish stocks in the Northern Hemisphere has led industrial countries to pay for access to southern exclusive economic zones. As global demand for fishery products grows, the competition for access to fisheries will grow stronger, with further negative effects likely for local fishing communities, who rarely share in the revenues from fishing concessions.

Promoting Sustainable Aquaculture

Aquaculture is the most rapidly growing area of global fish production. Production of fish and shellfish through aquaculture was valued at over $36 billion in 1995, up from about $10 billion in 1984. Traditional forms of aquaculture, such as integrated fish and rice farming in Asia, have offered an ecologically sound method of improving local food security and household nutrition. By contrast, the intensive aquaculture characteristic of much of the rapidly expanding industry has created a host of environmental problems.

The monoculture of primarily carnivorous fish species demands large amounts of feed, water, and fertilizers. Rather than contributing to resource reuse and recycling, as traditional aquaculture has typically done, modern systems are generally resource-intensive and high-polluting. Ecologically valuable coastal areas such as mangrove swamps are often destroyed to make room for intensive aquaculture. Raising carnivorous fish species such as shrimp and salmon actually contributes to further depletion of natural fisheries, since high-protein fishmeal pellets are produced from wild fish. Six million tons of wild ocean fish are used to feed farmed species each year. Waste and uneaten food pollute the aquatic environment, triggering eutrophication[2] and algal blooms. A further threat is posed by the escape of domesticated fish into rivers, lakes, and coastal areas. These escaped fish can dilute the wild gene pool and spread diseases.

To promote sustainability in aquaculture, it is important to encourage the use of species that do not require fishmeal—preferably native species, which do not pose genetic hazards to wild stocks. Conversion of ecologically valuable coastal areas must be halted, and resource recycling rates must be increased by adopting integrated systems on the model of traditional small-scale aquaculture.

Policies for Sustainable Fisheries

The 1982 Law of the Sea and the oceans chapter of Agenda 21, adopted at the 1992 Earth Summit, address the sustainable use and conservation of marine resources. Despite some successes, such as the banning of high-seas driftnets, efforts to promote sustainable fishing practices have generally had weak implementation. In 1995, more than sixty fishing countries agreed to a voluntary Code of Conduct for Responsible Fishing. Also in 1995, the Convention on Highly Migratory and Straddling Stocks was adopted. In theory, these commit governments to sustainable fisheries management and stock conservation. However, the convention lacks force because most major fishing nations have not signed or ratified it.

Policies that are needed to conserve fisheries include

- A shift from maximum yield to more conservative catch limits.
- Effective data collection and reporting systems.
- Reduction or elimination of fishing subsidies.
- Prohibition of indiscriminate and destructive fishing practices.
- Integrated coastal management and marine sanctuaries.
- Individual transferable quotas to limit entry to fisheries.
- Fishing permits, licenses, and user fees to recapture resource rents.
- Retraining and alternative income opportunities for displaced fishers.

- Consumer education and boycotts to promote sustainable practices.

The adoption of such policies can help both to conserve fish stocks and to create a healthy fishing industry that will provide stable employment for millions of people in traditional as well as properly managed commercial fisheries.

Notes

1. Matteo Milazzo, *Subsidies in World Fisheries: A Reexamination.* World Bank Technical Paper No. 406, Fisheries Series. Washington, D.C.: World Bank, April 1998.
2. Eutrophication occurs when water contains an excess of nutrients, causing algae and other simple forms of plant life to proliferate, reducing dissolved oxygen and killing off other aquatic life.

Summary of

The World's Forests: Problems and Potentials

by Norman Myers

[Published in *Environmental Conservation* 23, 2 (1996), 156–168.]

Although greater attention and resources are now being directed toward increasing the sustainability of forests, recent deforestation, especially in the humid tropics, constitutes the fastest land-use change of its scale in human history. In the absence of greatly expanded efforts for better management, many of the world's forests appear likely to decline at even more rapid rates.

This article appraises the forest situation from both a natural and social science standpoint, arguing that deforestation results from a lack of sufficient scientific and economic understanding of forests' contribution to human welfare, as well as from a lack of recognition by policy-makers that deforestation is principally driven by nonforestry factors. The recently established World Commission on Forests and Sustainable Development and the Intergovernmental Panel on Forests constitute institutions that can work to bridge these gaps in understanding and policy.

What Is at Stake

Forests offer an "exceptional array of goods and services" that "should be reckoned amongst our most valuable stocks of natural resources." [157] But forests are among the least developed of all natural resources insofar as they are not often managed in sustainable ways to serve the long-term interest of all human communities concerned. Many forests are exploited for only a few products, with "disregard and adverse repercussions" for their many other potential and actual outputs.

Among the goods and services supplied by forests are

- *Commercial timber*—Commercial timber products are worth over U.S.$400 billion annually. This income, however, is expected to decrease as timber stocks, especially in tropical moist forests, are increasingly overharvested.
- *Fuelwood*—Three billion people worldwide depend on fuelwood for almost all their household energy, but to do so, half of them must overcut tree stocks, which greatly curtails regeneration.
- *Nonwood products*—These include wild fruits, latexes, essential oils, waxes, and medicinals, among many others. If subsistence and nonmarketed items are included, the value of the world's nonwood products may amount to as much as $90 billion annually.
- *Biodiversity and genetic resources*—Tropical forests contain about 50 percent and possibly 80 percent of the earth's species. Biodiversity has scientific, aesthetic, and ethical value, and biodiverse tropical forests are a source for many medicinal resources. Preliminary estimates of the potential worth of all tropical forest plants range from $420 to $900 billion per year. Additionally, insects, which are by far the most abundant form of biodiversity, assist material welfare by facilitating essential ecosystem processes such as pollination and control of pests.
- *Environmental services*—Forests stabilize landscapes, protect soils by helping them to retain their moisture and to store and cycle nutrients, and serve as buffers against the spread of pests and diseases. They also preserve watershed functions, helping to regulate the quality and quantity of water flows, and to modulate climate by regulating rainfall regimes.
- *Climate regulation and global warming*—Forests account for 65 percent of net plant growth and carbon fixation on land (over half occurring in boreal forests), which prevents carbon from entering the atmosphere and contributing to global warming. When forests are burned, they release their carbon, accounting for approximately one-fifth of all carbon released. Ironically, global warming may also trigger increased decomposition and die-off of forest biomass and may cause boreal forests to become more vulnerable to fires, which would also lead to further release of carbon dioxide. This could result in a decline of 10 percent of all carbon held on land in plants and soils. However, reforestation would increase carbon sinks in forests.

The Problem of Forest Decline

Two-thirds of deforestation in tropical forests, where deforestation is most rampant, is due to slash-and-burn agriculture by displaced peasants. However, forestry policy tends to focus on the minor players in tropical deforestation, such as commercial loggers and cattle ranchers.

Policy-makers need to focus more on the sources of the problem and examine the motivations behind peasant action. The peasant cultivator is driven to deforest by population growth, poverty, and inequitable land-use systems, among other forces. Thus, it can be seen that the sources of the problem are often nonforestry related.

Temperate forests are more or less in equilibrium. However, boreal forests are also starting to decline in expanse, more from long-term degradation than from outright destruction. Clearcut logging, burning, acid precipitation, and industrial pollution affect boreal forests. Degradation in Siberian forests totals twice as much as recent annual deforestation in Brazilian Amazonia, although in contrast to the situation in the tropics, boreal forests can usually regenerate themselves over a period of decades.

Ultimate Sources of Deforestation

The ultimate source of deforestation is the fact that councils of power, including governments and international agencies, marginalize the forest issue and treat forests as dispensable. To illustrate, the forestry budget of the U.N. Food and Agriculture Organization, the U.N. agency in charge of forestry, has dropped from 5 percent in 1975 to 3 percent today. This stems from the fact that policy-makers lack a full scientific and economic understanding of the value of forests.

Two New Institutions: Their Scope and Scale

The World Commission on Forests and Sustainable Development works to propose policy reform and institutional changes, and assembles and strengthens scientific research related to the sustainability of forests. The Intergovernmental Panel on Forests aims "to pursue consensus for action toward sustainable development of forests through international cooperation." [161] Both institutions should build on the achievements of other forestry initiatives, such as the Declaration of Forestry Principles at the Earth Summit in Rio de Janeiro, and the Bandung Initiative for Global Partnership in Sustainable Forest Development, among others.

The challenge for the two institutional initiatives is to formulate a vision for forests' future, focusing on six questions:

- How much forest and of what type do we want in the twenty-first century?
- What environmental and socioeconomic purposes should forests serve?
- How can forests' development contribute to society's sustainable development?
- How far can existing institutions go to serve the long-term purposes of forestry, and how necessary are new institutions to promoting joint responses to joint problems?
- How can governments, international agencies, scientists, NGOs, and others work together to ensure that forests play full parts in the world of the future, and how can these participants bring about consensus between the scientific communities and political leaders?
- How can concrete proposals be formulated to support governments and institutions that seek sustainable forestry?

There is still inadequate recognition of what sustainable forestry entails. The forestry profession tends to focus on logging technology, industrial processing, timber markets, wood engineering, plantation genetics, and the like, and downplay external factors such as poverty that play the largest role in the fate of forests. Thus, the two new international bodies should seek to expand the policy horizon of traditional forestry.

Policy Reform

To reform policy to successfully achieve sustainable forestry, the following proposals should be considered:

- *Change the definition of sustainable forestry*—The definitions of sustainable forestry advanced by many international organizations do not view forests in the context of their wider physiobiotic, socioeconomic, or political-legal landscapes, thus ignoring the outside factors that play such a large role in forests' fates. To withstand the myriad pressures and threats overtaking forests, we need an approach that enables forests to make their full contribution to socioeconomic advancement for all communities concerned.
- *Enhance the institutional status of forests*—Basic forestry policies are effectively, though unwittingly, set by powerful bureaucracies in charge of non-forestry economic sectors such as agriculture or employment. Policy planners need to view the forests with an eye for their actual and potential goods and services. Additionally, forests ultimately benefit communities in all parts of the world, a factor that must be taken into account in policy formulation. One response would be to compensate forest countries that supply global-scope benefits through mechanisms such as the World

Bank's Global Environment Facility. Also useful are debt-for-nature swaps in which wealthy countries pay off part of the national debts of poorer countries in exchange for their agreement to preserve their natural resources.

- *Increase scientific understanding of forests*—A full scientific evaluation would include forests' character, extent, make-up, mechanisms, and dynamic linkages to the rest of the biosphere. This need is to be partly met by the formation of the Scientific Advisory Council under the World Commission.
- *Increase economic understanding of forests*—Economic analyses should be undertaken to evaluate the entire range of goods and services that forests provide. This is difficult because many of these products are consumed outside the marketplace. Where quantitative assessments fail, qualitative assessments can highlight actual and potential benefits. The analytic methodologies underlying these assessments should be geared to social equity as well as to economic efficiency.
- *Remove perverse subsidies*—Subsidies that encourage overlogging make it profitable for concessionaires to overexploit forests and to cut trees otherwise uneconomical to harvest. In addition, governments often receive only a fraction of the natural resource rent possible had these resources been sold at their true value. Subsidies to cattle ranchers also promote deforestation. The Brazilian government has spent $2.5 billion to subsidize ranchers, who often don't even bother to sell the trees felled to make pastureland, collectively torching $5 billion worth of timber per year. "Virtually every ranch has been a financial success for the individual entrepreneur while an economic disaster for the national economy." [164]
- *Calculate the costs of inaction*—The costs of not acting to preserve valuable ecosystem goods and services should be calculated, including the impact on watershed functions, fisheries, fuelwood, genetic resources, climate stabilization, and overall environmental values. For example, at an estimated value of $20/ton, the carbon storage function of tropical forests alone amounts to U.S.$3.7 trillion.
- *Include all goods and services in valuation*—Valuation methodology for all ecosystem services in forests should include direct-use values such as timber and medicinal plants, indirect-use values such as soil conservation and watershed protection, and existence values conferred by assuring the survival of a resource and option values of potential future use.

A preliminary review of ecosystem services in several dozen tropical forests indicates that the hypothetical overall value of sustainable use of one hectare of forest is about $220 per year, comprising $69 from forest products, $12 from recreation, $10 from watershed functions, $5 from hunting and fishing, $16 from option and existence values, and $110 from timber. It should also be noted that as the incomes of people rise, these people are willing to better recognize and pay more for ecosystem services. Additionally, as forests continue to disappear, the value of those remaining will increase.

- *Promote forests as a global commons resources*—There is need to reconcile the fact that the forests' environmental services benefit the global community yet fall within the sovereign jurisdiction of individual nations subject to the policy discretion of their governments. The two new forestry institutions should serve to "foster a coalition of interests as a basis for an eventual international instrument or set of instruments on forests," and should establish a consensus about the world's forests and their values so that the effort to sustain them is truly global.

PART VI

Materials, Energy, and Climate Change

Overview Essay

by Frank Ackerman

In an industrial economy based on fossil fuels, is sustainable development imaginable? And if so, is it achievable? Environmental advocates and policy-makers face one of their sharpest challenges here: Is there, even in theory, such a thing as sustainable industry? It is comparatively easy to describe the possibility of organic agriculture integrated with nature; it is much harder to picture "organic" steel mills or chemical plants.

Yet the invention of more ecologically friendly industry is essential if sustainability is to become meaningful in modern economic life. Economic development inevitably involves widespread industrial activity; a vision of utopia without manufacturing would be unlikely to win many adherents in the lower-income countries of the world. And industry, in its present form, constantly causes critical environmental impacts. Many of the most serious contemporary environmental problems, such as the threat of global climate change, are direct results of the industrial use of energy and materials. There may indeed be viable alternatives, but they require substantial changes in the shape and scale of existing industry.

A sustainable society will still use materials and energy for mass production—but it will have to do so more moderately than rich countries do at present, and in a qualitatively different manner. It is clear that the current patterns of production and consumption are unsustainable. On the production side, extractive and manufacturing industries are overexploiting both renewable and nonrenewable resources while causing substantial air and water pollution. On the consumption side, ever-escalating standards for new cars, homes, and assorted luxuries, conveyed worldwide by American-influenced or -dominated media, create a consumer demand for unsustainable levels of production. A very different approach will be needed in order to avert the threat of climate change and to create a society based on sustainable resource flows.

There is an extensive body of research related to these topics, including many detailed, applied analyses. This essay reviews the literature in three related areas: material use, energy, and climate change. Looking beyond the technical details, the discussions reviewed here display a striking divergence in tone. There are some differences about the urgency and magnitude of the environmental problems to be solved. There are also important differences, even among those who see substantial problems facing us, about the feasibility and cost of the needed changes. Some see signs of change in the right direction and can describe technologically feasible, affordable alternatives. Others point out how far we are from solving basic environmental problems, how much we will have to spend, and how rapidly we need to change in order to prevent ecological crisis. Sorting out the evidence for these contradictory perspectives is one of the challenges in this area.

Materials

There are inescapable natural limits to the use of materials in a sustainable society. For renewable resources, there is a sustainable annual yield—for example, the annual growth in a forest—which harvesting cannot exceed in the long run without causing depletion. For nonrenewable resources, there is no such thing as a sustainable long-run "yield" of virgin material; it is important to seek alternatives and move away from dependence on finite, exhaustible supplies. Are we close to bumping up against those limits, or are they so far away that other problems are more urgent to address? Ominous predictions of material shortages, often heard in environmental circles in the 1970s, have fortunately not been borne out by events to date. Yet the logic of the underlying argument remains unassailable over a longer time frame. Sustainable material use is still ultimately essential, and still far from reality.

A number of overviews of material use are available, such as the comprehensive study by Gary Gardner and Payal Sampat (1998). The use of all materials other than food and fuel reached 101 kilograms per capita per day in the United States in 1995; in that year, total U.S. material use was eighteen times higher than in 1900. Some progress has been made toward efficiency of material use (or, equivalently, toward declining material intensity of consumption), but much more needs to be done. Gardner and Sampat cite studies suggesting that equitable, sustainable worldwide material use might require a 90-percent reduction in per capita material consumption in industrialized countries—far beyond the level that can be easily achieved through gradual efficiency gains. Thus they conclude with a call for new programmatic initiatives, such as elimination of subsidies to virgin material production, imposition of environmental taxes and fees, and support for both established and new forms of recycling. Other recent overviews and studies along similar lines include Cutler Cleveland and Matthias Ruth (1999), Ruth (1998), and Friedrich Hinterberger and Eberhard Seifert (1997).

The theoretical perspective embodied in many recent studies is called "industrial ecology," stressing the parallels between the inter-industry flows of materials, energy, and wastes and the similar flows that occur between species and environments in nature. Just as there are nutrient cycles in nature, with wastes of one organism providing food for another, so, too, there can be material cycles in industry, with wastes and byproducts from one firm providing valuable inputs to another firm. A linear economy, in which everything is used once and then discarded, requires huge resource inputs and generates huge wastes in the process of satisfying consumer demand. In contrast, a cyclical economy, in which one industry's waste is another industry's resource, can achieve the same levels of production and consumption with far lower levels of material "throughput." For an introduction to industrial ecology, see Robert Socolow et al. (1994).

The best-known, and perhaps the gloomiest, writer in the field of industrial ecology is Robert Ayres (see Robert and Leslie Ayres 1998, Ayres and Paul Weaver 1998, Ayres and Ayres 1996, among many others). Ayres paints a bleak picture of the environmental damage and unsustainable resource consumption caused by current practices, filled in with encyclopedic knowledge of specific industrial processes. He advocates an "industrial metabolism" perspective, in which the circulation of materials and waste products throughout society is viewed holistically, as analogous to the metabolism of a biological organism. For Ayres, this perspective implies that environmental impacts should be viewed on an integrated or life-cycle basis, rather than in isolation. Thus he calls for the creation of closed materials cycles, in which the wastes of one process are always useful inputs into another, as the only long-run strategy that can lead to sustainable industrial production. Sweeping changes in industrial technology would be required to substitute closed materials cycles for the open, waste-generating processes now in use (on industrial futures see also Duchin and Lange 1994).

Counterposed to Ayres are more upbeat writers who find evidence of what they call "dematerialization"—that is, declining material use per constant dollar of GDP, or in some cases per capita, in the developed world. The computer industry provides a familiar, and extreme, example: the material required to produce the same amount of computing power is steadily and rapidly declining. The summarized article by **Iddo Wernick et al.** provides a clear explanation of the argument for dematerialization. There have been long-term declines in the intensity of use (kilograms of material per constant dollar of GDP) of many common materials, including timber, copper, and steel.

If, in fact, social and technological change are steadily reducing the intensity of material use, there is more room for sustainable economic growth before we run into resource constraints. However, the implications of global equity should be considered in this context: raising several billion people up to developed-country levels of material use would create a surge in demand and production, outweighing the reductions achieved by many years of dematerialization. If an industry is growing rapidly enough, declining material use per unit of output may coexist with rising total material use; the computer industry is again a good example.

Dematerialization stems in part from the shift from goods to services, and within manufacturing from the increasing role of technology (as industry relies on more skillful or knowledge-based fabrication rather than on increases in the bulk of materials). In part it also reflects the rise of recycling of waste materials, which lowers the required input of virgin materials per unit of output. The discussion of sustainability therefore intersects the discussion of recycling.

The second summary (**Frank Ackerman**—the concluding chapter of my book on recycling) addresses the connection between recycling and sustainability. A sustainable future economy will have to move increasingly toward reliance on renewable resources for both materials and energy; the rate of recycling is one of several key factors determining how hard it will be to reach and remain at that goal. Emphasis on material conservation and recycling seems to fly in the face of the current abundance of cheap materials; free marketeers unfortunately have a point in suggesting that the freedom to use and discard materials is intrinsic to the feeling of affluence. Yet only in the short run is it imaginable that we can afford to waste materials at current American levels. The act of recycling is a response to environmental concern at least as much as to market signals; it promotes an ethic of conservation and increases the chances for learning-curve effects that will lead to environmentally desirable future technologies.

The techniques of life-cycle analysis, and the field of industrial ecology in general, have promoted a helpful focus on the comprehensive impacts of material use from cradle to grave. However, most studies show that the principal environmental impacts of nonhazardous material life cycles occur in raw material extraction and in the first stages of purification and processing of the resulting materials (see the discussion in Ackerman 1997, Ch. 5). These are by far the dirtiest and most energy-intensive branches of industry. Impacts are much smaller in the later stages of refining, fabrication, transportation, use, and (in most but not all cases) disposal of the resulting materials. Thus the focus on industrial ecology, while considering the entire material life cycle, can also be read as a study of how to minimize the use of (or the impacts of producing) virgin raw materials.

The industrial ecology of paper production has been particularly controversial. The widespread belief that recycling is environmentally beneficial has been challenged by claims that paper recovery and recycling may be no better than garbage collection and disposal. This is of enormous importance for recycling efforts, since paper and cardboard account for the great majority of recycled materials collected in municipal programs, whether measured by weight, volume, or dollar value.

Controversy is possible because two major life-cycle analyses of the paper industry have reached somewhat contradictory conclusions. However, neither is entirely critical of recycling. A study by the Environmental Defense Fund, sponsored by several major paper-using companies and institutions, finds that recycling is almost always environmentally preferable to any form of waste disposal (for an article-length summary, see Lauren Blum et al. 1997; see also several related commentaries in the same journal). In contrast, a study performed by the International Institute for Environment and Development, and partially funded by the paper industry, finds that incineration is on balance environmentally preferable to recycling for many but not all grades of paper (Grieg-Gran et al. 1997). Despite this disagreement, both studies agree that recycling is almost always preferable to landfilling of paper, which is the only available alternative for most of North America.

Energy

Energy is one of the biggest and dirtiest industries of all, centrally implicated in problems such as acid rain and climate change. The threat of exhaustion of nonrenewable resources has been widely discussed since the oil crises of the 1970s. Early analyses by Amory Lovins, Barry Commoner, and others identified plausible alternatives based on conservation and renewables—the "soft energy path." Low-cost opportunities for increasing energy efficiency were routinely ignored before the first oil crisis in 1973; for years, total U.S. energy consumption had grown at essentially the same rate as economic output. Yet after 1973, in the atmosphere of crisis, energy conservation suddenly appeared to be a bargain; from 1973 to 1986 there was essentially no change in U.S. energy use, while real GNP grew by 40 percent (Brower 1992). Still more can undoubtedly be done, since energy use per capita in the United States remains about twice as high as in Germany or Japan.

Meanwhile, renewable energy sources are becoming steadily cheaper. Wind power and solar water heating are already competitive with conventional energy sources in many areas, and other technologies such as photovoltaics (rooftop solar electricity generation panels) and fuel cells are promising for the long run. Thus advocates of the "soft path" call for using the remaining supplies of fossil fuels to ease the transition to a renewable energy future. For overviews of energy production and consumption see John Holdren (1990), Brower (1992), and Christopher Flavin and Nicholas Lenssen (1994). The challenge at this point is not primarily to develop new alternatives, but rather to overcome the massive political and economic obstacles to moving toward the soft energy path.

The problems of economic development and energy use appear quite different in developing countries, ex-Soviet nations, and the industrialized world. However, the summarized article by **John Byrne et al.** argues that China's energy problems and solutions have many parallels to those in the United States. A gradual decline in China's energy/GDP ratio has been swamped by rapid economic growth, so that energy use has climbed sharply. In light of the country's abundant coal reserves, the path of least resistance seems to be to burn more of it, with dreadful environmental effects. As in the United States, increased efficiency of energy use is often a cost-effective short-run alternative. In the long run, China, like the United States, has ample potential to develop wind, solar, and geothermal energy.

The problem in Russia is almost the opposite. There the energy/GDP ratio remains high, reflecting the legacy of Soviet technology, which evolved in isolation from world markets and in a context of abundant fossil fuel supplies. Yet despite persistently high energy intensity, Russia's long economic slump has held down total energy use. If economic growth were to revive in Russia or other ex-Soviet nations, the prevailing patterns of inefficient energy use would cause severe economic and environmental problems (Martinot 1996).

The hope of achieving efficiency by removing market distortions is a major theme of the World Bank's many studies of energy in developing countries. Robin Bates (1993) argues that developing countries frequently subsidize and control energy prices, tolerate or unintentionally create barriers to investment in energy efficiency, and fail to provide adequate financing or management for energy-producing firms. However, Bates remains optimistic about the ability of market mechanisms to solve these and related energy sector problems. Mohan Munasinghe (1995) offers a broader view from the World Bank, emphasizing issues of sustainability and the need to incorporate externalities and nonquantifiable objectives into developing country energy policy. His analysis points to a few high-priority technical fixes but otherwise relies on the assumed ability of market-based policies to solve a wide range of problems.

Debate about energy policy in the United States increasingly centers on the role of the market. One goal that receives widespread support, at least in theory, is the elimination of subsidies to nonrenewable energy production. Several studies have identified numerous subsidies to fossil fuel and nuclear power, worth $15 to $35 billion annually depending on the study (see citations and discussion in Ackerman 1997, Ch. 2, and in Roodman 1996). The "Green Scissors" report from Friends of the Earth targets twenty egregiously wasteful federal energy expenditures that are ripe for repeal, worth several billion dollars annually in total (Friends of the Earth 2000). Unfortunately, many of these expenditures are protected by powerful special interests and have survived public criticism unscathed. While repeal of wasteful energy subsidies is obviously desirable, the amounts involved are too small, relative to the huge size of the industry, to have a significant effect on prices, on consumer demand for energy, or on the competitive position of alternative fuels.

In energy modeling, there is a long-standing debate between those who examine detailed energy supply and end-use technologies—so-called "bottom-up" studies—and those who rely on "top-down" or aggregate econometric techniques. Bottom-up studies routinely find ample opportunities for no-cost or low-cost conservation. Top-down models based on economic theory often rule this out a priori, sometimes casting the issue metaphorically as a debate about whether there are any $20 bills (free conservation opportunities) to be found lying on the sidewalk.

Examining U.S. energy use from the perspective of sustainability is a frustrating exercise. Many sensible opportunities to reorganize transportation, for example, are easy to describe in theory—such as smaller, fuel-efficient autos and more mass transit—but apparently politically impossible to implement in practice. One area where rapid institutional change is occurring is in the electric utility industry, where deregulation of generation has begun and will be adopted in most states over the next several years. Will this help or hurt the environment? For an overview that favors deregulation, see Keith Kozloff (1997). By breaking down utility monopolies, Kozloff argues that deregulation could allow market access for new, greener producers.

In contrast, the summarized chapters by **Peter Fox-Penner,** an experienced utility analyst, present a more skeptical view. Deregulation pursued without any attention to the environment could undercut the opportunities for conservation and renewable energy sources that were built into older regulatory systems, and could allow existing coal plants to capture a greater share of the national market for electricity. (These coal plants gain an important economic advantage by being "grandfathered" under regulations such as the Clean Air Act; see Ackerman et al. 1999.) Fox-Penner argues that deregulation has potentially positive implications for the environment only in those cases, such as in Massachusetts, where strong environmental provisions are incorporated into the law. Such provisions demonstrate that it is not the operation of the market per se, but rather the political and regulatory context within which the market operates, that creates the potential to move toward sustainability.

Climate Change

Among the most ominous effects of energy use (along with selected other activities) is the risk of long-term climate change. The basic science is by now familiar, and a formal international consensus of the experts has been reached (see the three volumes of IPCC 1996). Unfortunately, there is no comparable consensus in the political arena about how to proceed. The Kyoto Protocol, an international treaty proposing moderate initial steps toward climate change mitigation, was negotiated in late 1997. Although hailed by many countries and by independent observers, the Kyoto Protocol has thus far been stalled by U.S. congressional opposition, with conservatives arguing that it asks too little of developing countries and too much of the United States.

The debate over modeling strategies and assumptions, mentioned in connection with energy analyses, seems even deeper and harder to reconcile in the economics of climate change. Huge, intricate models run by differing researchers yield strikingly inconsistent results. Are there ample opportunities, as some studies find, to reduce carbon emissions at little or no cost? Or is any such "free lunch" implausible, as other studies report, since any substantial carbon reduction would be exorbitantly expensive? In some cases, zero-cost reduction is said to be impossible as a matter of economic theory, not empirical research. This view, quite common among conventional economists, implies that there must be "hidden costs" blocking the implementation of apparently costless opportunities. However, those hidden costs are rarely identified.

Many climate change analyses inappropriately apply very abstract economic models, as argued by Irene Peters et al. (1999). The general equilibrium framework of traditional economic theory is mathematically convenient but typically rests on unrealistic assumptions associated with models of perfect competition. (For a more extensive theoretical critique, see Ackerman 2000a.) These assumptions rule out a priori some of the most important features of energy and environmental technologies, such as increasing returns to scale and learning curve (or learning-by-doing) effects. The unrealistic assumptions of general equilibrium models lead to the unrealistic conclusion that there is a unique, optimal path for energy development, rather than a choice between multiple options—for instance, soft vs. hard energy paths.

The essay by **Robert Repetto and Duncan Austin** summarized here demonstrates that it is possible to reconcile the estimates of the costs of climate change mitigation from numerous different economic models. A handful of key economic assumptions, described in the summary, turn out to account for virtually all of the differences in cost estimates. Under the most favorable assumptions, substantial reduction in carbon emissions could have a net positive economic impact; under the least favorable case, it would have a very large net cost. Repetto and Austin conclude, on balance, that policy proposals such as carbon taxes would do no damage to the economy and would bring long-term environmental benefits. However, many uncertainties surround several of the key assumptions, and further analysis is needed. A helpful agenda for further research on the economics of climate is proposed by Michael Toman (1998a).

Some of the dilemmas in the field are matters of theory, not of empirical research. The summarized article by **Robert Lind and Richard Schuler** attacks one of the controversial theoretical foundations of climate change modeling, namely the use of discounting for events far in the future. The greatest benefits of climate change mitigation will be enjoyed by generations long after those who pay the costs. As discussed in Part I of this volume, there is no logical basis for adopting a numerical discount rate for intergenerational calculations; rather, there is a need for careful examination of costs and for public debate over alternatives. (Agreeing that the conventional approaches to discounting do not make sense, Cédric Philibert [1999] reviews the theoretical debates on these questions and proposes a different solution: a gradually declining discount rate combined with gradually rising prices for nonreproducible environmental assets.)

Lind and Schuler also raise the fundamental question of the equity implications of climate policy. Investment in climate change mitigation involves equity between rich and poor, as well as between present and future generations. Is it logically contradictory, and/or politically self-defeating, to recommend substantial investment on behalf of the future without doing the same on behalf of today's poor?

As important as these larger theoretical debates may be, it is also worth remembering that numerous studies do find practical steps that can be taken today at little or no net cost. Even if these steps do not yet lead all the way to a comprehensive long-term strategy, there is little harm, and potentially a lot to be gained, by starting as soon as possible. For a widely discussed economic analysis advocating immediate action, see William Cline (1992). More recent studies that reach similar conclusions include Stephen Bernow and Max Duckworth (1998), and Union of Concerned Scientists and Tellus Institute (1998).

Detailed engineering analyses of emissions of methane, the second most important greenhouse gas, also identify low-cost/no-cost options for reduction (United States Environmental Protection Agency 1999). Familiar "low-technology" policies such as recycling of municipal waste have substantial climate benefits, due to the combined effects of the reduction in landfill methane emissions, the reduction in industrial energy requirements, and (in the case of paper recycling) the increase in forest carbon sequestration (Ackerman 2000b).

Greenhouse emissions per capita are currently lower, but are growing faster, in developing countries. This is due both to the rate of economic growth and to structural change, for instance, from agriculture to manufacturing (Xiaoli Han and Lata Chatterjee 1997). Thus it is particularly important to develop strategies for emission reduction in developing countries. Options for technology transfer oriented to climate change mitigation are discussed in Martinot et al. (1997). There is no iron law of development requiring a fixed ratio of carbon emissions to GDP; in practice these ratios vary widely, suggesting that even with existing technology there is room for improvement almost everywhere (Moomaw and Tullis 1994).

Conclusion

Are we already moving toward solutions to the long-run problems of energy, materials, and climate change? Or are we far from doing enough to create a sustainable material world? The answer is undoubtedly yes to both questions. There are encouraging signs of change, and researchers have identified numerous technologically feasible solutions to serious environmental problems. The image of relentless and unmitigated ecological disaster, all too common in speeches and fundraising appeals, is both misleading as description and counterproductive (because it is so demoralizing) as advocacy.

At the same time, the magnitude of required changes is enormous, and the existing pace of environmental gains is only a start toward what will ultimately be needed. The Kyoto agreement on climate change, which to date has proved too controversial for U.S. congressional approval, still proposes to do noticeably less than what is called for in most scientific analyses of global warming. To create a world of renewable energy, sustainable material use, and low, limited carbon emissions—the world that we can and should leave to future generations—we will have to move much farther and faster.

In short, great things have been done, and greater things still need to be done, to address the problems of sustainable materials and energy use, and to respond to the threat of climate change.

Summary of

Materialization and Dematerialization: Measures and Trends

by Iddo K. Wernick, Robert Herman, Shekhar Govind, and Jesse H. Ausubel

[Published in *Daedalus* 125, 3 (Summer 1996), 171–198.]

Is the "dematerialization" of human societies under way? Is there, in other words, a tendency toward a decrease in the quantity of materials required to serve economic functions? The concept of dematerialization is analogous to energy conservation; both refer to achieving the same output with reduced resource inputs. "Dematerialization matters enormously for the human environment. Lower materials intensity of the economy could reduce the amount of garbage produced, limit human exposures to hazardous materials, and conserve landscapes. . . . A general trajectory of dematerialization would certainly favor sustaining the human economy over the long term." [172]

This article reviews the evidence for dematerialization in the United States in the late twentieth century, finding a mixed picture of successes and failures. It examines four economic stages: resource extraction and primary materials, industrial production, consumer behavior, and waste generation.

Primary Materials

In 1990, the United States consumed 1.9 billion (metric) tons of hydrocarbons, almost entirely for fuel, and 2.5 billion tons of nonenergy materials (excluding air and water)—almost 8 tons per capita of fuel and about 10 tons per capita of other materials. Construction materials, such as the huge quantities of crushed stone used for road building and other purposes, accounted for 70 percent of the nonenergy materials; while there may be local environmental impacts associated with excavation, the available resources of stone are immense. Greater problems are associated with the use of smaller quantities of other, scarcer materials.

There is no upward or downward trend in the total weight of materials consumed in the United States since 1970. However, there has been a reduction of about one-third in material use per constant dollar of GDP, much of it associated with the oil shocks of 1973 and 1979.

The intensity of use has changed dramatically for individual materials. Timber, the most heavily used material of the early twentieth century, has declined steadily in intensity (measured in kilograms per constant dollar of GDP), as have copper, steel, and lead. In contrast, plastics and aluminum have shot upward in intensity, as have phosphorus and potash, key ingredients in fertilizers. Despite advances in electronic communication, the intensity of paper use has showed little change since 1930. Today more than 25 percent of timber cut in the United States is used to produce pulp and paper.

Small quantities of exotic, newly exploited materials (such as gallium, platinum, vanadium, and beryllium) have come into use in electronics, in the production of steel alloys, and in other "designer materials." Mining and processing of these materials is often environmentally damaging, and the small, widely distributed quantities of rare elements are sometimes difficult to recover and recycle.

The explosive growth of plastics production was indirectly made possible by the rise of the automobile; when oil is refined to produce gasoline, refinery byproducts are available for use as plastic feedstocks. However, moves toward decarbonization of the energy system will reduce oil and gas use, encouraging plastics recycling, over the next few decades.

Industry and Industrial Products

Several individual products exhibit dematerialization. Beverage containers have become lighter as steel cans and glass bottles have been replaced by aluminum cans and plastic bottles. The aluminum can itself has gotten lighter over time.

Cars were becoming lighter on average, prior to the recent growth in sales of light trucks and sport vehicles. Cars have also become more materially complex, with increasing use of plastics, composites, and specialty steels and decreasing use of carbon steel. One kilogram of the new materials replaces about three of carbon steel, but the complex mixture of materials causes difficulty in disassembly and reuse of scrapped vehicles. New high-performance materials, some of which show up in cars and other products, are constantly being developed in the aerospace industry, where the payoff for weight reduction is immense.

Dematerialization in industry is based both on downsizing, or "light-weighting" of products, and on reuse of materials. Secondary materials recovery depends on the ease of isolation of the used materials and on consumer demand for the materials. Some hazardous wastes such as cadmium and arsenic are very difficult to isolate and therefore are rarely recovered. In contrast, lead, now used mainly for automobile batteries, is readily recovered, and secondary lead supplies more than 70 percent of demand. Steel, other common metals, and wastepaper are also easily recovered and have secondary markets supplying a significant fraction of demand.

Dematerialization and Consumers

One study suggests that the size of new houses in the United States has grown steadily since World War II, while the average residential plot of land has actually shrunk. "Our hankering for a domicile in idyllic settings was what drove us to suburbia. Contrary to conventional belief, once we get there, we do not seem to care about how small the plot area is. Notwithstanding professed tastes for open space, we seem to build, enclose, and accrete steadily." [187] Today's enlarged homes house fewer people; the average number of residents per housing unit has declined from five in 1890 to fewer than three today. Thus floor area per person has almost doubled since 1945.

Other data corroborates the growing material intensity of consumption. The average weight of household goods transferred in intercity moves increased by 20 percent from 1977 to 1991. The number of pieces of mail per capita has more than tripled since 1940. The amount of food packaging has grown rapidly. Previously saturated markets have begun new waves of expansion, as with the sales of telephones. Each new phone is smaller and lighter, but there are many more of them; it is uncertain whether the total mass of the telecommunications system, including cables and equipment, has changed much since the early twentieth century.

Wastes

Data on wastes are incomplete and inconsistent. A comprehensive review of wastes in the United States in 1985 found that industrial wastes dominate the picture, but 90 percent of industrial wastes can be water, making comparisons with other wastes potentially misleading. Total waste in 1985 was 10 billion tons, but a large and unknown fraction of this was water.

Sewage sludge almost doubled between 1972 and 1992, due to population growth and increased treatment of waste. Ash, now produced largely by coal-burning power plants, has slumped along with the use of coal; if past statistics included estimates of coal and wood ash, the growth trend in waste would be flattened. Long-term figures are not available for hazardous wastes; in this category the quantities that are environmentally harmful are often minute. Municipal solid waste has been growing slightly in per capita terms but decreasing when measured per constant dollar of GDP. United States levels of municipal waste generation per capita remain far above those of other leading industrial countries.

Conclusions

Is there an overall trend toward dematerialization? With regard to primary materials, there is some evidence of reduction in the weight of input per constant dollar of output. This change has been driven by the substitution of new, scientifically selected and designed material for old, familiar ones. In industry, there are encouraging examples of more efficient material use in particular products as firms seek to economize on material and nonmaterial inputs alike. However, the taste for complexity and high performance may intensify other problems and lead to growing use of exotic new materials.

The trends are least encouraging in consumer behavior, where there is no significant evidence of net dematerialization or saturation of material wants. In terms of wastes, spotty data make it difficult to assess trends, but international comparisons suggest that substantial further reductions can take place.

A logical next step is to develop a detailed scenario for a dematerialized economy and to explore the changes in technology and behavior needed to achieve it. The decoupling of material use from economic growth, like the decoupling of energy from economic growth, can make a substantial contribution to the creation of a sustainable future.

Summary of

Material Use and Sustainable Affluence

by Frank Ackerman

[Published in *Why Do We Recycle? Markets, Values, and Public Policy* (Washington, DC: Island Press, 1997), 173–188.]

What would sustainable patterns of material use look like? Will those who live in a sustainable future society feel affluent, or will they be constantly struggling to conserve resources? What is the role of the present-day practice of recycling in creating a sustainable future? These and related questions are the subject of the chapter summarized here, the conclusion to a recent book on the economic and environmental meaning of recycling.

Play It Again, and Again

A sustainable economy must include patterns of production and consumption that can be repeated, generation after generation, without cumulative or worsening environmental damage. Current consumption patterns in the developed countries rely heavily on nonrenewable resources, particularly fossil fuels and metals. Worldwide demands for these resources will intensify as incomes rise in developing countries. While exhaustion of nonrenewable resources is not a short-run problem, it is inescapable in the long run. No nonrenewable resource can be used forever; even the best recycling systems never recover 100 percent of any material, due to losses in collection and processing.

Therefore, the post-oil, post-metal world of the future will have to rely on the renewable products of the land: wood, plant and animal fibers, paper—and plant-based plastics. The first plastics were made from plants; celluloid was based on cellulose, derived from cotton. In general, the hydrocarbon chemicals found in plants are similar to those found in fossil fuels, the current feedstock for plastic production. The vast and growing technological sophistication of the modern plastics industry could, in theory, be redirected to making the same materials from plants on a sustainable basis.

Let a Hundred Fibers Bloom

Although it is technologically possible to rebuild the material world with plant-based products, there is no guarantee that it is economically possible. The use of land to grow industrial materials will be in competition with production of food and fuel, expansion of areas of human settlement, and the desire for recreational and wilderness land. Will there be enough land and other resources to provide a comfortable material existence for all? The answer depends on three factors: the volume of total consumption; the productivity of agriculture and industry; and the rate of recycling and recovery of used materials.

The first factor, total consumption, depends on both population and per capita consumption. Demographic forecasts suggest that the world's population may level off during the twenty-first century. While per capita consumption is still growing, it is difficult to imagine this continuing indefinitely. Sustainability clearly requires that both population and per capita consumption of materials eventually be stabilized.

The second factor, the productivity of resource use, has been an area of rapid innovation. New technologies will be required to develop a sustainable plant-based economy, often regionally differentiated to make optimum use of successful local crops. China's extensive production of paper from nonwood fibers and Brazil's production of ethanol fuel from sugar cane illustrate the potential for innovative plant-based industry (although neither example is free of problems). The ongoing tendency toward "dematerialization"—reduction of the material required to create a product or end-use service—holds out hope for achieving sustainability with fixed resource inputs.

The third factor, recycling of used materials, plays an essential role in stretching our finite resource base. Recycling of metals, a mature technology, conserves energy and increases the time available for the transition to an all-renewable economy. Recycling of paper, though widespread, has ample room for improvement; and recycling of plastics remains quite limited and technologically underdeveloped. Further progress in this area will greatly ease the transition to sustainability.

Affluence, Abundance, and Scarcity

In the short run, however, low and declining prices for energy and materials are taken by many free-market advocates as signs of abundance. New resource discoveries and improvements in extractive technologies are said to be increasing the availability of materials, ushering in a new era of unprecedented affluence. This is not entirely wrong as a description of the past, but there is no reason to think that it can continue indefinitely. Eventually, all the available resources will be discovered, and extraction of the remaining ores and fuels will become more and more expensive.

However unsustainable, the image of free-market abundance is an alluring one. It poses a crucial question: will a sustainable society feel affluent, or will there be constant pressure to scrimp and struggle to conserve materials? When materials are expensive and labor is cheap, market forces push people into undesirable roles such as landfill scavenging. A less extreme but still troubling image is the painstaking recovery, repair, and reuse of ordinary material goods that consumed so much of the effort of nineteenth-century American housewives.

In contrast, affluence consists in large part of being able to act as though materials are cheap. The time required for the average urban worker to earn the price of several common materials is less than one-tenth of what it was in the 1830s. Economic theory might suggest that the failure to recycle is a natural response to the current relative prices of materials and labor. If and when materials become scarce again, greater levels of recycling and conservation will become cost-effective and will be implemented by market forces.

What's Wrong with This Picture?

From this perspective, contemporary recycling is an anomaly. Modern recycling programs are designed for people in a hurry who believe their own time is scarce. Yet despite the hurry, they evidently believe that materials are scarce as well. People who recycle are acting as if materials (or landfill space) were expensive, despite the absence of market signals telling them to do so. The commitment to recycling is not confined to small numbers of activists and advocates; rather, recycling is one of the most widespread and popular environmental initiatives.

Is there a role for this anomalous institution? Does recycling materials "before we have to" play a part in moving toward sustainability? There are two quite distinct answers, one involving technology and the other involving human behavior and motivation.

The celebration of free-market material abundance rests on simple economic theories in which there is no need to worry about choice of technologies: price signals automatically guide producers toward use of the most efficient techniques. However, in reality many industries experience increasing returns to scale and learning-curve effects, meaning that the more that a technology is used, the more efficient its users become.

The result is the phenomenon of "path dependence." An initial head start for one technology may lead to snowballing advantages of accumulated engineering knowledge, production skills, and consumer acceptance. The choice of technologies, therefore, is not made automatically by the market, but depends on small events that give one or another technology the crucial initial boost.

In this context, recycling pushes industry toward the adoption of technologies to process and use recycled materials. In a path-dependent world, waiting for the market to recognize that materials are scarce runs the risk of allowing further development of virgin-material–based technologies. By acting as if materials are more valuable than the market now thinks they are, recycling helps select which learning curves industry will slide down next.

Frugality and Participation

The popularity of recycling also provides a hopeful counter to the standard economic theories of consumer behavior. *Homo economicus,* the acquisitive individualist who inhabits economics texts, would never participate in recycling programs unless given a market incentive. However, contemporary recycling largely cannot be interpreted as a response to market signals; it is, rather, encouraging evidence that people are motivated by social and environmental concerns. A study of one municipal program found that recycling was correlated with desires for frugality and public participation, motivations that have a vital role to play in the creation of a sustainable society.

> The practice of recycling pushes us in the right direction, toward the development of the technologies of sustainable material use and toward the creation of less materialistic, more socially and environmentally engaged ways of living. There is no greater hope in any other direction. Indeed, in the long run there is nowhere else to go. [187]

Summary of

Balancing China's Energy, Economic, and Environmental Goals

by John M. Byrne, Bo Shen, and Xiuguo Li

[Published in *Energy Policy* 24, 5 (1996), 455–462.]

China's economy has expanded rapidly, with annual growth of real GNP exceeding 9 percent throughout the 1980s. As a result there has been a rapid increase in energy use, largely relying on coal, the country's most abundant fuel. Combustion of coal has caused severe environmental degradation, particularly in urban areas, and the problems will only worsen if the same style of growth continues. This article suggests that an alternative energy path emphasizing energy efficiency and renewable energy development is in China's long-term economic and environmental interest.

Energy, Environment, and Development

Economic growth in developing countries is typically energy-intensive. China is no exception: from 1980 to 1991, China's per capita energy consumption grew by 3.8 percent annually, compared to less than 0.8 percent annual growth worldwide. Chinese industry is particularly energy-hungry, requiring three to four times as much energy input per unit of output as industry in developed countries. Energy-intensive economic growth often leads to mounting air pollution problems; in China, these problems are particularly serious because three-fourths of all commercial energy there comes from coal.

In one respect, China's energy intensity is no surprise. The country's industry is concentrated in processing raw materials and producing infrastructure and durable goods, all of which are highly energy intensive activities. The developed countries, which are now comparatively energy efficient, historically had rapid increases in energy intensity as they moved through the early stages of industrialization, followed much later by decreases (i.e., efficiency gains). But there are both economic and environmental obstacles to countries like China taking the same path today.

The globalization of markets is far more significant for countries industrializing today than it was for countries that industrialized earlier. China's energy-intensive industries, such as chemicals and steel, are forced to compete with more energy-efficient producers in other countries. Chinese producers may find themselves at a competitive disadvantage due to their high energy costs.

Moreover, there is not enough coal in China to continue energy-intensive growth indefinitely. While the country's total supply is large, China's per capita coal reserves are only two-thirds of the world average; China's proved recoverable coal reserves per capita are only half the global level. Thus energy efficiency and diversification of supply will be essential for continued growth.

In the developed countries, economic growth is becoming de-linked from energy use. Energy intensity in high-income countries is low and has declined steadily since the oil crises of the 1970s. In China, by contrast, the ratio of energy use to GDP rose until 1978, followed by a modest decline as new policies began to promote energy efficiency. Much more, however, will need to be done in the future.

Coal combustion has led to high levels of air pollution in Chinese cities. Much of the coal has very high sulfur content, and often it is not sorted or washed before burning, which only adds to the problems. Levels of both suspended particulates and sulfur dioxide are far above World Health Organization standards for healthy air. Among the world's large cities, five of the ten with the highest particulate levels are in China, as are three of the ten with the highest sulfur levels. Urban air pollution is worst in winter and spring, when coal is burned for heat as well as in industry and electric power plants. Sulfur emissions create another environmental threat: acid rain, which is now a problem in rural as well as in urban areas.

The Potential of Energy Efficiency and Renewables

Promotion of energy efficiency began in the Sixth and Seventh Five-Year Plans (1981–1985 and 1986–1990). As in developed countries, investment in energy conservation is often more cost-effective than investment in new energy supplies. More than a billion dollars was invested in energy efficiency under the Sixth Five-Year Plan, saving energy at less than three-fourths of the cost of new supplies. However, this made only a small dent in China's energy system as a whole. Energy use remains highly inefficient, and there are still many opportunities for cost-effective investments in efficiency.

China also has a number of renewable energy alternatives. The nation's geothermal reserves are equivalent to 3 billion tons of coal, only 0.01 percent of which is being used. China's wind power potential is estimated at 1600 GW, or eight times current total electricity generation. At least in some regions, tapping this resource could provide users with relatively low-cost power without adverse environmental impacts. China also has strong prospects for photovoltaic power, with high levels of solar radiation in most parts of the country. Photovoltaics can bring electricity to some remote agricultural areas at a lower cost than extension of the conventional power grid.

Renewable energy sources have two important economic advantages. First, they reduce the risk of future fuel-price variability, since they do not depend on fossil fuels. Second, economies of scale are generally much less significant for renewables than for conventional energy sources. This allows modular, small-scale expansion of energy supply when and where it is needed.

Equally important are the environmental benefits of energy efficiency and renewables. Just as energy efficiency is more economical than new supplies in meeting energy needs, it can also provide pollution reduction at a lower cost than retrofitting existing power plants. Analyses conducted in the United States have found that investment in high-efficiency refrigerators and lighting can reduce sulfur emissions from coal-burning power plants far more cheaply than emission controls.

In China, retrofitting the numerous coal-burning facilities with emission controls is an urgent environmental priority, but investment in efficiency and renewables can provide complementary benefits at a lower cost. If China used 3 percent of its wind power potential for electricity generation, it could reduce its annual sulfur emissions by more than 20 percent. Thus, greater use of renewables can offer sizeable environmental advantages to China.

Policies for Developing Energy Efficiency and Renewables

To enable energy efficiency and renewables to compete on a level playing field, China needs to reform several features of its institutional and economic structures. National energy planning should set goals and timetables for increasing the use of renewable resources in areas where grid extension is too costly. The country also needs to create or strengthen national and provincial institutions that promote energy efficiency and renewables, and it should encourage government-industry partnerships in this area.

Specific regulatory measures are needed, including comprehensive national air quality standards and energy efficiency codes. Changes in economic incentives should include a phase-out of government subsidies for fossil fuels, evaluation of unconventional energy resources based on avoided costs, utility rebates for investment in efficiency and renewables, cost-based electricity pricing, and favorable tax treatment for energy equipment expenditures.

Along with such incentives, market transformation strategies could encourage more rapid development of efficiency and renewables. Provincial and local governments could adopt renewable energy set-asides, establishing targets for increased use of renewables. Policy collaboratives involving governments, industries, communities, and research organizations could identify local opportunities for the growth of energy efficiency and renewables markets.

Finally, China could enhance its cooperation with the international community, seeking capacity building and institutional support in energy and the environment from the World Bank, the United Nations, and other multilateral organizations. China should work with developed countries to transfer energy efficiency and renewable technologies, and should participate in worldwide information exchanges in this area.

> Economic development is now and will remain a dominant goal for China. However, it is possible to achieve the country's goals in a sustainable way. Pursuing an alternative energy path that emphasizes efficiency and renewables can be in China's long-term economic and environmental interest. . . . China can move quickly in this direction because it can invest in efficiency and renewability from the outset rather than having to rebuild its energy system as industrial countries must do. [461]

Summary of

Environmental Quality, Energy Efficiency, and Renewable Energy

by Peter Fox-Penner

[Published in *Electric Utility Restructuring: A Guide to the Competitive Era* (Vienna, Va: Public Utility Reports, 1997), 333–369.]

Electric utilities were, until recently, controlled by extensive government regulation, including direct rate regulation. Now a movement toward deregulation and competition is rapidly transforming the industry. These chapters offer a brief overview of the environmental implications of deregulation of U.S. utilities. The regulatory system that is being replaced had important strengths as well as weaknesses; in its final years it incorporated significant environmental initiatives. The impact of deregulation depends on the extent to which these initiatives survive in the competitive industry of the future or are replaced by other means of achieving environmental protection.

Environmental Quality

At the end of the nineteenth century, the first electric utilities led to an improvement in urban air quality, replacing countless gas-burning lamps with smokeless electric lights. Yet by the late twentieth century electric power plants had themselves become major sources of air pollution, accounting for two-thirds of the nation's sulfur dioxide emissions and about one-third of nitrogen oxides and carbon dioxide, as well as smaller amounts of air toxins and other pollutants.

Both federal and state regulations have addressed the problems of utility air pollution. The principal federal legislation is the Clean Air Act, which sets strict standards for new facilities. Unfortunately, much of the pollution comes from older facilities that were "grandfathered in" and thus held to looser emission standards. As restructuring has proceeded, these plants have increasingly been targeted for increased pollution control efforts, partly to offset feared increases in sales from these low-cost plants.

Restructuring has also changed the ways state utility and air quality regulators have addressed power plant emissions. In some states, regulators first tried to determine the monetary value of environmental damages from emissions and added this cost into utilities' estimates of the costs of power supply. Traditional regulation required utilities to build and operate power plants that could meet their customers' needs at the lowest possible cost; this was amended to require production at the lowest social cost—in other words, the utility's actual costs plus the value of the environmental damages.

This approach has largely been discarded for a variety of reasons. Estimates of environmental damage costs were uncertain and controversial, application of environmental cost "adders" to new plants alone (the usual regulatory practice) increased the relative attractiveness of older, dirtier plants, and in general it became too difficult politically to monetize and inject nonmarket costs into a single portion of a single industry.

The risk of increased reliance on older, dirty facilities was also a concern in the first moves toward competition. In 1996, federal regulators ordered transmission lines to grant open access to wholesale transactions between utilities. Many large, low-cost coal plants in the Midwest were operating well below capacity; increased transmission access meant that they could sell more power to other regions. Due to prevailing wind patterns, increased coal combustion in the Midwest threatened to worsen pollution problems such as ozone and acid rain in the Northeast and in eastern Canada. Estimates of the magnitude of this effect differ widely, but the problem underscores the need for broad regional responses to air pollution.

Another aspect of environmental changes triggered by deregulation involves siting, zoning, and new technologies. The new plants being built by unregulated producers are almost always gas-fired plants, are highly fuel-efficient, and incorporate state-of-the-art emissions controls.

Additionally, many plant developers are using innovative siting strategies to reduce plant impacts. On the other hand, there are also reports that some developers are too cost-conscious to maximize environmental compatibility, perhaps choosing sites a regulated or local utility would not.

Perhaps the most interesting environmental development is a new emphasis on "combined heat and power" (CHP) systems and small-scale or "distributed" power technologies. By decentralizing energy production using advanced noncombustion technologies such as fuel cells and using the waste heat produced by power plants large and small, these new approaches promise a revolution in power generation and its environmental impacts. However, significant issues remain, including whether CHP will flourish under deregulation and the rate at which distributed generation technologies penetrate the market and solve (rather than cause) environmental problems.

Finally, it should be noted that restructuring is credited with greatly expanding the market for renewable or "green" energy. Most renewable energy groups now strongly support restructuring because it increases their ability to sell power to a small but loyal group of power customers who want to "buy green."

"It is possible—perhaps even likely—that restructuring will lead the nation down a path that ends in clean energy and a cleaner environment. It is also likely that the nation will first enter a substantial transition period in which some emissions will increase, perhaps significantly." [345]

Energy Efficiency and Restructuring

Is energy just a private good, or is there a public interest in using energy efficiently? Arguments for the latter position rest on two grounds. First, the price paid for energy does not reflect the total environmental cost of energy use to society as a whole. Second, there are large economies of scale in the provision of energy-saving programs and technologies; lack of information, or high transaction costs, may prevent customers from making profitable investments in energy-conserving techniques such as insulation.

These arguments have led in the past to utility-sponsored energy efficiency, or demand-side management (DSM), expenditures. In many states, regulators ordered utilities to provide DSM programs, including rebates or reduced prices for energy-efficient equipment, low-interest loans for approved purchases, information and design assistance, cooperation with appliance manufacturers in promoting energy-efficient models, and solicitation of new DSM proposals from customers. Some states required utilities to compare the costs of DSM programs to new power plants and to invest in DSM whenever it was cheaper than electricity supply. Starting on a small scale in the late 1970s, DSM programs grew rapidly; by 1994 DSM reduced U.S. peak electricity demand by 25,000 megawatts—the equivalent of fifty large power plants or about 3 percent of nationwide generating capacity.

Some economists claimed that most DSM programs cost more than the value of the energy they saved, while others vigorously disagreed. The heated controversy over this subject was never fully resolved. There is little disagreement, however, about the fact that utility DSM programs entered a period of decline in the mid-1990s as the industry began to prepare for competition and restructuring. DSM was made possible by the regulated utility's status as a monopoly supplier of electricity; in a competitive electricity market, there is no easy way to make a profit selling energy conservation to small customers.

To prevent the loss of energy efficiency programs under restructuring, some states are incorporating mandated minimum levels of DSM expenditures into their deregulation laws. New institutional arrangements are required to fund and administer DSM programs in a competitive industry; one approach is to collect a system benefits charge from all electricity customers to be used for DSM programs by a state agency or by the local electricity distribution companies. For example, Vermont has created a statewide "energy efficiency utility" that will operate statewide efficiency programs, even after restructuring.

Renewable Energy

Renewable energy, aside from well-established hydroelectric plants, remains more expensive than conventional generation for most purposes. In the mid-1990s, only about 2 percent of electric generation came from renewable generators connected to the power grid (other than large hydropower). Costs for new technologies such as wind energy and solar photovoltaics have dropped rapidly as the infant industries have expanded; renewables have become competitive for providing service to some remote locations or for use in particular (e.g., very windy) areas. Advocates have proposed many initiatives to increase production in the hopes of achieving further cost reductions.

Recognizing the environmental benefits of renewables, many state regulators sought to promote their adoption. The use of environmental damage costs for planning purposes, as described above, gave a boost to renewables (since, unlike combustion plants, they cause little or no environmental harm).

Several proposals to protect renewables under deregulation have been discussed and in some cases have been incorporated into state restructuring laws.

One popular notion is a renewable portfolio standard, requiring each electricity producer to obtain a minimum percentage of its generation from renewable sources. Another option is a system benefits charge that could be used to fund new investments, similar to that discussed for DSM. Some proposals for restructuring, in fact, contain stronger mandates for renewables than exist today.

The simplest, but perhaps least effective, alternative is to encourage renewable energy producers to offer "green energy" at a premium price. Renewable energy with its environmental benefits is a public good, with a social value greater than its market price; there is no reason to think that voluntary individual purchases will produce the optimal level of investment. While many customers say they would pay more for green energy, early tests have found a small minority actually signing up for it. Nevertheless, direct (unregulated) sales of green power are sufficiently brisk to have encouraged significant new investment and activity in some states. Coupled with continuing technology cost reductions, this is causing many traditional power plant builders to also begin investing in renewable plants. Indeed, many renewable technology companies are now part of larger energy conglomerates, including gas and oil companies. The net impact of restructuring on renewable energy, as on DSM and on air pollution, is uncertain. Restructuring may propel environmentally sound alternatives forward into a new era of customer-driven, distributed resources, or backward into smaller niche applications.

Summary of

The Costs of Climate Protection: A Guide for the Perplexed

by Robert Repetto and Duncan Austin

[Published as *The Costs of Climate Protection: A Guide for the Perplexed* (Washington, D.C.: World Resources Institute, 1997).]

How much will it cost to reduce carbon emissions? Answers to this crucial question frequently involve simulations performed with complex economic models. More than a dozen different models have been used, leading to widely divergent forecasts. There are hundreds of variables and numerous, intricate relationships in the leading models, making it appear all but impossible to explain why forecasts differ. However, this analysis finds that eight key assumptions largely determine the predicted economic impacts of reaching CO_2 abatement targets. Using reasonable choices about these assumptions, the predicted costs of substantial reduction in carbon emissions are quite low or even negative—that is, carbon reduction may even stimulate economic growth.

Models, Assumptions, and Conclusions

An economic model is just a set of assumptions about the structure and functioning of the economy. It inevitably simplifies reality in order to make the model easier to analyze. Economic modelers try, with mixed results, to develop simplified yet still realistic representations of economic relationships.

Just as atmospheric models of climate change have improved through debate among practitioners, so have economic models. So-called "top-down" models, based on aggregate representations of the economy as a whole, provide a focus on overall balances and constraints, often drawing heavily on economic theory; these models often, though not always, lead to pessimistic conclusions about the costs of energy savings and carbon reduction. "Bottom-up" models, on the other hand, begin with a disaggregated examination of the potential for energy savings and emission reduction in individual sectors of the economy, often leading to more optimistic conclusions about the potential for low-cost or no-cost savings. Debate between the advocates of these two approaches has led to some models adopting features of the other approach; as the assumptions converge, so do the forecasts.

In either style of model, two kinds of assumptions are critical: those that determine the predicted costs of abating carbon emissions and those that estimate the value of the environmental benefits from reducing fossil fuel combustion. Abatement costs depend in part on assumptions about possible substitutions among fuels and technologies, the expected future rate of technical change, and the availability of non-fossil-fuel alternatives. Abatement costs also depend on assumptions about markets and institutions, such as the extent of market distortions and low-cost savings opportunities that exist at present, the potential for international trading in emission reduction credits, and the use that will be made of carbon tax revenues.

Most models are not constructed in ways that can take into account the environmental benefits of reduced fossil fuel consumption. However, a few models do factor in these benefits, including both the avoidance of climate change damages and the reduction in other air pollution damages such as acid rain and human health hazards. These factors, along with the abatement cost assumptions, largely explain why models predict such different economic costs of stipulated emission reductions. "Under a reasonable standardized set of assumptions, most economic models would predict that the macroeconomic impacts of a carbon tax designed to stabilize carbon emissions would be small and potentially favorable." [8]

How the Assumptions Determine the Predictions

To clarify how the key assumptions shape a model's economic predictions, the authors analyzed 162 different predictions from 16 of the leading models. Each of the models has been used repeatedly, with differing input assumptions leading to differing forecasts. Each forecast includes a predicted percentage change in GDP in some future year (relative to the expected baseline GDP if there were no new climate-change policies) and a corresponding percentage change in CO_2 emissions in the same year. Forecasts involving a 35 percent reduction in CO_2 emissions relative to baseline, for example, ranged from a 1.5 percent increase in GDP above the projected baseline to a 3 percent decrease.

A statistical analysis of the 162 predictions shows that just eight assumptions, plus the level of reduction in emissions, account for 80 percent of the variation in predicted economic impacts. (Emission reduction alone accounted for only 35 percent of the variation.) Four of the eight assumptions stand out as having the greatest effects:

1. *Does the model assume that the economy always adapts efficiently to changed conditions, at least in the long run, or does it assume that there can be persistent inefficiencies?* Efficient adaptation leads to lower predicted costs of emission reduction.
2. *Will international "joint implementation" of emission reduction, such as trading emission rights between countries, be achieved?* Predicted costs are lower with joint implementation.
3. *Will government revenues from a carbon tax or from auctioning emission permits be "recycled" in the form of reductions in other distortionary taxes?* Costs are lower with revenue recycling.
4. *Does the model include nonclimate economic benefits from air pollution abatement?* According to the models, the nonclimate benefits of reduced air pollution are much more valuable, in the near term, than the climate benefits.

The other four assumptions have a smaller but still noticeable impact on predictions of the GDP change associated with a given level of emission reduction:

5. *How much scope for inter-fuel and product substitution does the model assume?* The more substitution is possible, the lower the predicted costs of emission reduction.
6. *Does the model assume that "backstop" non-fossil energy sources are available at a constant cost?* The availability of a backstop energy source (such as solar power) limits future energy price increases, thereby lowering the costs of emission reduction.
7. *Does the model include economic benefits from avoiding or reducing climate change?* Inclusion of climate change benefits makes the net cost of emission reduction lower.

8. *How many years does the model assume it will take to reach a CO_2 reduction target?* Slower reduction is less costly.

With best-case assumptions in all or even in most of the eight areas, economic models predict that reduction of carbon emissions will actually increase GDP in 2020 relative to baseline. Conversely, with worst-case assumptions in each of these areas, economic models predict that substantial reduction in carbon emissions will impose a loss in GDP relative to the business-as-usual case. For example, consider the target of reduction of carbon emissions to 1990 levels by 2010 and stabilization at that level thereafter. Under the least favorable assumptions in the eight key areas, meeting this target would reduce U.S. GDP by 2.4 percent, relative to baseline. Under the most favorable assumptions, meeting the same target would increase U.S. GDP by 2.4 percent. Either way, the impact on U.S. economic growth over the next two decades would be negligible.

Further Modeling Issues

Many additional issues have been raised (and are discussed in the original essay) concerning long-run economic modeling of energy use, carbon emissions, and climate change. Top-down models typically assume that all cost-effective improvements in energy efficiency have already been made, an assumption that is repeatedly contradicted by actual experience. On the other hand, bottom-up models have sometimes identified vast inefficiencies in current energy use, perhaps understating the costs of converting to new technologies.

International joint implementation, the subject of one of the key assumptions, has been tried only in experimental pilot programs to date. Many details remain to be ironed out before this is a workable policy; still, the modeling results highlight its importance. Similarly, the models emphasize the significance of the recycling of carbon tax or emission permit revenues via cuts in other taxes. A lively theoretical debate among economists has addressed the exact nature and magnitude of the benefits from recycling environmental tax revenues, but revenue recycling clearly should be part of a climate change mitigation strategy.

> Predictions that a carbon tax or a cap-and-trade policy to reduce CO_2 emissions would seriously harm the economy are unrealistic. They stem from worst-case modeling assumptions. Under more reasonable assumptions and preferable policy approaches, a carbon tax is a cost-effective way of reducing the risks of climate change and would do no damage to the economy. More likely, taking the environmental effects into account, it would bring long-term benefits. [36]

Summary of

Equity and Discounting in Climate-Change Decisions

by Robert C. Lind and Richard E. Schuler

[Published in *Economics and Policy Issues in Climate Change*, ed. William D. Nordhaus (Washington, D.C.: Resources for the Future, 1998), 59–96.]

Climate change mitigation policies typically call for very long-term investments, with current costs justified largely by the benefits to future generations. When evaluating such policies, economists use the techniques of cost-benefit analysis, including discounting of future costs and benefits. This essay reviews the standard economic approach to discounting as it applies to climate change, arguing that because there is no simple way to justify the choice of a discount rate for investments whose benefits and costs span several generations, collapsing those effects into a single number is highly misleading.

In the authors' view, the question of discounting these intergenerational impacts is closely connected to problems of fairness and distribution; cost-benefit analyses frequently rest on hidden and controversial assumptions about equity. The answer is to use the most sophisticated methods of economic analysis that display the distribution of costs and benefits both between nations and between generations.

Discounting and Intergenerational Equity

Climate change involves questions of risk and uncertainty, which pose unique problems for economic analysis. Many of the crucial outcomes are not only far in the future, but are also subject to inescapably great uncertainty. If the degree of risk associated with each future possibility were known in advance, then potential outcomes could in theory be risk adjusted, or converted to "certainty equivalents," for purposes of cost-benefit analysis. However, this theoretical device cannot be used in practice, since the risks are rarely known. The modern literature on investment under uncertainty (real-options analyses) suggests a different approach: recognizing that additional information will become available over time, a sequential decision process can reject once-and-for-all choices in favor of single steps that reach only to the next decision point.

Cost-benefit analysis has traditionally relied heavily on a hypothetical compensation test. If the present value of net benefits from a project is positive, those who receive the benefits could hypothetically compensate those who incur the costs. There is a strong economic argument for rejecting most or all projects that fail this test.

The logic of the compensation test runs into difficulty when costs and benefits are separated by several generations. If we decide against investment in climate change mitigation now on the grounds that other investments have higher returns, there is no way to compensate the future generations who will suffer as a result. Should we establish a global trust fund for future compensation, in lieu of climate change investments? Even if we could overcome the political obstacles to creation of such a fund, it would be impossible to commit intervening generations to maintain the trust fund for the far-future generations who will need it the most.

Conversely, if we do decide to invest now for the benefit of future generations, there is no way for them to compensate us today—as might seem equitable if we expect per capita incomes to be higher in the future. Intergenerational transfers happen all the time, but there is no mechanism that allows a planned reallocation across generations to offset the costs of an investment. In short, climate change investments involve redistribution of resources, a question that cannot be settled by cost-benefit analysis.

Optimal Growth and Discounting

In the selection of a discount rate for climate change analyses, economists have used two methods, which have been called the prescriptive and the descriptive approaches. Both approaches are invalidated in this application by flaws in their logic.

The prescriptive approach argues that, in a theoretical model of optimal economic growth, the discount rate would be the sum of two different components. One is the discount rate for utility—that is, the satisfaction produced by consumption: how much less is the same amount of happiness worth if it occurs in the future? There is widespread agreement that this should be zero; that is, an equal amount of human satisfaction should be valued equally regardless of when it occurs.

The other component of the discount rate reflects the expected growth of per capita income and the accompanying expected change in the marginal utility of income: if we are richer in the future, will a dollar of additional income produce less satisfaction then than it does now? Empirical estimates of this component of the discount rate generally range from 0.5 to 3.0 percent.

One problem with the prescriptive approach is that it is derived from an economic theory of optimal growth, which clearly implies that the discount rate should equal the rate of return on capital. Yet the return on capital is 5 percent or more, well above any of the prescriptive estimates of the discount rate. Therefore, contrary to the theory's assumptions, we are far from being on an optimal growth path.

Another problem is implicit in the assumption that human satisfaction is equally important in all time periods (i.e., the assumption that the first component of the discount rate must be zero). This assumption embodies a radical egalitarianism that is not commonly applied in other contexts. People do not act as if everyone's satisfaction is equally important in the present; they do not give the same weight to the welfare of their own distant descendants as they do to themselves and their children; and even less do they act as if people living in distant countries many generations in the future should count equally with themselves in cost-benefit calculations. In short, a controversial, frequently violated ethical premise is buried in the analysis.

The alternative, descriptive approach takes the position that the observed marginal rate of return on capital should be used as the discount rate. This approach assumes that society's preferences about intertemporal transfers are revealed by current market rates of return, an assumption that avoids many of the problems of the prescriptive method. However, the discount rate in standard economic models is based on an individual's short-run trade-off between consumption now and at a moment in the near or medium-term future. There is no way to interpret this as the rate at which someone would trade consumption today for someone else's consumption two hundred years from now—since, as we have seen, there is no way to arrange compensation for such a trade.

Thus neither market data, in the descriptive approach, nor economic principles, in the prescriptive approach, leads to a justifiable discount rate that applies to intergenerational trade-offs. What role does this leave for economic analysis in determining climate change policy? In view of the difficulties with discounting, the costs and benefits of climate change policies should be displayed as time profiles rather than collapsed into present values. Such profiles provide important information for decision-making, especially when contrasted to similar calculations for alternative investment scenarios. When comparing two scenarios, "any economist . . . will obviously feel a strong urge to discount the difference in the consumption streams to a present value. But given the previous discussion, the corresponding value will be essentially meaningless." [80]

A focus on time profiles rather than present values can also help address the issue of risk and uncertainty. The goal is not to make a definitive yes-or-no decision on a multicentury endeavor. Instead, the sequential approach to decision-making under uncertainty suggests that we should gather the information that is relevant to a tractable planning horizon, perhaps ten to thirty years, and plot a course of action that will allow us to move ahead with additional mitigation investments in the future if we choose to do so.

Concepts of Equity

Numerous issues of equity arise in connection with global climate change policies, either between nations, between generations, or both. Discount rate calculations mask the impacts of an investment on across-nation or intergenerational equity. However, agreement on contemporaneous equity between nations may be a prerequisite for effective, future-oriented international action.

Since current conditions are viewed as inequitable by poor countries, wealthier nations may have to accept a greater share of the costs of forestalling climate change in order to gain broad participation in an accord. That distribution of the burden of climate change mitigation is consistent with one important concept of equity: benefits and costs should be shared in similar proportions. The idea of proportionality echoes the theory of economic efficiency, in which factor payments are proportional to marginal costs; it is also consistent with legal concepts of responsibility and the philosophical principle of "fault-based equity," and with the ecological "polluter pays" principle.

Equity in the present does not guarantee equity over time. Consistent behavior across generations cannot be guaranteed; future generations may choose different values. Once we act, posterity is in the driver's seat. With free choice and no intertemporal enforcer of values, contradictory behavior across generations is certainly possible.

The requirements for sustainability compound this dilemma, particularly for poor societies. To ensure sustainability we must both leave the resources needed by future generations and provide the resources needed by the current generation. Ecologists emphasize the importance of slowing population growth as a way to guarantee sustainability. But this also poses a difficult equity question in the present: whose population growth will be curtailed? The major point here is that intergenerational equity has implications for intragenerational equity, and vice versa.

> The central conclusion of this chapter is that the mechanical application of the discounting apparatus to large-scale economic models cannot automatically lead to policy prescriptions for generation-spanning, global climate change–mitigating strategies that are equitable and therefore likely to be adopted. Nevertheless, economists' tools can provide tremendous insights into forging fair and efficient methods, policies, and institutions for dealing with global climate change. [94]

POLICIES FOR SUSTAINABILITY

PART VII
Globalization and Sustainability

Overview Essay

by Kevin P. Gallagher

Globalization is fast becoming a fact of life—too fast for many people to accept. It is transforming trade, finance, employment, migration, technology, communications, the environment, social systems, ways of living, cultures, and patterns of governance. The main features of the current phase of globalization that are discussed in this section are the accelerated integration of the world's economies through the liberalization of trade and investment regimes; the adoption of structural adjustment programs for many of the less-developed countries; and the diminishing role of the state in developed and developing economies alike.

At the global level the 1990s witnessed a new round of negotiation under the General Agreement on Tariffs and Trade (GATT) that resulted in the creation of the World Trade Organization (WTO). At the regional level, free trade and investment agreements were initiated in Europe, Asia, Africa, Latin America, and North America. Also during that time many structural adjustment programs were adopted in the developing world. While the pace of globalization appears to be increasing, a growing number of scholars and activists, some of whom are represented in this section, are making the case that the current form of globalization is not compatible with sustainable development.

Theory and Reality

Economic arguments for trade liberalization are based on David Ricardo's theory of comparative advantage. Ricardo argued that different countries with different technologies, customs, and resources will have different costs to produce the same product. If each country produces and exports the goods for which it has comparatively lower costs, then all parties will benefit. The mutual benefits from trade are said to include greater efficiency in production and higher worldwide rates of consumption. In this model, some groups may lose as a result of trade. However, the net national gains are predicted to exceed losses.

The effects of comparative advantage on factors of production are dealt with in the Heckscher-Ohlin model. This economic model assumes a framework in which all countries have perfectly competitive economies and can make the same diversified mix of products, with perfect factor mobility between industries. In this framework, the Stolper-Samuelson theorem predicts that international trade increases the prices of products in which a country has a comparative advantage. This in turn has similar effects on the country's demand for, and prices of, factors of production: demand and prices rise for those factors in which the country is relatively abundant.

Under this theory, free trade would lead rich countries to specialize in capital- and skill-intensive products, while poor countries would specialize in unskilled-labor-intensive products. As a result, wages would rise for skilled workers in rich countries and for unskilled workers in poor countries. Conversely, in this simple model, trade should cause employment and wages to fall for unskilled labor in rich countries and for skilled labor in poor countries.

Criticisms of this model are of two sorts. One approach is based on analysis of the costs to the "losers" in liberalized trade, the other on the model's assumptions. The former criticism was addressed in Volumes 4 and 5 in this series, which covered the ongoing debate over to what extent trade liberalization is responsible for job losses and inequality in the developed and developing world alike. This volume examines the latter criticism: what happens when key assumptions in trade theory break down?

Externalities such as pollution and resource depletion are assumed not to exist in conventional trade models. Since externalities are, in fact, pervasive, trade liberalization can have unintended consequences: nations with lower environmental standards could gain a comparative advantage in "dirty" production, relative to nations with higher standards. This could create an incentive to lower environmental regulation worldwide, an effect that is sometimes referred to as the "race to the bottom." (Daly 1996, 1999).

A separate problem is that trade theory assumes that every country is capable of moving into any industry in which it could potentially obtain a comparative advantage. Some countries are at a stage of development where they do not yet have the networks of infrastructure, suppliers, and so forth that would allow them to develop the export industries necessary to accrue full benefits from trade. In such cases, the optimal solution may be to protect industries temporarily until a certain level of development is achieved. Import substitution, a development strategy based on protection of key industries from trade pressures, has long been out of favor with economists; recently, however, it has received a new and at least partially sympathetic hearing (Bruton 1998).

Race to the Bottom or Climb to the Top?

While there are a growing number of analysts who share the view that the underpinnings of trade policy do not reflect the goals of sustainability, there is a wide range of opinion regarding the impact of globalization. Some see global economic integration as a crisis for environment, labor, and community; others see it as an opportunity.

Paul Streeten argues that for globalization to be successful, markets should be channeled in a way that allows their energies to flourish. Unfortunately, in Streeten's view, the decreasing role of the state, in addition to the unprecedented spread of world markets, without a countervailing authority to regulate them, is making the goals of social and environmental well-being harder to reach. While Streeten acknowledges the significant achievements of globalization in promoting economic growth and development, he stresses that it has also contributed to increased inequality and has weakened important social structures and institutions.

Herman Daly argues that there is a clear conflict between the goals of free trade and environmental protection. Daly worries about creating free trade relationships between developed nations with stringent environmental policies and developing countries that do not internalize environmental and social costs through environmental regulation. When free trade partnerships like this are created, he argues that the developing countries will become "havens" for highly polluting industries. Through their lack of environmental regulation, such nations could gain a comparative advantage in pollution-intensive industry. This is particularly alarming to Daly, because as an ecological economist he is concerned that the resulting increases in overall output will push local and global ecosystems beyond their limits.

Like Daly, many of the analysts concerned with the detrimental effects of trade liberalization assume that *developed* country environmental practice will be jeopardized. The summarized article by **James Boyce** shows how trade liberalization can also adversely affect sustainable practices in *developing* countries. Using the same logic of externalities articulated by Daly and others, Boyce illustrates how the North American Free Trade Agreement (NAFTA) has impacted Mexican maize farming. Boyce shows that the higher cost of Mexico's labor intensive maize farming is indirectly protecting the genetic diversity of Mexican maize—some of the most genetically diverse cropland on earth. Trade liberalization in Mexico, however, has brought dramatic increases in maize imports from the United States that do not include internalization of externalities in their prices. Boyce and others have gone on to document how these imbalances have affected local production, the environment, and poverty in the Mexican countryside (Nadal 2000).

The empirical story on labor and social standards, extensively covered in the previous two volumes in this series, is mixed. "There is a strong case for trade effects as one of the important causes of the declining prospects for low-skilled labor in developed countries, but there is also a strong case for the view that other effects must be of equal or greater importance. Perhaps the most surprising conclusion is the difficulty of precisely defining and measuring the effects of trade and the extent to which the evaluation of these effects rests on a series of subtle technical judgments" (Ackerman 1998). One effect that is difficult to pick up in empirical analysis is the use by employers of the threat of moving overseas as a means to win concessions from workers, even when no actual movement of jobs occurs (Belman and Lee 1992). A more recent article by **James Crotty, Gerald Epstein, and Patricia Kelly** calls this the "magnification effect," and adds that stagnant wages and contractionary monetary policies increase the need for communities to bid to attract employment.

Although much of this discussion has focused on ways in which nations can derive comparative advantage through economic and institutional means that are harmful to society and the environment, there are also a number of scholars who focus on how comparative advantage can be associated with positive externalities that will result in a "climb to the top" rather than a race to the bottom. These analysts contend that the globalization process, especially in the realm of private capital flows, can be harnessed to achieve sustainable development. Since 1990, Foreign Direct Investment (FDI) has increased from $44 billion in 1990 to over $650 billion in 1998, while official development assistance continued to hover at close to $50 billion (UNCTAD 2000). Private capital flows that are moving to areas where they will enjoy a new comparative advantage often originate from developed countries where environmental and social standards are more stringent. Thus, it is argued, these practices can be transferred to the developing countries where these firms will now concentrate, therefore improving the prospects for sustainable development in the developing world. This more positive view of trade impacts is presented in the article by **Daniel Esty and Bradford Gentry,** summarized here (see also Gentry 1996).

A well-known advocate of this view is businessman Stephan Schmidheiny, who cites the Mexican privatization of its steel industry as an example of environmental progress through privatization. Altos Hornos de Mexico (AHMSA), the largest Mexican steel plant, was sold to a group of Dutch and Mexican investors in 1991. The Dutch investor supplied international and European environmental certification of its manufacturing standards and installed a state-of-the-art environmental management system (Schmidheiny and Gentry 1997). This approach has promise but falls far short of being the key to sustainable development. Of the FDI flows in 1998—$657 billion—only 25 percent were located in the developing world. Moreover, three nations—China, Mexico, and Brazil—receive almost half of the developing world's share (UNCTAD 2000).

What about the rest of the developing world that receives relatively little private capital? Moreover, massive capital flows to developing countries are not a sustained guarantee; such flows have proved to be erratic and volatile over time. In addition to receiving relatively little investment, most developing countries are minor players in the trade picture as well. Although the value of world trade tripled from 1980 to 1997, the least-developed countries only imported and exported one-sixteenth of all imports and exports in 1997[1] (World Bank 2000).

Dharam Ghai, director of the United Nations Research Institute for Social Development (UNRISD) adds that in addition to the effects of global integration on the environment, working conditions, and job creation, globalization is harming the social fabric in the developing world as well. Expanding on issues raised in Part II of this volume, Ghai emphasizes how the marginalization that is caused from the globalization process can cause internal political divisions and international migration. Similar concerns have been voiced regarding the developed world. Herman Daly and John Cobb, for example, have argued that economic integration is undermining communities on both sides of the comparative advantage divide (Daly and Cobb 1994).

Reassessing Structural Adjustment

Structural adjustment programs (SAPs) are the policy instruments that are most frequently relied on to bring the developing world into the world economy. SAPs are often applauded for achieving their desired macro-economic changes. From a sustainability perspective the question is: At what cost?

The structural adjustment programs that have been prescribed by the International Monetary Fund (IMF) and the World Bank have the stated goals of achieving higher output growth and rising real incomes in the developing world. Structural adjustment policies have sought to bring developing countries into the world economy by adopting a development strategy based on the following: the promotion of export-oriented growth; the privatization of state-owned industry; the elimination of barriers to international trade and investment flows; the reduction of the role of the state as an economic agent; and the deregulation of domestic labor markets.

Even the harshest critics of SAPs acknowledge that it is erroneous to think that developing nations can develop in the context of hyperinflation, ballooning deficits, and extreme poverty (MacEwan 2000). Macroeconomic stability is a key ingredient for sustainability. From a macroeconomic point of view, there is little doubt that adjustment programs are having some positive effects. In most cases, per capita GDP, agricultural exports, and revenues from extractive industries are all on the rise, while budget deficits and inflation have been brought under control. But this is not the whole picture. In two comprehensive book-length studies, authors **Lance Taylor, Ute Peiper**, and **David Reed**, show that the social and environmental costs of these programs may outweigh the short-term economic benefits. Taylor and Peiper's work, while including a chapter on environmental impacts, highlight the social impacts of SAPs. Specifically, they focus on the effects of SAPs on poverty and inequality, gender relations, and education and human health.

SAPs can create significant distributional conflicts and worsen inequality. Where this has occurred they have run a strong risk of failing. In cases such as Turkey and Mexico, these authors found that the beneficiaries of SAP reform have been households in the top 20 percent of the income distribution, while the losers were people in the bottom 80 percent. Such ramifications have provoked extreme political backlash. The result has been corruption, accelerating inflation exacerbated by distributional conflict, high interest rates, and overvalued exchange rates—all problems that economic integration is supposed to overcome.

Taylor and Peiper especially emphasize the impact of SAPs on women's productive roles. Other studies have also concluded that SAPs leave women at a clear disadvantage, and, because women don't own the necessary factors of production, they are often "spectators" of SAPs rather than participants (Haddad et al. 1995).

While some analysts argue that it is still too early and too difficult to fully assess the environmental impacts of SAPs, David Reed's work draws on case studies of SAP experience in Cameroon, Mali, Tanzania, Zambia, El Salvador, Jamaica, Venezuela, Pakistan, and Vietnam. Reed acknowledges that although the short-term economic problems were often solved in these countries, environmental problems were often, though not always, exacerbated. He finds that the social and economic benefits of SAPs have been unequally distributed, with the heaviest costs being borne by small farmers, low-income families, and workers in the informed sector.

Reed's study also points out that there are a number of impacts on the environment that are clearly positive. In some cases, exchange rate reforms have led to a shift away from "erosive" and toward "nonerosive" crops, and price corrections have created new agricultural incentives that have stimulated expansion and diversification of tradeable crops and other commodities. These are positive effects in that they increase relative returns to the agricultural sector and raise the incomes of some farmers, thus encouraging on-farm investment.

Theodore Panayotou argues that since environmental improvement has not been the aim of SAPs, it is not fair to use an environmental measuring rod to assess such policies. However, he notes that to the extent that SAPs lead to natural resource depletion and environmental damage beyond the economic optimum, they are defective in economic terms by failing to meet their own objectives. Panayotou also argues that SAPs should treat the natural resource base in the same way that man-made capital is treated (Panayotou 1996).

Conclusion

While Esty and Gentry have shown how globalization can promote sustainability, the other authors in this section demonstrate significant ways in which globalization in its current form is unsustainable. However, they also note that we are faced with many choices regarding how to globalize the world economy. Many of those choices can promote sustainability, but not all of them can be addressed on the global level. A sustainable global economy will require the efforts of many different actors. These include corporations, local and national communities, as well as global institutions. In the sections that follow, we will focus on roles that these actors can play in shaping sustainable systems at the local, national, and global levels.

Note

1. "Least-developed" refers to "low-income" countries as classified by the World Bank, and include countries such as Nigeria, Vietnam, Nicaragua, Armenia, and Mali. "Middle-income" countries, ranging from Mexico, the Czech Republic, and South Africa, to Egypt, Bolivia, Indonesia, and the Russian Federation, only imported or exported one-fifth of total imports and exports.

Summary of

Globalization: Threat or Salvation?

by Paul Streeten

[Published in *Globalization, Growth, and Marginalization,* ed. A.S. Bhalla (New York: St. Martin's Press, 1998), 13–46.]

Is globalization a threat to humanity, or its salvation? Globalization is transforming trade, finance, employment, migration, technology, communications, the environment, social systems, ways of living, cultures, and patterns of governance. For globalization to be successful, markets have to rest on a framework that enables their energies to flourish and to be used for socially and ecologically sustainable development. However, the reduced power of national governments combined with the spread of worldwide free markets and technological innovation without a corresponding authority to regulate them is leading us in another direction.

Integration and Interdependence

We hear everywhere that international integration is occurring rapidly as a result of increased trade and capital, technology, and information flows. The four components of an integrated international system to promote development are today fragmented: generation of current account surpluses; financial institutions that convert surpluses into loans and investments; production and sale of producer and consumer goods and up-to-date technology; and the military power to keep peace and enforce contracts. These components were relatively centralized, first with Great Britain and then with the United States, in the first three-quarters of the twentieth century. But today we live in a schizophrenic, fragmented world without effective coordination.

Some important qualifications should be made about international integration. Because international trade has been increasing rapidly, it is often used as an indicator. From 1820 to 1992, world population increased 5-fold, income per head 8-fold, world income 40-fold, and world trade 540-fold. However, as a proportion of GDP, trade is more similar to pre–World War I levels, and many developing countries contribute only minimally to global trade. Indeed, much trade is intra-industry or intra-firm. Transnational Corporations (TNCs) have to a certain extent replaced states as the major agents in trade, although most TNCs are actually centered in one or a few countries.

Similarly, globalization of financial flows, while talked about a great deal, is only partial. Most financial flows occur among rich or upper-income countries; private flows to low-income countries are minimal and development assistance to these countries has stagnated. Such trends pose the danger of instability, and, while the system as a whole is unlikely to collapse, there is a need for the reregulation and harmonization of legislation. The more oriented toward free enterprise a nation is, the greater its need for official supervision.

Technological diffusion and communication occur much more rapidly, but here also vast areas of the developing world have been left out. Moreover, such diffusion has not led to more-rapid economic growth. Comparing annual growth rates of GNP per head in two periods 1965–1980 and 1980–1993, we find that for all developing countries the rates were 4.6 percent and 4 percent respectively, and for OECD (Organization for Economic Cooperation and Development) countries, 3.9 percent and 1.67 percent respectively. Real wages have converged in the United States and Europe as a result of increased trade and some labor mobility.

Uneven Benefits and Costs of Globalization

Globalization has helped to achieve enormous improvements in human development indicators but has also contributed to increased impoverishment, inequality, insecurity, and a weakening of institutions and social systems. Social services have been cut back in both developed and developing economies. The share of global wealth for developing countries has shrunk. In the poorest countries, poverty, malnutrition, and disease have increased. With a decline in the extended-family system, there is a need for social safety nets and better access to employment and training.

Globalization requires firms to compete internationally, and international competition has reinforced globalization. However, cost reductions, greater efficiency, and higher incomes have been achieved at the expense of growing uncertainty, unemployment, and inequality. In Europe, unemployment rates have remained high. In the United States, employment insecurity and the persistence of an underclass are major problems. Mass unemployment is also a problem in rapidly growing China because many state enterprises have reduced employment to become more efficient. Other formerly socialist economies have similar problems, although often with less growth than China's.

Globally, the interest of consumers in cheaper goods has been elevated over the human need for secure employment. Marginalization and social exclusion worsen social problems of crime, drug abuse, domestic violence, and suicide. Promoting employment in health care, education, childcare, and environmental protection is important, but this requires action by the public sector, which now lacks funding and is in disrepute.

How does globalization affect income distribution within rich and poor countries and between them? Economic theory suggests that with globalization wage differentials should widen in the North and decrease in the South. But only the first is empirically supported. Globalization appears to lead to growing inequality between capital and labor incomes, and also between skilled and unskilled workers.

Institutions are lagging behind rapid technological growth and diffusion. This can lead to prisoner's-dilemma-type situations in which apparently optimal strategies for nation-states are self-destructive (e.g., arms races, environmental damage, competitive protectionism or exchange rate movements, destructive bidding to attract investment, overfishing, and forest and species destruction). To avoid these traps, policy coordination, cooperation, and enforcement are needed, but the international political order is having trouble coping with these needs.

It might be expected that globalization would go hand in hand with a decreasing role of the state—that is, the more open the economy, the smaller the government. This would seem logical because liberal trade policies reflect a preference for markets over government and because globalization makes monetary and fiscal policies less effective. However, Rodrik (1996) has shown that the scope of government has grown in countries that take advantage of world markets. He suggests that an increased role of government is needed as an insulator against external shocks, a kind of insurance against risk and income volatility. Thus, globalization might require bigger, not smaller, government.

Conclusion

Partial international integration can lead to national disintegration manifested in unemployment, poverty, inequality, exclusion, and marginalization. In addition, the requirements of social harmony and stability are in conflict with the pressure for increased international competitiveness. Destruction of communities and the environment are social costs of globalization that are unmeasured by most economic and market metrics. The public good suffers at the national level, but institutions for defending the public good are weak or nonexistent at the global level. National democratic institutions are weakened, but global democratic institutions do not exist. The challenge is to preserve positive aspects of globalization while taking action against the harmful effects. This will require delegation of state functions both upward (to global institutions) and downward (to the local or regional level).

Summary of

From Adjustment to Sustainable Development: The Obstacle of Free Trade

by Herman E. Daly

[Published in *Beyond Growth: The Economics of Sustainable Development* (Boston: Beacon Press, 1996), 158–167.]

"Adjustment" is frequently used to describe economic policies prescribed by international authorities such as the International Monetary Fund (IMF) and the World Bank. By narrowly defining "adjustment" in terms of the doctrine of free trade, mainstream economists are charting a course for the world economy that is not sustainable. For the world economy to make the transition to a sustainable society, distribution and scale considerations have to take center stage in economic thinking. This chapter describes five reasons why focusing solely on free trade as an economic strategy is unsustainable and outlines a vision of a world where "development," not "growth," is society's goal.

The Neoclassical View

In neoclassical economics, the adjustment of an economy to an efficient one involves three main objectives: the internalization of social and environmental costs into prices; the adjustment of macroeconomic conditions to attain monetary stability; and the integration of national markets into the world trading system so as to increase global productivity. While the first two goals are essential to sustainable development in national economies, the third goal of economic integration can undercut both of them.

We have three fundamental economic problems: allocation, distribution, and scale. Neoclassical economics deals especially with allocation but gives short shrift to distribution and ignores scale entirely. Allocation refers to the use of resources to produce goods and services. Distribution is the apportioning of goods and services produced among different people, and scale refers to the physical size of the economy relative to the ecosystem. Allocation can be efficient or inefficient, distribution can be just or unjust, scale can be sustainable or unsustainable, but relying on free trade alone can lead to unjust and unsustainable outcomes.

Why Free Trade Conflicts with Sustainable Development

There are at least five reasons why free trade conflicts with national efforts to develop sustainably. Free trade impinges on nations' ability to "get the prices right" by internalizing social and environmental costs; distribute resources in a just manner; foster community; balance and control the macroeconomy; and keep the scale of the economy within ecological limits. Each of these reasons will be discussed in turn.

First, there is clearly a conflict between free trade and a national policy of internalization of external costs. To take the simple case of trade between two nations, if one nation embarks on an effort to internalize environmental and social costs, and enters into trade with a nation that does not, the latter nation can enjoy an advantage in goods that incur high amounts of such costs. What is necessary, albeit very difficult, for arguments for free trade to have some salience, is for all trading nations to agree on common rules for internalizing external costs.

Even if the condition of multilateral agreement on rules for cost internalization was met, achieving a just level of distribution through free trade may be insurmountable. Wage levels vary enormously among the world's trading partners due to different conditions of labor supply and population growth rates. For most traded goods, labor is still the major cost item and is therefore a very significant influence on prices. Cheap labor therefore means an advantage in trade. When both capital and goods are internationally mobile, capital will migrate to low-wage countries, creating a tendency for wages to equalize worldwide. Neoclassical economists do not fret over this outcome. They believe that wages will eventually equalize at a higher level because the gains from free trade will be so enormous. Such a thought can only be entertained by those who ignore the issue of scale. It is ecologically impossible for a world population of over 6 billion people to consume resources at the same per capita rate as Americans and Europeans.

The third reason why free trade conflicts with sustainable development is that it tends to break down community. Free trade increases "the separation of ownership and control and the forced mobility of labor which are so inimical to community." [163] With free trade, life and community can be made subject to distant decisions and events over which communities have no control.

Fourth, free trade interferes with macroeconomic stability by allowing nations to run up excessive debt. In an attempt to repay these debts there is an incentive for nations to embark upon unsustainable rates of exploitation of exportable resources and to take out yet more loans to get foreign exchange to pay old loans. These efforts often spiral into crisis:

"Efforts to pay back loans and still meet domestic obligations lead to government budget deficits and monetary creation with resulting inflation. Inflation, plus the need to export to pay off loans, leads to currency devaluations, giving rise to foreign exchange speculation, capital flight, and hot money movements, disrupting the macroeconomic stability that adjustment was supposed to foster." [164]

The fifth reason, alluded to earlier, is that free trade violates the criterion of sustainable scale. The world economy is an open subsystem of a closed, nongrowing, and finite ecosystem. Sustainable development means living within the limits of this finite ecosystem. Free trade allows nations, regions, or localities to live beyond their own ecological capacities by importing those capacities from abroad. Within limits, this is reasonable and necessary. But carried to extremes, it becomes destructive to the global ecosystem. Trade does not remove carrying capacity; it just guarantees that nations will hit that global constraints simultaneously rather then sequentially—in effect it converts differing local constraints into an aggregated global constraint.

Development, Not Growth

Underlying the five reasons why free trade conflicts with sustainable development is the concept of growth. The term "growth" means a quantitative increase in physical size by assimilation of materials. As has been demonstrated, the growth that often accompanies free trade eventually increases environmental and social costs faster than it increases production benefits. Growth is quite distinct from "development," which means a qualitative change, a realization of potential, or a transition to a better state.

What needs to be sustained is development, not growth. Sustainable development is "development without growth in the scale of the economy beyond some point that is within biospheric carrying capacity." [167] Such a perspective does not put an end to economics; on the contrary, it requires a more subtle and complex economics of better, not bigger.

Summary of

Ecological Distribution, Agricultural Trade Liberalization, and In Situ Genetic Diversity

by James K. Boyce

[Published in *Journal of Income Distribution* 6, 2, (1996), 265–286.]

Most of the discussion surrounding the environmental effects of trade liberalization between developed and developing countries assumes that liberalization may threaten strong environmental standards in developed countries. Conversely, this paper focuses on how liberalization can undercut more sustainable practices in developing countries by inducing them to import goods from them that have not internalized the negative externalities associated with those goods.

This notion is illustrated in the context of genetic diversity. Genetic diversity in crop plants is essential for long-term world food security. Such diversity is sustained "in the field" by poor farmers in developing countries. Agricultural imports from developed countries that do not include the internalization of externalities in their prices, can displace local production in centers of genetic diversity. Such a displacement can threaten both rural livelihoods and the continued provision of crop genetic diversity. The paper shows how the North American Free Trade Agreement (NAFTA) has impacted Mexican maize farming in this manner.

Ecological Distribution and the Power-Weighted Social Decision Rule

There are distributional issues associated with environmental externalities. Those who impose external costs on others benefit from negative externalities by avoiding internalization of the environmental costs of production. In effect, they are subsidized by those who bear those costs. Those who produce positive externalities cannot internalize those benefits as well. They, in effect, pay a tax. To identify the most economically efficient environmental policies, conventional economic theory holds that analysts should be guided by use of cost-benefit analyses. This rule requires that, constrained by available resources, net benefits summed over all individual members of society should be maximized. This formula treats all individuals the same in that costs or benefits to one person count as much as the costs or benefits to another.

In practice, social decisions may give greater weight to the costs or benefits to some people than to others. Such outcomes can be described as following a "power weighted social decision rule." That is, social decisions maximize net benefits "weighted by the power of those who receive them." [267] Put another way, the extent to which a government intervenes to correct market failures in the economy is governed by the balance of power between winners and losers. The effects of trade liberalization between Mexico and the United States with respect to genetic diversity and maize production can be evaluated with respect to these principles.

In Situ Genetic Diversity

By providing the raw material necessary for future crop adaptations to changing pests, pathogens, and environmental conditions, genetic diversity in the world's major food crops is a key to maintaining long-term food security around the globe. Modern agriculture is associated with less genetic diversity than traditional agriculture. While such uniformity can bring high land productivity, it can also increase vulnerability to large-scale crop failures due to plant disease and pest epidemics.

The centers of origin and diversity of the world's major food crops are located in the developing world. To respond to the long-term threats posed by genetic erosion, national and international agencies now collect and store seed samples in what are called *ex situ* (off-site) germplasm banks. These banks give plant breeders access to genetic diversity while providing insurance against losses of *in situ* (on-site or "in the field") genetic diversity.

Ex situ banks are not substitutes for but are rather complements to in situ genetic diversity for three reasons. First, gene banks are not completely secure due to financial constraints, human and mechanical failures, and the delicate nature of the seeds themselves. Second, many genetic attributes can only be ascertained by growing plants in habitats very similar to those from which they originated—costly and difficult to reproduce in the laboratory. Finally, gene banks can only store the existing *stock* of genetic diversity at a point in time—not the evolutionary *flow* of new varieties that can only occur by the processes of mutation and natural selection of genes in the field.

NAFTA and Mexican Maize

Maize originated in what is now southern and central Mexico and Guatemala. Today, central and southern Mexico remain the global center of genetic diversity in maize. Scientists refer to the Mexican campesino farmers' maize plots as "evolutionary gardens" because they not only maintain and produce a vast stock of maize varieties, they also manage an ongoing evolutionary flow of new varieties.

While maize is the leading crop in both Mexico and the United States, use and production techniques differ greatly in the two countries. The United States has average yields of 7.4 metric tons per hectare and produces roughly 200 million metric tons of maize each year on 300,000 farms. Mexico has average yields of 2 metric tons per hectare and produces roughly 14 million metric tons on 2.7 million farms each year. U.S. maize is used mainly as animal feed; Mexican maize is consumed by people in the form of tortillas.

Where the contrast is most stark, however, is in production techniques. Only six varieties of maize account for almost half of total U.S. acreage, leaving U.S. maize genetically vulnerable to insect and disease epidemics. In addition, 96 percent of U.S. maize acreage is treated with herbicides and one-third with insecticides. These inputs have contaminated groundwater supplies in a number of U.S. states.

By standard market prices, U.S. maize is more "efficient" than its Mexican counterpart. When NAFTA was negotiated, U.S. maize was priced at roughly $110 per ton whereas Mexican farmers received $240 per ton. The price advantage of U.S. maize has four sources: natural factors such as better soil and rainfall conditions; farm subsidies; the exclusion of environmental costs such as groundwater contamination from market prices; and the failure of market prices to capture the value of the maintenance of genetic diversity by Mexican maize farmers.

Estimates of the number of Mexican farmers who will eventually be displaced by U.S. imports vary widely. NAFTA will not totally eliminate Mexican maize production, but it is likely to hurt the most genetically diverse areas cultivated disproportionately by poor campesinos. Much of the abandoned maize land is likely to be converted into cattle pastures, which require less labor. Relatively conservative estimates of the amount of campesinos who will migrate to Mexican cities number in the hundreds of thousands; upper-end estimates reach as high as 15 million.

The impact of NAFTA on ecological sustainability arises from both positive and negative externalities; the erosion of positive externalities in the form of conservation of the evolution of in situ genetic diversity in Mexico, and the proliferation of negative externalities associated with intensive agrochemical use in the United States. Referring to the power-weighted social decision rule discussed earlier, the clear winners from NAFTA in this case are U.S. maize producers and to some extent Mexican consumers. The clear losers are Mexican maize producers and the environment. The Mexican producers will lose their livelihoods. There are two environmental "losers": future generations whose food security will be diminished by the reduction of in situ genetic diversity, and those adversely affected by the agrochemical pollution resulting from the U.S. corn production.

Conclusions and Policy Recommendations

Industrialized nations have long protected their farm sectors to safeguard employment, political stability, cultural values, and food security. In the case of Mexican maize, there is an even more compelling case for trade protection: the need to sustain genetic diversity in one of humankind's most important food crops. In principle, the first-best policy in Mexico would be ecological subsidies in the form of payments to Mexican maize farmers to reward their contribution to the public good via in situ conservation of genetic diversity. Since a good part of the benefits accrue to the planet as a whole, there is also a case for international assistance in such subsidization. However, tariff policy—which brings in revenue for the government rather than requiring expenditures—may be more politically feasible.

Given the power-weighted social decision rule, it will take international political backing, as well as financial support, to protect genetic diversity in maize. Recent history has shown that the Mexican campesinos hold little leverage over Mexican government policy. But international cooperation to sustain in situ diversity may alter the political balance and help to devise ways to compensate farmers who perform this valuable service.

Summary of

Multinational Corporations and the Neo-Liberal Regime

by James R. Crotty, Gerald Epstein, and Patricia Kelly

[Published in *Globalization and Progressive Economic Policy,* ed. Dean Baker, Gerald Epstein, and Robert Pollin (Cambridge: Cambridge University Press, 1998), Ch. 5, 117–143.]

A number of theories about the socioeconomic effects of increased foreign direct investment (FDI) have emerged in recent decades. These include

- *The race to the bottom:* Increased mobility of multinational corporations (MNCs) allows them to play off workers and to play communities and nations against one another, hurting low-skilled workers and the unemployed while benefiting owners of capital and some professional workers.
- *The climb to the top:* Competition for FDI will promote improved infrastructure, education, and economic growth. This is compatible with "neoliberal convergence": the claim that capital and technology transfer will raise living standards in poorer countries and promote a lessening of international income gaps.
- *Uneven development:* Some regions will benefit at the expense of others. In the theory of imperialism, the North exploits the South. More recently, there are fears that export-oriented growth in the South damages the North through competition with cheap Southern labor.
- *Much ado about nothing:* FDI is still a relatively small proportion of national GDP and flows primarily among already developed nations and a handful of developing countries. It therefore has little effect on inequality, unemployment, or wages.

In this chapter, the authors argue that within the current neo-liberal context, the "race to the bottom," is the most likely outcome of FDI. Their analysis focuses on aggregate demand for labor, domestic and international "rules of the game," and the nature of domestic and international competition. In particular, they examine the effects of these three factors on the relative bargaining power of firms and workers and on the ability of governments to capture benefits from FDI.

Shift in Bargaining Power

In the high-employment, high-growth era of the 1960s, outward flows of U.S. FDI were of about the same magnitude relative to GDP as they are today. But at that time, foreign investment often translated into increased exports for domestic companies. When companies did go abroad, the domestic demand for workers was high enough that the move did not increase the bargaining power of firms.

In contrast, the 1990s are characterized by a neo-liberal regime that is conducive to chronic unemployment, coercive competition, and destructive domestic and international rules of the game. The key components of this regime are fiscal austerity; financial, trade, and investment liberalization; privatization; and increased labor market flexibility.

In this context, the impact of FDI is much larger than its size as a proportion of GDP might suggest. The threat of mobility and its spillover effects transform labor markets to the detriment of workers and communities in both sending and receiving nations. Whether or not capital is actually transferred is not as important as the impacts of the bidding process, which creates a "magnification effect" of FDI.

Countries like the United States that have large inward and outward investment flows are vulnerable to these effects. "It is not only the *net* mobility of capital but also the problems associated with the possible destructive impact of *gross* mobility of capital in a particular setting," that affects wages, income equality, and unemployment. [121] In addition, countries that receive little FDI may nonetheless suffer social losses as they engage in a bidding war to attempt to attract it.

Trends in Direct Foreign Investment

Global FDI flows have increased steadily since 1980 in gross and net terms, and as a proportion of GDP and capital formation in all major regions. While the distribution of FDI flows is uneven, with the bulk going to the most developed countries, the amount going to the developing world is also growing rapidly. MNCs also increasingly use licensing, joint ventures, and outsourcing, so that FDI flows understate the total global influence of MNCs. Technological change and computerization can partly account for the explosive growth of FDI. But equally important are changes in the "enforcement structure—the set of domestic and international institutions and rules that secure the property rights and enhance the prerogatives of multinational corporations." [122] Most of the changes in countries' regulatory regimes have been toward liberalization. New areas have been opened to FDI, approval processes streamlined, and firms' exit options strengthened. In addition, the collapse of communist economies, together with deliberate sabotage by the United States and international organizations of alternative development models, has led to a widespread perception among governments that there is no alternative to integration into the global economy.

An Alternative Framework for Analysis of FDI

Some mainstream models of FDI allow for the possibility of uneven development or a race to the bottom. But almost all of these models assume full employment. The model presented here does not assume full employment and focuses instead on bargaining power and threat effects. In this model, MNCs will move to a new location as long as the benefits outweigh the fixed costs of moving. Communities trying to attract MNCs will "bid" by offering costs (measured in terms of wage and tax levels) that are sufficiently low to cover the fixed costs of moving. In response, communities that currently host MNCs must lower wages to a level equal to that of the community trying to attract the MNC minus the fixed cost of moving.

If the host communities cooperate with MNCs, the firms will not move, "but there will be a decline in wages induced by the threat of moving . . . the 'magnification effect.'" [126] The effect is more pronounced when more than two countries are involved in the bidding process. Other bidders may receive little FDI yet end up with reduced tax and wage rates, altering their distribution of income and reducing their level of public services.

Current global conditions are conducive to this model rather than to a mainstream full-employment model. Global competition has broken down the oligopolistic control, capital-labor cooperation, and high aggregate demand characteristic of the "Golden Age" of the 1950s and 1960s. Thus firms have turned to conflictual rather than to cooperative labor relations, and have initiated a coercive process of leveling-down. Stagnant wages and contractionary monetary policies keep aggregate demand low, increasing the need for communities to bid to attract employment. The liberalized "rules of the game" also reduce the bargaining powers of governments that subscribe to bilateral or multilateral investment agreements.

Effects of FDI and MNCs on Wages, Employment, and Social Conditions

There is considerable specific evidence supporting the hypothesis that FDI has negative effects on workers and communities. In the United States, for example, manufacturing firms frequently threaten to close if workers unionize, and workers find these threats credible. Several studies indicate that increased FDI has reduced United States employment by parent companies. Within the United States, individual states commonly offer substantial increases in subsidies and tax breaks to attract investment. In developing nations, a review of empirical literature suggests that positive spillover effects from FDI on wage levels are small or nonexistent.

In East Asia, external forces pressured the state-led economies responsible for the "East Asian economic miracle" to adopt neo-liberal policies. Financial deregulation, open markets, and decontrol of capital flows created the conditions for a boom-and-bust cycle as massive inflows of speculative investment gave way to panic and equally massive outflows. The International Monetary Fund (IMF) further diffused neo-liberalism in Asia by conditioning the allocation of its rescue funds on the implementation of further liberalization policies and fiscal austerity. This allowed MNCs to use the IMF against labor unions and government to "win major labor concessions and to eliminate burdensome regulations." [131]

The successful "Swedish Model," characterized by a centralized bargaining structure and a set of traditional relations and mutual obligations connecting business, labor, and the state in the Golden Age, was also dismantled by the negative effects of FDI flows. A more free-market–oriented position in the 1980s encouraged Swedish MNCs to locate a substantial amount of their resources outside Sweden and to outsource high-value-added, high-skilled segments of the production process. Thus, high-growth industries with positive economic spillover effects were exchanged for low-skilled raw material processing and intermediate good production, and unemployment in Sweden rose to record levels.

Policy Suggestions

Measures that could be taken to reverse the negative impacts of the current neo-liberal regime include:

- Initiate expansionary macroeconomic policies by national governments and international institutions to restore the faster global aggregate demand growth and to lower unemployment levels that minimize the harmful effects of FDI. Control of short-term capital flows may be required to assure that these policies are sustainable.
- Impose a moratorium on all international agreements that liberalize FDI controls until an effective set of international rules governing FDI is implemented.
- End World Bank and IMF policies of pressuring developing countries to open their economies in exchange for credit.

- Establish international labor standards and corporate codes of conduct.
- Establish stronger national laws protecting union rights, improved social safety nets, and worker/community input into corporate governance.
- Regulate more strongly capital markets and Tobin taxes on short-term capital flows.

These and other measures to restrain coercive competition can alter the domestic and international context for FDI and thus enable communities to capture more of the benefits of FDI flows. Globalization is not an irreversible juggernaut; its impacts can be steered by constructive policy action to change the "rules of the game."

Summary of

Foreign Investment, Globalization, and Environment

by Daniel C. Esty and Bradford S. Gentry

[Published in *Globalisation and Environment,* ed. Tom Jones (Paris: Organization for Economic Cooperation and Development, 1997), Ch. 6, 141–172.]

Hundreds of billions of dollars of private capital annually flow from North to South in the form of foreign direct investment (FDI), portfolio equity investment, and debt finance. How this capital is spent will have a much more profound effect on the quality of the global environment than the few billion dollars of official assistance devoted to environmental investments each year. This chapter argues that a variety of factors that influence how private capital flows to developing countries also affect the environment. In particular, it reviews the incentives and conditions that determine who gets private capital funds and outlines ways environmental goals could be better integrated into foreign investment processes.

Diversity in International Capital Flows

Any evaluation of private capital flows and the environment must consider not only FDI (which accounts for 54 percent of total capital flows), but also portfolio equity investments and debt finance as well:

- *Portfolio investment:* Environmental performance may affect the value of a portfolio investment in overseas companies' shares; the price of a company adhering to high environmental standards may be bid up if investors believe this will positively affect performance and bid down if investors believe environmental corners should be cut for short-term profits.

- *Debt financing:* Because commercial lending to private companies gives debt holders a stake in the borrower's financial success, debt holders are concerned about the environmental risks and performance of borrowers as it relates to their ability to pay back loans. Thus, they may encourage the adoption of either lax or stringent environmental standards by their clients.

Qualitative Dimensions of Expanded Foreign Direct Investment

Private international finance has both positive and negative environmental consequences. As FDI encourages growth in the developing world, pollution may be exacerbated. However, as countries become wealthier, they can afford larger investments in pollution prevention and control. This may eventually lead to reduction in some pollution impacts—but not all. "Thus, some environmental harms (notably climate change) . . . may worsen *more slowly* as more countries develop, but never diminish in *absolute* terms." [160]

As more and more countries develop, the competition increases for limited FDI funds that speed the development process. The process of luring foreign investors sometimes includes a commitment to more lax enforcement of environmental standards.

Yet these competitive pressures differ for different industries. In more commodity-like industries where products are relatively undifferentiated and small cost differences translate into large market share gains and losses, investment flows are particularly prone to influence based on the level of environmental standards in each country and the standards of the specific projects to be financed.

Pressures to lower environmental standards may come from international investors or from host governments. North American and European companies have found themselves pressured to eliminate environmental components from proposed power plants in China in order to cut costs. Additionally, some U.S. furniture makers moved their operations from the United States to Mexico, where environmental regulations are more lax.

Competitive pressures may also operate in the opposite direction: foreign investors in Costa Rican banana production insisted that production be more environmentally sound, because their European customers wanted a better product. Some Asian lumber products have similarly improved environmental performance in response to European consumer sensitivities.

The traditional belief that multinational companies exploit weak pollution control programs in developing countries and despoil the environment is outdated and generally wrong. Foreign investors often set up operations with modern, less polluting, technologies and environmental management systems and training programs that are more advanced than those that exist locally. There are only a few isolated cases in which multinational companies dismantled outdated polluting facilities in their home countries and reassembled them in developing countries. This "technology dumping" usually involves non-Organization for Economic Cooperation and Development (OECD) countries such as Hong Kong, Singapore, and Taiwan and concerns sales of outdated equipment to companies in developing countries rather than FDI per se.

Environmental Aspects of FDI

- Because the environmental character of industries that receive FDI vary depending on the type of investment and the goal of the investor, it is necessary to distinguish three types of FDI:
- *Market-seeking FDI:* Pursued by investors seeking opportunities to sell in overseas markets that have a broad sales potential, such as China.
- *Resource-seeking FDI:* Pursued by investors seeking access to critical resources, such as raw materials and cheap labor.
- *Production-platform-seeking FDI:* Pursued by investors desiring to set up overseas facilities in a particular country as a platform for production and sales in a regional market.

Many multinational corporations (MNCs) adhere to their home country's high environmental standards for a number of reasons:

- The efficiency of adhering to a single set of management practices, pollution control, and training technologies outweighs any cost advantage of relaxing environmental regulations in an overseas facility.
- Because MNCs operate on a large scale, their visibility makes them prone to exposure by local investment officials.
- The threat of liability for failure to meet standards encourages the adoption of better environmental standards than are locally required.

However, even when MNCs follow strict environmental regulations in developing countries, the results are varied. The host country may not provide the same degree of pollution control infrastructure found in the corporation's developed-country home base. For example, many developing countries lack sewage systems that can handle the partially treated wastewater of MNCs. Additionally, local suppliers and service providers of MNCs may not adhere to sound environmental standards.

The increased scale of economic activity resulting from FDI may have serious environmental consequences. Forestry and agricultural activities at commercial scale for export markets may involve monocultures with ecologically damaging impacts. Further, companies from countries that lack strict environmental standards may ignore the requirements of environmental stewardship.

Economists have found little empirical evidence exists for the "race to the bottom" theory, which states that countries with low environmental standards attract dirty industries. This may be because environmental compliance costs are usually too small a factor in a company's overall production costs to affect its choice of location. There is some evidence, however, that companies with much higher-than-average pollution control costs *do* move to areas where environmental standards are more lax.

The race to the bottom model is more evident *within* countries with high environmental standards such as OECD nations (among which most international capital flows occur); companies in these countries move to jurisdictions with relatively lower environmental standards. This results in a "political drag" effect in which standards are not raised to optimal levels or existing rules are not enforced in the majority of jurisdictions.

On the other hand, there is scant evidence that developing countries set higher environmental standards for projects funded by foreign investors. So what drives FDI environmental performance? FDI improves environmental performance when it translates into better business. Four major categories of commercial benefits, which motivated corporate environmental improvements, were identified in a study of Latin America:

- *Improved access to export markets:* Perceived consumer demands for environmentally responsible products and a desire to keep up with environmentally friendly competitors.
- *Increased productivity:* Because pollution equals waste, foreign investors seek to achieve a profitable balance between increased production efficiencies and higher pollution control costs.
- *Maintenance of a "social license" to operate and expand:* MNCs face international pressures to be "good environmental actors" and home-country pressures to not "export" pollution.
- *Access to finance:* Many external financing sources, including government financing bodies such as the U.S. Export-Import Bank, make loans conditional on adherence to certain environmental standards. Even private financing often requires a minimum level of environmental "due diligence."

MNCs typically seek *consistent* environmental enforcement, not lax enforcement. They may be deterred from investing in a country that has a precedent of placing excessive burdens on companies for the cleanup of past contamination.

Integrating FDI Practices and Environmental Goals

"To the extent that environmental investments yield adequate returns, FDI source and host countries alike would benefit from finding ways to channel private capital flows into environmental infrastructure projects . . . as part of broader development efforts." [167] These rules should be developed and enforced multilaterally.

Research shows that countries that have straightforward, transparent, and efficient environmental regulations do not as a result lose, and may even attract, FDI (Gentry et al. 1996). For example, after Mexico recently increased its environmental enforcement, FDI in the Mexico City area expanded, and the air quality actually improved. OECD countries can support and encourage efforts by developing countries to create functioning environmental programs in the following ways:

- Move toward consistent implementation of minimum environmental standards for major development projects that have implications for the "global commons." This could be institutionalized through the proposed OECD Multilateral Agreements on Investments (MAI).
- Make foreign assistance conditional on specified environmental results.
- Offer funding to promote environmental compliance in developing countries.
- Provide specific incentives for cooperation (such as the joint implementation program in the Framework Convention on Climate Change).
- Increase the data available to companies and governments regarding environmental components of siting decisions and the role of environmental standards in such decisions. This data should divide up FDI into its components (such as market-seeking investments, etc.).
- Institutionalize eco-labels and standards, such as the ISO 14000 series, to empower consumers to bring pressure on producers.

Foreign investors are generally not "anti-environment." They are, however, profit maximizers. If it is to their business advantage to improve environmental performance, they will do so." [169] Government policy frameworks and consumer pressures can have a major influence on business calculations concerning the environment.

Summary of

Globalization and Social Integration

by Dharam Ghai and Cynthia Hewitt de Alcantara

[Published in *People: From Impoverishment to Empowerment,* ed. Uner Kidar and Leonard Silk (New York: UNDP Press, 1995), 301–318.]

Analyzing "social integration" is the process of examining the patterns of human relations and values that bind people together in time and place and that define their life opportunities. Patterns of social integration are shaped by trends in politics, economics, culture, and technology. The recent wave of globalization has radically changed the context of social integration across the globe. After discussing the overarching bases of social integration, this chapter examines how the process of globalization is altering existing values and behavior, institutions and governance, and relations among ethnic groups around the world.

Setting the Stage for Social Integration

A number of aspects of globalization are shaping the patterns of social integration. With the collapse of the Soviet Union there has been a great upswing in the degree of democracy and individual freedom across the globe. This has opened up new possibilities for participation and has created a new, wide-ranging set of voluntary associations and interests groups. While such organizations can widen and deepen the bonds of citizenship on one hand, they can also create new divisions or accentuate long-standing rivalries.

Coupled with the turn toward liberal democracy and the search for individual freedom has been the endorsement of market forces as the principal means to manage economic systems. This shift is manifest in changes such as the retreat of the state from intervening in economies, the deregulation and privatization of many industries, and, perhaps most importantly, the integration of different economies through the liberalization of trade and investment in goods and services.

This reliance on the market has markedly altered the economic and political context for social integration by changing the power relations among different social groups and countries. For example, it is clear that the power of organized labor has weakened, while transnational corporations, the owners of capital, and other groups have been strengthened. Changes in production systems and labor markets have also altered patterns of social integration around the world. Longer-term trends that give nations with highly trained personnel a comparative advantage over the unskilled will present difficulties for many of the developing economies.

These economic trends, paired with rapid technological change (especially in electronics, communications, etc.) are transforming the nature of work relations and job creation, altering patterns of leisure and consumption activity, and encouraging the creation of a global culture. The revolution in mass communications, while having the potential for promoting understanding and solidarity and enhancing knowledge throughout the world, also has the capacity to exalt consumerism.

Exclusion and Inclusion

The characteristics of globalization outlined above give the people of the world a contradictory picture of inclusion and exclusion. At the same time that images of wealth and leisure are widely disseminated by global media, the life chances of many are becoming more and more restricted and marginalized. Income and wealth are becoming more polarized and the ability of governments to provide social services is decreasing.

One of the most obvious responses by those threatened with exclusion or marginalization is to migrate, whether within countries or abroad. Migration can be a process both of integration and of division. For the affluent and educated migrants, life chances are often improved. Larger-scale migration by poor people, in contrast, can sometimes mean greater impoverishment and a disruption of existing forms of social organization in both sending and receiving regions.

Changing Values, Behavior, and Institutions

Globalization is creating profound changes in basic social institutions, modifying existing bonds and patterns of behavior, and creating new forms of interaction and integration between the household and the larger society.

While households around the world vary considerably, self-contained, two-parent families are no longer the norm either in developed or developing nations. Single-parent households, most often headed by women, tend to be the poorest in industrialized countries. This is true also in some areas of the developing world, but in others extended families provide support for female-headed households.

Globalization is also weakening communities and neighborhoods. The computerization of the workplace, the declining legitimacy of altruism, the necessity to accommodate refugees and migrants, in addition to exposure to the global economy, all stress core social norms and relationships.

One of the most positive changes has been the opening of civil society, but this must be looked at with caution. The strengthening of civil society should be based not only on citizens' initiatives and nongovernmental organizations, but also on the breadth and scope of a kind of "civic culture" that presupposes a basic adherence to an agreed-upon set of universal values and an acceptance of workable rules for the adjudication of interests and the protection of minority voices.

Governance and Social Integration

Developments in the global economy and society are posing increasingly complex problems for political and administrative institutions at both the local and national level. The basis of good governance is a package of institutions, laws, procedures, and norms that allow people to express their concerns within a predictable and equitable context. Structural and ethical changes are jeopardizing such contexts.

In developed and developing worlds alike, ideological and economic forces are sponsoring a deep reduction in public expenditure on social services. Business, labor, and political alliances are also increasingly confronting the threat of fragmentation, instability, and ineffectiveness. This reinforces public perceptions of ineffectiveness on the part of the state. These trends manifest in the growing numbers of those in the informal sectors of the world economy. Growing numbers of people provide for themselves without contributing through taxes for public purpose. Some of these informal groups are criminal organizations, which have gained significant power in countries such as Colombia and Russia.

Social Integration and Multi-ethnic Societies

Many of the problems discussed here are exacerbated within the context of multi-ethnic societies. Of course most countries in the world are multi-ethnic, but some have been more successful than others in attempting to forge tolerant and just societies. Drawing from the experiences of diverse countries such as Belgium, India, Lebanon, Malaysia, Nigeria, and Switzerland, it is possible to identify institutions and policies that can be effective in easing ethnic tensions during the integration process:

- Systems of governments with power-sharing arrangements between the center and the regions, as well as among various ethnic groups.
- Electoral systems that are tailored to the specific ethnic structures and problems of individual countries so that they guarantee a place for minority ethnic groups.
- The fostering of a network of advocacy groups concerned with humanitarian questions and human rights.
- Strong educational systems that promote understanding and tolerance in multicultural societies.

- The enforcement of tough policies directed against those who stir ethnic hatred and violence.

Conclusion

"There is at present a striking incongruence between patterns of social integration, that bind people around the world more closely than ever before, on the one hand, and the frailty of existing mechanisms for subjecting global processes to regulation and channeling them toward human welfare on the other." [318] This gives rise to a number of challenges that will require institutional reform at many levels of society. These issues include the following:

- How can the international community control the polarizing effects of the liberalization of trade and investment?
- How can new bases of solidarity be created during a period when capital and labor are much more mobile than ever before?
- How can new forms of personal worth and livelihood be created in the context of the secular decline of the need for human labor?
- How can we deal with the unprecedented degree of environmental degradation associated with new patterns of resource use?

"The international context assumes greater importance in this endeavor than ever before—not because it is possible to design universal solutions, uniformly applied around the world, but because global forces have created inescapable common problems of worldwide scope. The concept of an 'international community' is no longer a simple ideal. It is a fact of life." [318]

Summary of

Reconciling Economic Reform and Sustainable Human Development: Social Consequences of Neo-Liberalism

by Lance Taylor and Ute Pieper

[Published as *Reconciling Economic Reform and Sustainable Human Development: Social Consequences of Neo-Liberalism* (New York: UNDP, 1996).]

The structural adjustment programs (SAPs) that have been prescribed by the International Monetary Fund (IMF) and the World Bank have sought to achieve higher output growth and rising real incomes in the developing world. However, the framers of these programs have given little attention to their social and environmental costs. This report is an in depth review of the social effects of the SAP approach, with a briefer treatment of environmental repercussions.[1] Drawing from the report, this summary focuses on the effects of SAPs on poverty and inequality, gender relations, education, and human health.

The Washington Consensus

The package of instruments used to administer SAPs in the 1980s and 1990s is guided by a set of principles known as the "Washington Consensus," the principal focus of which has been market liberalization. The first stage of SAPs, the domain of the IMF, is the "stabilizing" period. This period is characterized by fiscal and monetary austerity, exchange rate adjustments, policy-induced relative price shifts, and the liberalization of foreign trade restrictions. This is followed by the "adjusting" period—the task of the World Bank. The World Bank urges nations to continue trade liberalization, lower barriers to foreign capital flows, deregulate financial markets, deregulate labor markets, rationalize taxes, and privatize public enterprises. While many of these programs can be seen as successes by traditional economic measures, they have come at considerable social cost.

Poverty and Inequality

Levels of poverty and inequality have changed during periods of SAPs. It is important to note that poverty and inequality are not the same, and a distinction is made between absolute and relative poverty. The concept of relative poverty is linked to distribution of income, for example, by the use of a benchmark of 50 percent of median national income in developed countries. Some nations that do not experience absolute poverty may experience relative poverty due to large amounts of inequality.

Trends in both median national income and inequality can thus affect poverty trends. For example, relative poverty has not increased much in formerly socialist economies of Eastern Europe that still maintain extensive social safety nets, but absolute poverty rose as real GDP declined from 1987 to 1993.

In general, there has been an overall decline in the rate of poverty incidence (percentage of households or individuals below a "given" poverty line) in the developing countries in the postwar period. This decline masks a number of significant interregional disparities. Latin America is the clearest example of an increase in poverty incidence during adjustment. The incidence of poverty in this region climbed from 13 percent in 1985 to 52 percent in 1990.

Poverty rates in Central and Eastern Europe rose dramatically during their transition, from 1 to 36 percent of the total population in the Baltics, and from 2 to 41 percent in the Slavics and Moldova. South and East Asia are the success stories from a poverty perspective; both regions achieved sustained output growth rates that are correlated with reduced poverty levels.

Official figures show no upswing in poverty incidence during periods of SAPs in Africa, but these appear to be relative poverty measures that fail to reflect the full degree of absolute poverty.

Two (less-well established) global patterns can also be identified. First is the feminization of poverty in developing countries, due to explicit gender biases, a drop in agricultural production and environmental degradation in sub-Saharan Africa, and family disintegration due to male desertion and migration. The second is an increasing poverty incidence among the elderly, particularly elderly women in the United States, due in part to rising health-care costs.

While the overall incidence in poverty is falling in developing countries, inequality is on the rise in all regions except for East Asia. Income inequality is most severe in Latin America, where Gini coefficients are usually in the 0.5 to 0.6 range.[2] In Eastern Europe Gini coefficients rose dramatically during adjustment as well, rising from 0.23 to 0.35 in the Baltics, and from 0.24 to 0.32 for the Slavics. In Central Europe the rise was not as sharp, but a rise nonetheless from 0.22 to 0.26. In East and South Asia the Gini coefficients are between 0.3 and 0.4. In the Middle East and Africa data on income distribution are scarce.

Impact on Gender Relations

A number of empirical studies have examined the effects of SAPs on gender relations. Based on this work, a number of general patterns can be observed:

- Men and women feel the effects of these programs in practically every aspect of their lives: food price increases, declines in family incomes, and reductions in public expenditures on social and health services. These changes directly affect women's role as principal homemakers.
- Many policies of SAPs alter the extent to which women enter the formal labor force. Examples of such policies are the general decline in employment opportunities and a withdrawal of credit and technical support to small farmers in a number of SAPs.
- The degree to which women's productive roles are affected by SAPs is partly governed by existing social rules and gender roles, as well as by the process of class differentiation.

  Anecdotal evidence suggests that females adapt unpaid labor time to changes such as declines in household income and reductions in social sector services.
- Employment increases for women in export manufacturing sectors as a result of relative price shifts through trade liberalization. However, in most

cases, this employment comes at a lower wage than that of male workers with similar skill levels.

- Depending on the public sector contraction of SAPs, women are subject to layoffs.

The evidence indicates that women are prepared to seize economic opportunities when they arise, but often such gains have arisen from previous public infrastructure investments such as services and transportation, rather than as a result of SAPs. This casts doubt on the World Bank's "investing in people" rhetoric, since macroeconomic policies are generally made without consideration for their impact on gender relations.

Education and Health

A clear pattern of deterioration in social well-being has occurred during adjustment in many developing countries. While some of those losses could be lagged results of economic recessions that led to adjustment, many SAPs either had negative effects on social well-being or they were ineffective in reversing negative effects caused by other factors.

In sub-Saharan Africa, where data availability is a problem, nutrition levels in many countries worsened during the early phases of adjustment. No evidence was found on infant and child mortality and morbidity rates. Educational achievement was found to be declining. School enrollment expanded rapidly in the region previous to 1980, but the average enrollment ratio fell from 59 to 51 percent by 1992.

In Latin America, the education of young people is on the rise. The proportion of urban youths dropping out of schools with nine or fewer years of schooling declined between the early 1980s and 1990s. However, schooling in Latin America is generally inadequate for the bottom part of the income distribution and for rural youth. Between 70 and 96 percent from the bottom quartile will leave home with a level of education deemed insufficient for modern life. These educational patterns magnify the impact of SAPs in increasing inequality. Increasing inequality is also a key issue in the health sector in Latin America. Health expenditures are inequitable in quality and coverage, with nonpoor segments of the population receiving most of the benefits of public expenditure and investment in health services.

Central and Eastern Europe have witnessed serious deterioration of their health and education indicators. A shift in dietary composition in the region has compounded a problem of a diet too rich in animal fats and starch and too poor in vitamins, minerals, and other essential nutrients. This decline, together with worsening sanitary conditions, has led to increased outbreaks of such diseases as tuberculosis and rickets. There has been an increase in overall mortality in the region, and a growing number of indicators show that social cohesion and personal safety are under serious threat. Pre-primary, primary, and secondary educational enrollment rates have fallen in most countries in the region (with the exception of Hungary).

As with gender relations, there has generally been a failure to consider the feedback effects between macroeconomic policy and social outcomes. In addition, feedback effects among different social indicators are significant and have not been taken into account in policy formulation. A more integrated approach is needed to examine the roles of income, technology, and behavior in relation to social outcomes such as health and education.

Policy Recommendations

Based on the full report, a set of recommendations emerge that can guide SAPs toward development paths that are sustainable:

- Develop policies to offset social shocks and promote social equity.
- Ease debt burdens and allow the state to play a more active role in capital formation.
- Recognize the importance of industrial strategies involving trade protection, financial market control, and corporatist deals with between the state, capital, and labor.[3]
- Invent appropriate micro-macro links such as the coordination of macroeconomic, industrial, and educational policy.
- Provide educational and health services as well as targeted income redistribution to create the conditions for successful and equitable economic growth.
- Link measures of social and gender equity, environmental quality, and similar indicators to output growth.
- Gradualism in adjustment and stabilization of capital and trade flows is needed.
- States must adopt a developmental role rather than following the neo-liberal guidelines of the World Bank and International Monetary Fund.

Notes

1. Environmental impacts of SAPs are dealt with in detail by Reed (this volume).
2. The Gini coefficient, or Gini ratio (the terms are used interchangeably) ranges from a minimum of 0 at perfect equality to a maximum of 1 at perfect inequality. A larger Gini coefficient indicates a society that is farther away from perfect equality, or in other words a more unequal society. Gini coefficients range from 0.23 for the most equal European

countries to about 0.60 for Brazil, the most unequal major country. For an extensive discussion of measures and trends in inequality, see Ackerman et al., *The Political Economy of Inequality* (Frontier Issues in Economic Thought, Volume 5; Washington, D.C.: Island Press, 2000).
3. Historically, such nonliberal intervention has been associated with rapid growth in South Korea and other successful capitalist economies.

Summary of

Impacts of Structural Adjustment on the Sustainability of Developing Countries

by David Reed

[Published in *Structural Adjustment, the Environment, and Sustainable Development*, ed. David Reed (London: Earthscan, 1996), Chs. 12–14, 299–350.]

Structural adjustment programs (SAPs) have been implemented in many developing nations to promote macroeconomic reform and integration into the world economy. In the concluding chapters of *Structural Adjustment, the Environment, and Sustainable Development*, the author evaluates the environmental and social impacts of SAPs in nine developing countries. These chapters draw on case studies of Cameroon, Mali, Tanzania, Zambia, El Salvador, Jamaica, Venezuela, Pakistan, and Vietnam. The question at issue is: Are such programs putting these countries on a development path that is sustainable? Both short- and long-term impacts on the environment and social equity are addressed.

Structural Adjustment and Sustainable Development

The object of structural adjustment policies is to bring developing countries into the world economy by adopting a development strategy based on the promotion of export-oriented growth; the privatization of state-owned industry; the elimination of barriers to international trade and investment flows; the reduction of the role of the state as an economic agent; and the deregulation of domestic labor markets.

The nine developing countries examined in this study share several commonalities. Their economies are dependent on the export of agricultural products or the extraction of natural resources; a large proportion of the population lives under poor social and economic conditions; all except Jamaica have high population growth rates; and all except Venezuela have a high percentage of their populations in rural areas. In order to achieve a sustainable result in these countries, the economic, social, and environmental dimensions of development should "converge in such a way as to generate a steady stream of income, ensure social equity, pursue socially agreed upon population levels, maintain human-generated and natural capital stocks, and protect the life-giving services of the environment." [336] To fully evaluate these effects, the short-term and long-term impacts of SAPs need to be examined separately.

Short-term Environmental Impacts

Short-term impacts of SAPs occur in the early stages of adjustment: currency devaluation and trade liberalization (together seen as price corrections), and changes in fiscal and monetary policy. These policies have varied impacts on the economic, social, and environmental dimensions of the adjustment process.

If measured by traditional economic measures, adjustment programs are having their desired effect in most of the nine countries. In most cases per capita GDP, agricultural exports, and revenues from extractive industries are all on the rise, while budget deficits and inflation have been brought under control.

In relation to distribution, the picture is less encouraging. Most of the benefits of SAPs accrued to wealthy, outward-oriented producers and merchants, commercial farmers, and investors in extractive industries. The heaviest costs of SAPs have been borne by small rural farmers, workers in the informal sector, urban consumers, redundant government employees, and women. Economic and social inequalities between these "winners" and "losers" have become exacerbated during the adjustment process as well. While the winners often respond more efficiently to price corrections, the losers often end up drawing down their productive assets to survive. This "accelerates the most intractable environmental problem facing many countries—that is, poverty-induced environmental degradation." [319]

Compounding these problems is the changing role of the state. Consciously limiting the state's role as an economic agent has been highly successful in eliminating inefficiencies, reducing government mismanagement, and correcting fiscal imbalances in these economies. However, in many areas the dismantling of government is undermining economic reforms, jeopardizing social stability, and weakening environmental sustainability as a result of reductions in social, environmental, and extension services. Paradoxically, it may end up being more difficult for the efficiency goals of price corrections to be achieved because the state's ability to correct policy and market failures has been virtually eliminated.

As with the economic and social dimensions of SAPs, there are a number of impacts on the environment that are clearly positive. In some cases, exchange rate reforms have led to a shift away from "erosive" to "nonerosive" crops, and price corrections have created new agricultural incentives that have stimulated expansion and diversification of tradeable crops and other commodities. These are positive effects in that they increase relative returns to the agricultural sector and raise the incomes of some farmers, thus encouraging on-farm investments. In other cases, however, nontraditional cash crops can be more erosive or generate other negative environmental effects such as deforestation during the processing stage (e.g., drying and curing of tea and tobacco).

The environmental impacts of agricultural sector adjustments are largely a function of the status of farmers in the countries being considered. Large, commercial producers can respond to the new playing field by diversifying crops, intensifying production, and introducing new technologies. Small farmers and rural workers, however, cannot absorb the increased costs of inputs (due to new external conditions and the removal of internal subsidies), nor are they flexible enough to respond to the new incentives stemming from economic integration. They may respond instead by agricultural extensification, deforestation, and intensified use of marginal lands. Adjustment can thus lead both to lower living standards for the rural poor and to increased environmental damage.

In countries with large extractive sectors, output growth and employment benefits have resulted from structural adjustment, but these benefits have been accompanied by a dismantling of government capacity to manage and regulate those industries so as to minimize environmental costs. In Tanzania, Zambia, and Venezuela, such policy failures are generating highly damaging environmental impacts. The failure to couple internal and external adjustment with adequate regulatory reform and institutional strengthening is generating high revenue and employment on the one hand and higher environmental costs to be absorbed in the future, on the other.

Long-term Impacts on Development Paths

The structural adjustment goal of "getting the prices right" may correct inefficiencies in these countries in the short term, but it raises two questions for the long run: Does such an approach guarantee economic security and a place in the future world economy? And does it provide continuing income for the majority of the citizens in these societies? The answer to these questions will depend on whether the new economic regime is equipped to give these countries the proper incentives to utilize their natural resources and environmental services on an enduring basis.

The country studies indicate that adjustment has accelerated the drawing down and overuse of natural resources and environmental services. Deforestation, soil degradation, and watershed disruption have been widespread. In some cases, environmental trends worsened with adjustment, in others pre-established trends have continued. "Without exception, however, the studies affirm the fact that the current environmental trends are serious, have long-term implications, and in many cases show signs of irreversible damage." [347]

There is a conflict between the short- and long-term costs and benefits in the agricultural sector also. The export value of primary commodities has been steadily declining relative to the export value of manufactures. This suggests that the more diversified economies will continue to benefit more from export-oriented growth policies than countries that are focusing on a small number of agricultural or extractive products. If these trends continue, then agricultural countries will see a decreasing relative share in global wealth over the long term. "In short, is this development path increasing the risks of mortgaging the economic futures of the countries for the prospects of gaining greater access to global markets in the short term?" [351]

It may be that the long-run benefits of adjustment will raise per capita incomes for small and large farmers alike. However, the studies suggest that in the short run, existing inequalities have been exacerbated. This in turn has accelerated the draw-down of natural capital. This poses critical problems in societies where poverty is pervasive and where population growth exceeds productivity gains. "What institutional force will mitigate the trend toward growing inequality and poverty-induced environmental degradation if the ability of governments to promote basic standards of equity and decency has been weakened?" [352]

The failure to correct social and environmental problems at the national level arouses concern for possible global ramifications. Global consumption of resources may cause large-scale environmental irreversibilities before relative resource scarcity is reflected in world prices. "If the designers of structural adjustment programs gave little thought to national-level environmental impacts of economic restructuring, they certainly gave no consideration to the global implications." [353] The lack of effective institutions for global environmental management means that there is no safeguard against cumulative problems arising at the national level but affecting planetary resources.

Policy Recommendations

While economic reforms are clearly necessary on a global scale, the process of how the costs and benefits of adjustment are distributed is in need of reform. The following recommendations call for changes in development priorities and a shift in function among the various actors within nations:

- Integrate environmental issues into macroeconomic reforms.
- Incorporate the needs of the poor in adjustment programs.
- Recognize the state's role in complementing the goals of adjustment by minimizing social and environmental costs.
- Enhance and expand the role of community groups and nongovernmental

organizations in designing and implementing aspects of the development process.

- Reform international financial institutions to incorporate "getting long-term social and environmental strategies right" in addition to "getting the prices right."
- Reform systems of national accounts by establishing and incorporating sustainability indicators.
- Reform the international trading system so that participation in the world economy is predicated on adherence to the social and environmental dimensions of development.

PART VIII

Taming the Corporation

Overview Essay

by Neva R. Goodwin

This essay, and the writings it introduces, is action-oriented. The thinking on corporate responsibility that is represented here is motivated by the perception of a problem, an image of a preferred state, and some specific ideas on how to move from where we are to where we would prefer to be. These ideas are discussed under three headings: internalizing externalities;[1] gaining a truer understanding of business costs (with an emphasis on environmental accounting); and lengthening the time horizons of corporations. The final section in the essay suggests (unsurprisingly) that, while some real progress is being made, not all corporations are going along; and it is by no means certain that the progress is sufficiently rapid or broad to counteract the unsustainable orientation of many businesses and the social and environmental problems that result.

The Problem: Corporate Power—Often Misused

Corporations control the vast majority of the world's productive assets. They are the major institutional forms through which raw materials are extracted, processed, and turned into products for sale. They are the exemplars and the upholders of the capitalist way of life, with its focus on sales, growth, and profits, often at the expense of other values, such as community, cooperation, and satisfaction of basic needs. Their economic power translates into political and cultural power that in many ways exceeds the power wielded by governments, educational and religious institutions, families, and other forces that shape society. Critics charge that corporations promote socially, ethically, and environmentally noxious consumerism, making special efforts to indoctrinate children and infants with materialist values.[2] The largest corporations—or collections of corporations in industrial coalitions—use their power to distort the economy, inducing governments to give them tax breaks, subsidies, and other favors.[3] Their lobbying efforts, and the funds they bestow on politicians, corrupt and corrode democracy. Given these and many other negative effects of corporations, some critics believe that the entire capitalist system is past reform; it must be replaced from the ground up.[4]

To other observers, the net effect of corporations is clearly positive, for they provide jobs, they generate technology and wealth, and they produce goods and services that are presumed to contribute to the well-being of consumers. A large part of humanity is, after all, dependent upon corporations for basic necessities—food, clothing, building materials—as well as for the materials and services we rely on for entertainment, information, and much else in our lives.

This essay will not attempt to present a balanced picture, or a "bottom line" of the beneficial and harmful effects of corporations. Nor will it weigh the arguments for systemic change.[5] Rather, starting from a relatively optimistic, reformist approach, it will ask: how can corporations be encouraged—or forced—to improve their social, economic, and environmental effects? The articles selected for summarization in this section all address that question. Starting from what they consider a realistic picture of what corporations do and what motivates them, they emphasize ways to improve the present reality.

The Goal: Corporate Responsibility

> Having organized their expansion based on globally-integrated efficiencies made possible by liberalized investment and trade regimes, TNCs (trans-national corporations) now confront a substantial challenge to this permissive regime. Globalization could bring about a serious backlash from unresolved societal needs. Considered within a global context, social responsibility therefore takes on immediate practical and political importance for an international business community whose operations are conditioned on continued globalization. In fact, there is a significant recent expansion of attempts to design newly cooperative ways for TNCs to respond, individually and collectively, to the evolving public expectations of a global social contract. (UNCTAD 1999, 355)

What is the social contract to which UNCTAD refers above? Ideally, business should serve people in society—rather than vice versa. At a minimum, each firm should bear the costs generated in its own processes of production. (Exceptions may be made for products of such generalized social value that it is deemed acceptable to start down the slippery slope of subsidies.) More positively, the business sector should produce goods and services that people intrinsically want (i.e., without having been manipulated through advertisements or other influences to want things that will not enhance their lives); it should provide meaningful and dignified work; and it should generate and distribute revenues so that workers and owners can purchase reasonable shares of society's output. More ambitiously, business should help to anticipate and plan for the future needs and constraints of society and of the natural world within which society—and its subset, business—are embedded.

These ideals can be summarized in the idea of *corporate responsibility.* Given a clear need, in today's world, for much more corporate responsibility than now exists, how can we promote change in the desired direction? Two concepts have emerged as centrally important to this goal: *accountability* and *transparency*.

Accountability is the idea that corporations must interact with and be answerable to all of their stakeholders. Lists of stakeholders have been drawn up in various ways; one such list has been drawn up using the principle that a firm's stakeholders are all those who are affected by, or who affect, its activities. These include workers, customers, suppliers, governments, creditors, investors, and neighbors—where the latter may be anyone "downstream" of the firm's environmental effects. They also include nonhuman entities often subsumed under the terms, "nature," or "the environment," as well as people of the future, and organizations who represent these otherwise voiceless stakeholders.

Corporate transparency is an idea that has, in a relatively narrow sense, long been promoted by the U.S. Securities and Exchange Commission (SEC). Since the Great Depression of the 1930s, the SEC has sought to make firms reveal all financial data that could be materially important to their investors. As the concept of stakeholders has expanded to include many other groups besides investors, the modern idea of transparency implies that all stakeholders have a right to all of the information that could have a bearing on how their interests are affected by corporate actions. The emphasis has been on issues in the areas of environment (e.g., adherence to the ISO [International Organization for Standardization] 14000 and 14001 standards);[6] human rights; and labor. While nongovernmental organization (NGO) activity in the last of these categories has included concern for wages and working conditions in both domestic and foreign factories, demands for transparency have most often come from the North, and pay less attention to some Southern development concerns such as technology transfer (UNCTAD 1999, 367).

Some of these issues have been on the business agenda for a long time. Unions and the International Labor Organization have insisted that corporations take some responsibility for social justice issues in the quality of their workers' lives. Starting more recently, environmentalists have waged similar battles. The social and economic agendas that promote communitarian and community values against the destructive impacts of corporation action have become more widely known since about 1980. Southern NGOs have protested many of the economic and cultural impacts of transnational corporations (TNCs), while pressuring TNCs to devote more of their profits to technology and skills transfer, wages, and the like.

An important, related agenda concerns the relationship between large corporations—especially TNCs—and the local companies that buy from and sell to them. On the one hand, this nexus may provide an opportunity for public relations pressure to be channeled through the large corporations to their suppliers; an example is the decision of Home Depot to purchase only from suppliers that provide environmentally certified wood products. On the other hand, when TNCs sell to smaller companies, they are often relieved of the pressures that broader consumer groups may exert.

Not many organizations are pursuing the full gamut of issues implied by the term "corporate responsibility." Some initiatives that will be examined in this essay are oriented largely toward environmental responsibility. This emphasis is not intended to suggest that the environment is the most important aspect of responsible corporations, but simply that some of the frontier writings lean this way.

Moving in the Right Direction

An image suggested by R. Buckminster Fuller will be helpful in dealing with the abstractions of responsibility, accountability, and transparency. Fuller proposed the image of a ship whose great size gives it so much momentum that a huge rudder is required to steer it. One or even several people cannot push hard enough to turn such a rudder; therefore a little rudder, called a trim-tab, is attached to the large one. It is within human strength to operate the trim tab, which then turns the rudder, which turns the boat.

Figure VIII.1a is a simple application of Fuller's image to our subject, showing the corporation as the ship we want to steer toward responsibility. The "rudder" is accountability and the "trim-tab" is transparency. This is, however, still a collection of abstractions. To make it more concrete, Figure VIII.1b uses some poetic license to suggest three rudders—the three groups in society that are most directly able to influence corporations. None of these groups is much

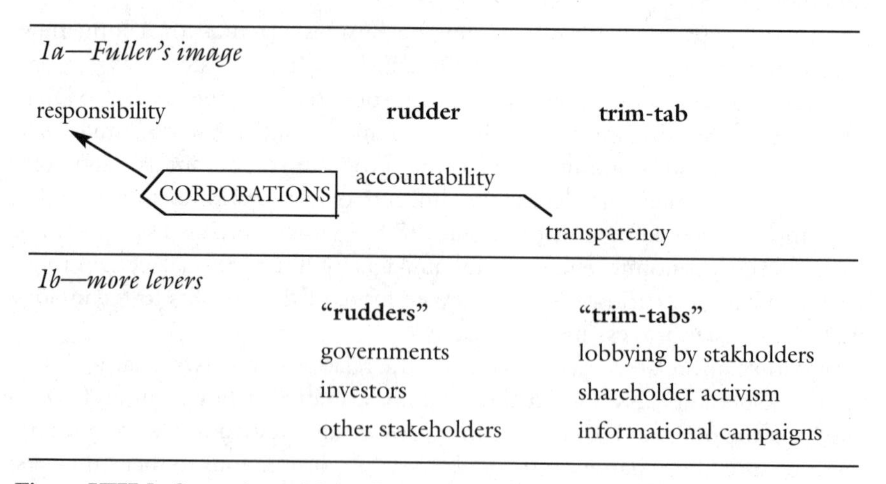

Figure VIII.1. Strategies to Move Corporations Toward Responsibility.

easier to move than the corporations themselves. Someone who sets out to move these rudders to affect corporations has a better chance of being able to take action through one of the suggested trim-tabs.

An awareness of the levers for action illustrated in Figure VIII.1 will help us understand the types of solutions that are commonly proposed to deal with failures of business responsibility. Solutions are apt to be proposed within two major categories: (1) regulate business ("command and control") and (2) change the goals and incentives that motivate business actions. The authors represented in this section generally start by focusing on the second approach, with government seen as essential to back up and enforce the desired regime of incentives. We are led, therefore, to ask what institutional, legal, or other changes or pressures can lead toward such a regime? The next three sections will explore three types of answers to this question.

Internalize the Externalities[7]

The first summary article in this section, of an article by **Robin Broad and John Cavanagh,** gives an overview of the corporate accountability movement, classifying its campaigns in terms of strategic goals, targets, methods, the initiating actors, and geographical scope. In the course of assessing the successes, challenges, and lessons learned from these campaigns, the authors survey a variety of approaches to internalizing costs externalized by corporations. These range from proposals to recharter corporations so that they can be dissolved if they do not act in the public interest, to direct actions (such as sit-ins) against individual firms or facilities.

This article draws our attention to the critical role of governments, which can impose specific obligations and restrictions on corporations. Unfortunately, the literature on corporate accountability gives relatively little attention to the political science issue of how governments themselves can be motivated to make the right laws and regulations, often simply assuming effective lobbying by other stakeholders. This is the only article summarized here in which we see any specific examples of this trim-tab.

Broad and Cavanagh also introduce a second category of direct influences on corporations: their investors. Investors can move corporations through their decisions on where to invest their money and also in the "voice" that they, as the ultimate owners of most corporations, can exercise on corporate decision-making. Religious groups have taken leadership in this area. An example is the United Methodist Board of Pension, with $9.7 billion in assets, which, after two years of filing shareholder resolutions with Delta Airlines, "finally won a commitment from the airline's management to publish a report on equal employment and diversity practices."[8] Other investor groups have successfully lobbied corporations to diversify their boards.

Investors are, however, a large and diffuse group: what is the trim-tab for this rudder? An important piece of the answer may be in what could most generally be described as empowerment. As Alinsky notes below, the investor responsibility movement offers to people who own stocks (a growing proportion of U.S. citizens, through pension and insurance funds, if not directly) a means to make their beliefs heard. As has been shown in relation to the anti-slavery movement of the nineteenth century, moral indignation does not turn into action until people believe that their action can have an effect.[9]

As described by **Peter D. Kinder, Steven D. Lydenberg, and Amy L. Domini,** shareholder activist groups such as the Interfaith Center on Corporate Responsibility (ICCR) have offered shareholder resolutions as ways to increase shareholder responsibility. After several decades of this prodding, investors are increasingly recognizing that they have the option—indeed, it may be regarded as a moral responsibility—to consider the issues that have been put up for shareholder votes. Whether the exercise of shareholders responsibility is increasing because investors are becoming better educated about the impacts of their actions, or because they are responding to fears of reduced share value following effective public relations, this movement is beginning to make itself felt in a number of ways (Krumsiek 1997; see also other articles in a special issue of the *Journal of Investing,* "Socially Responsible Investing," 6, 4, Winter 1997).

Examples of successful recent shareholder actions include the agreement of American Airlines to endorse the Coalition for Environmentally Responsible Economies (CERES) principles of environmental responsibility, and the decision of Baxter International (the world's largest health-care products manufacturer) to phase out PVC materials in its intravenous products. However, one lesson from the history of the movement is that significant victories may not come quickly. The impact of divestment from companies doing business with South Africa during apartheid is generally credited as significant in the dismantling of that regime; but more than three decades passed between the beginning of that movement in 1967 and its fruition in 1991 (see Massie 1997).

Saul Alinsky was the conceptualizer of an early move from the relatively passive strategy of divestment to the more active mobilization of shareholders to support social reform through their voting power. In his 1965 campaign against Eastman Kodak, Alinsky mobilized inner-city minority and church organizations in Rochester, New York, as well as major foundations and union retirements funds. Though the results were not earthshaking (Kodak allocated a few hundred jobs to poor, mostly minority, candidates), Alinsky felt that he had found a new tool that had potentially greater impact than investor decisions simply to buy or sell shares:

> Proxies can be the mechanism by which [the middle class] can organize. . . . Once organized around proxies they will have a reason to examine, to be-

come educated about the various corporate policies, both domestic and foreign, because they can now do something about them. . . . The way of proxy participation could mean the democratization of corporate America (Alinsky 1972).

Into the early 1990s business leaders generally regarded shareholder resolutions as nonsense engendered by nuts. Indeed, some individuals and organizations do put forward impractical, irrelevant, or unsustainable proposals. A stockholder who owns shares in many companies may be daunted by the task of sorting out the good from the nutty ideas. Fortunately, there is the beginning of a promising trend for portfolio managers themselves to offer to take a more socially responsible position on the resolutions that come up each year in relation to their clients' holdings. This sort of service, which is still in an early stage of development, could take several forms: for example, voting with ICCR on all issues; voting to support a defined set of proposals, such as workforce diversity, exclusion of child labor, or adoption of the CERES principles for environmental responsibility; or following an agreed-upon philosophy of corporate governance and behavior.

The Social Investment Forum estimated in 1999 that more than $2 trillion is now involved in socially responsible investing—an increase of 82 percent since 1997.[10] More than half of this is in portfolios whose exercise of responsibility comes through portfolio selection (to screen out "bad" companies—as in the South Africa divestment movement—or, more proactively, to screen in "good" ones). Kinder et al. describe other forms of socially responsible investing, including more active approaches such as community investing. In the latter, investors may accept below-market rates of return on loans to groups that sponsor housing, job creation, and other kinds of economic development in a given community. This differs in several ways from both guideline portfolio investing and shareholder activism: it may establish more direct involvement between the investor and the activity supported by his or her funds; and it is not deterred by the possibility of lower returns.

Aside from some community investors, most socially responsible investors have sought at least a reasonable chance of earning competitive returns. As reported by **George A. Steiner and John F. Steiner,** the evidence suggests that social and environmental screens average out to a neutral effect. Others claim that the average effect is positive. An interesting summary on this subject is provided by John B. Guerard, Jr., who concludes that portfolios that screen out negative environmental impacts, nuclear energy, alcohol, tobacco, and gaming derive on balance higher returns than those from unscreened portfolios. "The only social screen that consistently costs the investor returns is the military screen" (Guerard 1997, 31).[11]

Revealing Environmental and Social Feedbacks

The second category of ways to translate social and environmental harms and benefits into signals that will be felt by firms focuses on the reality that these harms and benefits are, in fact, already affecting the bottom line. When firms come to understand this—when they have a more sophisticated appreciation of their own operations and of the feedback loops between their own success and the health of their human and physical environment—they will be motivated to decrease the harm they cause, and increase the benefit.

Michael Porter of the Harvard Business School is a leading advocate for the idea that many environmental costs already are internal and will be properly dealt with when firms recognize this fact. Assuming a world in which citizens and governments are clearly trending toward ever-more stringent environmental standards and regulations, **Michael E. Porter and Claas van der Linde** argue forcefully that many of the environmentally destructive activities now being ruled illegal are also economically inefficient.[12]

There is an immense literature on how to design government laws and regulations so that they have their desired effects with minimum negative side effects. Porter and van der Linde, giving a taste of that literature, stress that government regulations have often in the past been designed as if to minimize the potential beneficial effects. Suggesting how this can be done better,[13] they offer a dual message: environmental regulations may be good for firms, or at least not as bad as they think; and, in any case, environmental regulations, as society's most obvious means of internalizing environmental externalities, are here to stay.

Ditz et al. (discussed below) note that "As firms come to terms with current environmental costs, they will appreciate that the boundary between private costs and social costs is porous and moving. Other environmental costs, now borne by society, will exert a growing influence over the decisions made within companies" (Ditz et al. 1995, 44). Corporations, and some writers about corporations, would like this shift to occur through voluntary self-regulation, as an alterative to government regulation—the latter being, in most cases, impossibly costly without industry cooperation. This option is explored in depth in a recent book by sociologist **Severyn T. Bruyn.** Bruyn describes how corporations and industry groups can and do self-regulate, what this possibility means for the development of a more "civil economy," and how it can be fostered. Self-regulation includes a number of the voluntary initiatives that have been welcomed by many as softening the differences between corporations and, especially, environmentalists: initiatives such as ISO 14000, and various labeling and certification schemes.

Bruyn emphasizes what Ditz et al. call the "porous and moving" boundary between what has been traditionally defined as the public realm, where society-wide interests are represented, and the private realm of business, with profit-maximization assumed as its only goal.[14] At the same time, Bruyn is in agreement with virtually all other authors quoted here on two key points: the profit motive is not going to go away, and appropriate government regulation will continue to be an essential spur to keep corporate interests converging with broader interests.

A focus on regulation, whether it is initiated by government or carried out by firms, is effectively complemented by the focus offered by **Allen White** on codes of conduct. Examples include the codes developed by CERES (Coalition for Environmentally Responsible Economies); International Labor Organization (ILO) conventions on international labor standards (ILO 1998); the Keidanren Charter for Good Corporate Behavior developed by Japan's Keidanren organization in 1996; and the Social Accountability Standard, SA 8000. Such codes, backed up by accounting, auditing, and reporting, are the keys that can lock in transparency and accountability.

An accounting system requires a company to gather and organize data on its performance. Financial accounting systems have been developed in the United States as a system of communication among three groups: investors, who need information about corporations' financial performance; the SEC, which protects the interest of the investors by setting standards for accounting as well as for auditing systems that check that the accounting data is accurate; and firms, which collect the required data and report it out to the SEC and to their investors, and which also benefit internally from improved self-knowledge. With an enlarged definition of corporate stakeholders, goals for accounting, auditing, and reporting have expanded significantly—most notably in the area of environmental accounting.

A good description of what is involved in environmental accounting may be found in the overview to a World Resources Institute publication, *Green Ledgers: Case Studies in Corporate Environmental Accounting* (Ditz et al. 1995; see also Owen 1993). As described by these authors, the concept of "full cost accounting" has been adapted to a specifically environmental meaning: "the practice of introducing environmental costs once considered external into corporate decision making." (Ditz et al. 1995) This conception encompasses all of the private and social costs generated throughout a product's life cycle, from raw material extraction to product disposal. Managers who adopt this approach may find that products with lower environmental costs have been subsidizing those that generated higher costs at some point along their lives.

It is important to note the distinction between environmental costs actually borne by the firm versus those that are externalized onto society. In traditional accounting only (but all of) the former are supposed to be noted. However, conventional accounting practices can overlook significant environmental costs, for these are frequently indirect or dispersed throughout a business, or can appear long after decisions are made. SEC-type standards require firms to account

Box VIII.1. Green Accounting

According to a study by the Institute for Environmental Management and the accounting firm KPMG, 35 percent of the world's 250 largest corporations now issue environmental reports. Companies are voluntarily embracing "green" reporting because it makes good business sense. Not only does public reporting push companies to be more disciplined about their environmental performance, which, in turn, reduces their environmental risk, it also creates positive PR. Good green reporting can serve as a differentiator in the war for talent—people like working for socially responsible companies—and it can make a company more attractive to customers and investors as well. Moreover, because green reporting puts all business practices under scrutiny, it often helps managers identify cost savings, and even new business opportunities.

"Trend: Green Reporting,"*Harvard Business Review* January–February 2000.

for those costs that would appear if the firm operated in a social vacuum. Full cost accounting has a better chance of including costs that are, or will in be the future, brought back onto the firm's ledgers, as public relations and legislation increasingly cause firms to take responsibility for the costs they had previously externalized.

In the areas of accounting and reporting there is a longer history and more sophisticated development with respect to the environment than in relation to the other parts of what Elkington (1998) calls the triple bottom line of environmental quality, economic prosperity, and social justice. It may turn out that the business benefits for responsibility in the last two areas are not as evident as, or show up in different ways from, the benefits to environmental responsibility cited by, for example, the *Harvard Business Review* or by Porter and van der Linde. However, for all aspects of corporate responsibility, advances in accounting, reporting, and auditing are critically important for providing information both to firms and to stakeholders. As depicted in Figure VIII.1, this transparency trim-tab will in turn promote accountability. When these standards become accepted practice—whether through laws, codes, or widespread business norms—they help to lock in other types of progress in corporate responsibility.

Making the Future Matter to Corporations

Responsibility cannot be imposed entirely from the outside; if the goals of corporate responsibility are to be achieved, people in business—owners, managers, workers—must make some kind of moral commitment, accepting responsibility for their firm's impact on the world. This implies an ethos—a widely and deeply held belief—that corporations should bear the costs they generate, to the extent of not causing harm to any entity outside of the corporation, and undoing the harm they have caused in the past. Hoffman (1997) provides an excellent treatment of the evolution of corporate norms of environmental responsibility. He emphasizes that the set of options that corporations consider in responding to environmental demands is largely determined by what other corporations are doing, and that, over the last three decades of the twentieth century, waves of industry responsiveness tracked the ups and downs of public concerns about environmental issues more closely than they followed either their own cost structures or trends in government regulations. Nevertheless, a positive corporate ethos, though essential, cannot be the only solution. As we have noted earlier, it must be stimulated and reinforced by an environment in which firms will perceive their interests to coincide with broader social interests.

The complex issues involved in encouraging corporations to transcend the short-horizon, next-quarter view of the world are explored in *As If the Future Mattered: Translating Social and Economic Theory into Human Behavior* (Goodwin 1996). In that book Michael Porter stresses the problems created by an institutional environment that encourages short-term thinking. Noting that "private and social returns will tend to converge more in the long term than in the short term" (Porter 1996, 19), he lays out an ambitious agenda for altering the legal and normative framework for business, to encourage longer time horizons. This approach is strengthened and extended by a package that includes internal accounting, reporting to interested outside parties, and external audits and monitoring.

Our summarized section of the business textbook by Steiner and Steiner contrasts the business norms of the nineteenth century with those that are emerging in the twenty-first. It associates with the obsolete norms an economic theory, prominently advocated by Milton Friedman, that regards the concept of business responsibility to stakeholders (other than stockholders) as "a fundamentally subversive doctrine."[15] The reality that Steiner and Steiner see as setting the stage for current and future norms is that "a manager operates within a set of economic, political, cultural, and technical constraints. They are powerful, and as societal expectations change corporate actions must conform. This is the equation of legitimacy" (Steiner and Steiner 1997, 116).

The legitimacy equation explains some of the strength of the rudder, in Figure VIII,1b, that was labeled "other stakeholders." Corporations often perceive these as the collection of forces that make up public opinion. Corporations pay attention to public opinion for a variety of reasons, including how it may affect sales, how it may affect the regulatory environment, and its personal impact on corporate decision makers (to the extent that public opinion filters into the social circles in which they and their families live).

Steiner and Steiner emphasize the convergence between the long-run interests of business and of society, noting that firms cannot thrive in a badly degraded environment or in a society riddled with violence, poverty, and ignorance. This point, though not in much dispute, does not necessarily change corporate behavior, for it leads to a classic free-rider situation. If some large corporations support the health or education of workers and potential workers, why should not the rest sit back and enjoy the benefits? John Elkington does, indeed, note that

> [m]ost companies continue to enjoy a "free ride" at the expense of both those pioneering companies that have made a start at internalizing costs and of the wider environment. However transparent the operating environment in which it does business, a company ultimately must face financial reality. If internalizing a range of triple bottom line costs starts to threaten stock market valuations, salaries or jobs, then only regulations will swing the argument. So the challenge is not simply one of making the costs imposed more visible, a task which accountants can certainly help, but also of forcing all economic players to internalize their fair share of costs. (Elkington 1998, 183)

Bruyn has noted that corporations recognize the advantages of enforceable common standards that will preserve their social legitimacy without encouraging free riders. When they cannot achieve this through their own efforts, they sometimes quietly ask for assistance in the form of government regulation. There is, at the same time, a record of corporate opposition to standards imposed on them from the outside.

If we focus on the reasons for corporate opposition to regulation, we can note a difference between objections to regulations that might hurt a firm's competitiveness vis-à-vis other firms, versus objections based on a fear of absolute cost increases. The first kind of concern can be assuaged when it is shown that all competing firms are affected equally. The level playing field argument is especially persuasive to those who expect to be affected "more equally" than others; these include firms that have more technical capacity to respond in an innovative manner to the new demands or that have more financial capacity to absorb the additional costs of monitoring and reporting. However, the large firms that possess these capacities are precisely those that are most likely to be engaged in international competition, and their appreciation of possible domestic advantage is often overshadowed by their fear of losing ground against foreign competitors that face less stringent standards. These—led by the TNCs—are the firms that have the clout and the funds to lobby most effectively against increased regulation. Yet the picture is very mixed: "most of the companies responding effectively to the transparency revolution operate internationally" (Elkington 1998, 164).

Sailing Upstream, Toward Responsibility

Given the great power of corporations to shape the social and physical world, optimism or pessimism about humanity's future depends on where in the corporate world we look. On the one hand, there are indications that pressure from governments, investors, and other stakeholders has caused real movement toward corporate responsibility. As examples, some energy companies are starting to take seriously the need to convert to renewable sources; some automobile manufacturers have ceased denying the reality of global warming,[16] and some apparel manufacturers are accepting in-depth, third-party evaluations of their Third World suppliers' social and environmental responsibility.

On the other hand, we might see the future in the corporations that, facing a choice between profits versus the triple bottom line, ignore the latter, or simply try to "greenwash" their image. Agribusinesses continue to develop technologies and sales strategies with little or no regard to the viability of small farms or threats to the health of humans or other biota; many private producers or operators of public goods such as health, education, waste disposal, and prisons lobby for government regulations that will increase their revenue stream (including the number of prisoners) rather than solving the social problems with which they are charged; and a stunning cascade of mergers and acquisitions poses an ever-greater threat to the balance of power between for-profit activities and the rest of society.

There are two areas in which it will likely be most difficult to get corporations to take responsibility for their impacts. One of these areas concerns the social and cultural effects of their products, both in themselves and in the advertising that supports them. The other concerns the myriad ways in which the economic power of huge, especially transnational, corporations, translates into political power. Given the existing, often pernicious political power being wielded in what the corporations see as their interest, it is hard to take seriously Milton Friedman's concern that the efforts of firms to be socially responsible would put them into the political arena *de novo,* or his tenacious faith that competition is sufficient to keep them out of it.[17] For too long, economists have worried only about economic power (especially monopoly power, with its cost to the consumer in setting high prices), ignoring the rise of corporate power in the cultural and political arenas.

Steiner and Steiner, addressing future corporate managers, stress the need for a new paradigm in economic theory: one that could accommodate the historical changes under way in the area of corporate responsibility. Increasing transparency continues to make corporations more accountable to a broader group of stakeholders. The paradigm change is, in fact, under way; corporations are—however slowly and unevenly—accepting increased responsibility for social, economic, and environmental sustainability. Now the critical questions are:

How fast can this change gather momentum? Can it successfully contend with growing corporate size and power? And what will come of the deepest of all conflicts between corporate profits and social and environmental health—the fact that much corporate production is intrinsically unsustainable? The answer to these questions will lie in the efforts of stakeholders and the groups that work to mobilize them to make business serve society—rather than the other way around.

Notes

1. A negative externality incurred by a firm is a cost that it generates but does not pay for, shifting the burden to others in society or to the natural environment (in which case the cost is likely to be born by people in the future, if not in the present). To give an example of what it means to internalize an externality: if a government mandates that firms are responsible for the environmental effects of their product during its entire life, from production through disposal, then costs of waste handling that had been paid by municipalities or by consumers become "internalized" as part of the firm's total production cost. The firm now has a cost incentive, previously lacking, to redesign the product for disassembly and recycling or reuse.
2. "More than $2 billion is spent annually on advertising directed at children, over 10 times the amount spent just 10 years ago. At three years of age, before they can even read, most American children start making specific requests for brand-name products." *Co-op America Quarterly*, No. 50, Spring 2000, 17.
3. Obvious examples in the United States are the beverage industry, using its power to defeat bills for bottle recycling, or the automobile industry, using its ability to direct transportation planning for over half a century. *Cf.* Blumberg and Gottlieb 1989; Adams and Brock 1987.
4. Severe critics of corporations include Derber (1998); Dugger (1988); George (Appendix and Afterword to *The Lugano Report* [1999]); Greider (1997); Heiman (1997); Korten (1995, 1999); and Mokhiber and Weissman (1999). Most of these authors tend to lean toward the end of the spectrum that suggests that radical change is necessary because there is little hope in the various kinds of reform from within that are the focus of this essay. Critics of corporations, perhaps because they find a readier audience in the broad public than in academia, sometimes adopt a populist style that can make it difficult to compare this side of the argument with the academics who express more favorable or optimistic views.
5. Among the many alternative ideas that could not fit into this essay, an especially important one is the movement for employee share ownership. A good overview of this topic is Kelly et al. 1997.
6. The members of ISO are national standards-setting bodies, including both government agencies and NGOs. Although the standards are voluntary, compliance with ISO 14001 (guidelines for the creation of environmental management systems) can be certified by outside auditors. Some 5,000 companies have adopted ISO 14001, mostly in Europe and the Far East. (UNESCO 1999, 365.)

7. See Note 1 for a definition of externalities.
8. The *United Methodist Reporter*, February 13, 1998.
9. See Haskel 1985.
10. This was reported in Co-op America's *Financial Planning Handbook*, 2000 edition, which adds that "nearly $1 out of every $7 under professional management in the U.S. is part of a responsibly invested portfolio" (including screened portfolios, which exclude, for example, tobacco, gambling, weapons, alcohol, and corporations with bad human rights or environmental records). Additionally, "an estimated $922 billion is controlled by investors with social goals who either sponsor shareholder resolutions, vote their proxies on the basis of socially responsible goals or communicate with problem companies" (25).
11. Without going into this intriguing exception in detail, two likely explanations concern the technology stocks that might be eliminated in a military screen, and the profitably cozy relationship between large military suppliers and governments.
12. For a variety of views on the "Porter hypothesis," see, for example, "The Challenge of Going Green" in the *Harvard Business Review* (1994); see also Gallarotti 1995 and Portney 1994.
13. Suggestions for improved government regulations are offered in the summarized article by Porter and van der Linde, as well as in Porter (1996). *The Civil Economy*, by Severyn Bruyn (2000), of which two chapters are summarized in this section, provides additional suggestions for a lighter but more effective regime of government regulations.
14. See Bowles and Gintis (1986) (summarized in Part 4 of Ackerman et al., *The Political Economy of Inequality*, Volume 5 in this series). Charles Derber, in an extended comparison between the present era and the Gilded Age, a century ago, notes how "corporations became private governments with quasi-public powers, while government itself became a servant of private interests" (Derber 1998, 25).
15. Friedman, 1962, 133; quoted in Steiner and Steiner, 118.
16. The Global Climate Coalition, which tried to discredit concern over climate change, has been unraveling; at the time of this writing, for the past few months the GCC has lost a member a week—including giants like Ford and GM.
17. For descriptions of the political activity and impact of corporate lobbying and political donations, see Part IV of Ackerman et al., *The Political Economy of Inequality* (Volume 5 in this series), and especially summaries of work by Dan Clawson, Alan Neustadtl, and Denise Scott; Walter Adams and James S. Brock; and Jerome L. Himmelstein.

Summary of

The Corporate Accountability Movement: Lessons and Opportunities

by Robin Broad and John Cavanagh

[Published in the *Fletcher Forum of World Affairs*, 23, 2 (Fall 1999).]

In recent decades thousands of campaigns have been mounted to encourage or force corporations to improve the social, environmental, and economic impacts of their activities. This article, emphasizing especially the attempts to affect transnational corporations in developing countries, identifies these activities as part of a global corporate accountability movement. The article seeks to encourage the various nongovernmental actors in this movement to interact more, to learn from one another, and to build on each other's strengths.

Goals and Actors in the Movement

Nongovernmental organizations (NGOs) in the corporate accountability arena include environmental and religious groups, labor unions, farmers, consumers, and investors. They derive their ability to make a difference from various sources. Some, like unions and religious and environmental groups, are able to mobilize millions of constituents. Investor groups, by contrast, wield power as shareholders. These different actors pursue a variety of goals through a variety of methods.

One pole on the spectrum of strategic goals is a desire to change the nature of corporations fundamentally, or even to get rid of them entirely. Near this pole, for example, we find the rechartering movement in the United States, which seeks to revive the possibility of revoking the charters of corporations that fail to advance the public good.

In the middle of the spectrum are efforts aimed at changing the rules that govern corporate behavior. A successful effort along these lines was the Organization for Economic Cooperation and Development (OECD) Convention to Outlaw Foreign Bribery, led by an international NGO, Transparency International. The convention, which went into force in 1999, requires the signatory governments to impose criminal penalties on companies that bribe foreign officials while soliciting business. Since the world's most dynamic economic actors are transnational corporations (TNCs), effective rules for corporate behavior need to be designed and enforced internationally. For this reason, efforts in this realm often seek to involve the United Nations or else to create a new global agency, such as a global antitrust authority.

At the other pole are efforts aimed at reforming abusive practices of individual corporations. Thousands of such campaigns have been organized, some on a local scale, such as that mounted by the Southwest Network on Economic and Environmental Justice against Intel in Albuquerque. Other protests against specific corporations have attained a global scope, such as the campaign against Union Carbide after the Bhopal accident, or against Shell in Nigeria.

The nongovernmental groups that have taken on these goals often see themselves as a necessary counterweight to government-established protections for corporate business as usual, such as those implemented by the World Trade Organization. "Indeed, current attempts to influence corporate behavior may arguably be categorized as a subset of a larger universe of citizens' organizations working to stop, slow down or reshape the path of economic globalization in ways that promote democracy, equity and sustainability." [153]

Methods Used to Achieve Accountability

The methods used vary as widely as the initiating actors or their goals. They "run the gamut from confrontation (kicking Coca Cola or Kentucky Fried Chicken out of India) to engagement (the Environmental Defense Fund convincing McDonalds to change its packaging materials)." [153] Greenpeace is especially well known for its direct-action campaigns, but other organizations also have picketed, occupied corporate offices, lain down in front of bulldozers, or plugged effluent pipes. Lawsuits may be seen by corporations as equally aggressive, and are sometimes even more effective.

Efforts to inform or alarm consumers may be a little less direct than a physical approach to a corporation's headquarters, but they can be extremely effective. The most immediate way in which consumers can make known their disapproval of a company's practices is to boycott its products. At the same time that they wield the stick of citizen disapproval in the form of boycotts, organizations seeking corporate accountability also use carrots such as labels that will steer consumers toward products produced in a responsible way. Examples are Europe's green seal program for products passing an environmental test, and labels that announce a product's recycled content.

While some campaigns seek leverage through consumers, others use the influence of investors and/or shareholders. Pressure was applied at many levels to persuade companies not to do business in South Africa under apartheid. "Participants used the pressures of selective investment, the power of government procurement contracts, divestment, and other measures to put pressure on the apartheid regime. . . . The U.N. concluded that these actions ultimately convinced two-thirds of U.S. companies to sell off equity shares in their South African operations." [154]

For a long time corporations regarded shareholder resolutions as strictly adversarial; however, in contrast to the early years of shareholder activists when corporations automatically fought their proposals or dismissed them as absurd, firms are increasingly likely to enter into dialogue with such organizations as the Interfaith Center for Corporate Responsibility, which may in a given year submit up to several hundred resolutions on a wide variety of issues.

The purpose of dialogue is, of course, to persuade corporations to behave more responsibly. Among the possible results, an important category is acceptance of a corporate code of conduct. In some cases codes are urged on industry associations (recent industry targets, for example, have been the apparel and footwear industries globally, and, in Europe, the toy industry). In other cases they are accepted by individual firms.

Governments can also affect corporate behavior through a variety of incentives or disincentives, including taxes or other influences on the corporate environment, and the creation of institutions to review products, processes, investments, transactions, and services before or after they appear in the market. The Generalized System of Preferences (GSP), operated by many governments with respect to import tariffs, can be used to express national preferences for responsibly produced products. For example, in 1984, labor, human rights, and religious groups convinced the U.S. Congress to amend the GSP system so that reduced tariffs would only apply to countries taking steps to ensure internationally recognized worker rights.

On an international level, there has been at least some progress in attaching labor and environmental issues to such trade agreements as NAFTA. Continuing campaigns involving governments on an international level include efforts to enforce the Basel Convention, which restricts international trade in hazardous wastes; and the effort by trade unions, with the assistance of some governments, to include a "Social Charter" in the regulations governing the World Trade Organization.

Effectiveness Criteria

"A major void in recent initiatives to make corporations more socially and environmentally responsible is that there have been shockingly few, if any, comprehensive attempts to assess the levels of effectiveness (or ineffectiveness) of different strategies." [161] In attempting to draw lessons from the record to date, the following criteria are useful.

Change in corporate behavior can be assessed on three levels. First, has the corporation stated an intention to behave more responsibly? Second, is there evidence that it has followed through on its stated intentions? And, third, have the changes in corporate behavior had real, positive effects on communities, workers, or the environment? Improvements in oversight by governments or NGOs can be similarly assessed in terms of new legislation or new mechanisms of nongovernmental oversight, along with the real impact of these changes.

With regard to changes in public attitude, it is important to ask whether a given campaign has had an educational effect. "Has broader public awareness or media attention led to actual changes in purchasing or consumption patterns?" [162] With regard to the actors seeking to affect change, campaigns may also be assessed with respect to their introduction of new actors or creation of new working alliances that can strengthen the corporate accountability movement.

Lessons and Challenges

Synergies can be created when different groups employ different tactics toward shared goals. In particular, "[c]orporations are much more likely to engage in dialogue or negotiation when they face large-scale negative publicity from direct actions, well-targeted media campaigns and other more confrontational tactics. Greenpeace and Public Citizen's confrontational opposition to NAFTA, for example, created more space for groups attempting to link environmental issues to the agreement." [163–164] There appears to be the greatest chance of succeeding with a corporation that already makes a claim to social responsibility, but there is some danger of sending the wrong message to corporations when public campaigns hit hardest at those that are readiest to engage.

Good use of media is often critical to the success of a campaign. Since simple, dramatic, easy-to-empathize-with issues are those that get most media play, "it might be advisable to open the door to the campaign with the most graphic issue and use the opening to educate around more difficult issues." [165] Campaigns that aim to ameliorate the impact of TNCs in the Third World work best when they achieve good North-South cooperation. At the same time, it is easiest to succeed with issues in countries that are of least strategic importance to powerful nations—especially the United States.

There are important disagreements between Northern and Southern advocates, including attitudes toward the inclusion of labor and environmental standards in trade agreements. Some Third World groups perceive linkages between trade and labor practices as a sell-out to the North. However, the existence of the International Labor Organization (ILO) has helped to keep workers' rights on the global agenda for eight decades and has achieved wide agreement, even across the North-South divide, over what constitutes basic worker rights. "By contrast, there is no governmental body where internationally recognized environmental rights and standards are negotiated." [165] Hence, while there is greater public awareness of environmental concerns, their inclusion in corporate codes lags labor issues by about a decade.

The work of the corporate accountability movement—fragmented and unselfconscious though it is—has achieved major changes in industries such as tobacco, arms, apparel, and footwear and has changed the way many corporations treat the environment and their workers. It is important that those involved in this movement be open to the possibility that their goals may need to evolve as some measure of success is achieved, or as possibilities and constraints shift. As corporations increasingly do sign on to standards of conduct, including agreements to report on their social and environmental impacts, such flexibility assumes a greater urgency.

Summary of

Socially Responsible Investing: Doing Good While Doing Well

by Peter D. Kinder, Steven D. Lydenberg, and Amy L. Domini

[Published in *Investing for Good: Making Money While Being Socially Responsible* (New York: HarperCollins Publishers 1993), Ch. 1.]

The History of Socially Responsible Investing

Humanity has had only a century plus a few decades to figure out how to deal with the extraordinary new force represented by the modern corporation. "Apart from armies, no one had ever put together human organizations the size of, say, U.S. Steel (now USX), which was created in 1901. Nor had anyone ever tried to concentrate so many functions—so many human relations, both internal and external—in a single structure." [15] Government and civil society is still reorganizing and redefining social and political relationships to respond to the challenge posed by the existence of these gigantic new economic forces.

The period from 1890 to 1917, called "the age of reform," generated mass movements that attempted to redress the balance between individualism and social needs. Government control had, in fact, begun to be exerted through the creation, beginning in the 1870s, of such regulatory agencies as the Food and Drug Administration. Civil society in the age of reform initiated a variety of crusades, fads, political movements, and other proposals for individual and collective action. One of these new directions was the beginning of the socially responsible investing (SRI) movement.

An even earlier phase of the SRI movement might be traced to the seventeenth-century Quakers, who refused to profit from war or the slave trade. However, it was not until the early twentieth century that this impulse confronted corporate power in its modern U.S. form. The Social Creed of Churches, adopted by the Federal Council of Churches in 1904, stressed most of the issues of workers rights that are still in the foreground today, along with a broad appeal for "the abatement of poverty" and "the most equitable division of the products of industry that can ultimately be devised." [14] However, from then until the 1960s the movement did little more than encourage churches and individuals to exclude "sin stocks" from their portfolios.

The late 1960s combined the fervor of the civil rights movement and a new sense of urgency about the environment with the unrest associated with the Vietnam War. Following the death of Dr. Martin Luther King, the divestiture movement—the effort to disengage university and other portfolios from companies seen to support South African apartheid—became an obvious outlet for a general sense of moral outrage. Then Ralph Nader took on General Motors, following up his book, *Unsafe at Any Speed,* with Campaign GM, which managed—for the first time in history—to put social issues on stockholders' proxy ballots. "U.S. churches and religious orders grasped the importance of what Nader had done. In 1971 they formed the Interfaith Center on Corporate Responsibility (ICCR), which has filed a torrent of shareholder resolutions ever since (17–18)."

Types of Socially Responsible Investors

Socially responsible investment may be carried out in several ways. One is guideline portfolio investment. Guideline portfolio investors may be individuals working through brokers, mutual funds, or other kinds of investment managers. What they have in common is the use of ethical guidelines, or screens, which affect their choice of investments. These may range from simple negative screens (e.g., eliminating all tobacco companies) to more proactive screens that seek out companies with, for example, good labor relations records. Social screens do not replace financial screens. While most investors prefer that their SRI portfolios perform reasonably close to the market average (a requirement that many have met or exceeded), others are willing to accept a below-market return for a worthy cause.

"Done quietly and individually, this type of investing has no effect whatsoever on publicly traded corporations." [3] However, the growth in the number and size of SRI mutual funds has not escaped the notice of corporations. Working with others to bring pressure for accountability, SRI investors have contributed to a social environment that, perhaps most importantly, has forced corporations to disclose more of their workings to an informed public.

Another type of socially responsible investor is the shareholder activist who, at minimum, votes on the resolutions for responsible corporate behavior that are put forth at annual meetings,[1] and, at maximum, writes and sponsors such resolutions. Religious groups such as the Interfaith Center on Corporate Responsibility and specific Christian orders have led the way in using shareholder resolutions to lobby corporations. Large pension funds such as TIAA-CREF or the New York City Employees Retirement System have also worked to convince corporations to divest from South Africa, to adopt the CERES (Coalition for Environmentally Responsible Economies) principles on the environment,[2] and so forth. So far, however, the pension funds have hewed to a narrower agenda than the religious groups, stressing especially the governance issues around the relationship between shareholders and managers.

There are many complementarities between guideline portfolio investors and shareholder activists. The issues developed by the latter often become screens used by the former. The two approaches together may be needed to get the attention of corporations. Their effectiveness is less in winning proxy votes—in fact, this is rarely achieved—than in getting corporations to pay attention to the issues they are raising. Even a small percentage of votes against management can make a company realize that it faces a potential public relations problem. Corporations see it as in their interest to prevent proxy fights, and increasingly often will agree to come to the table with the SRI group before a resolution goes to vote.

The third kind of SRI activity involves community investors. These groups deposit their money in community loan funds or credit unions, which in turn make loans that other credit institutions might turn down. These loans may be used to support small businesses in depressed areas, to support housing in districts considered poor risks, or to create jobs. Accion International, which loans U.S.-generated funds to microenterprises in Central and South America, is an example of the first of these approaches. South Shore Bank in Chicago, which channels funds from wealthy communities to support housing in poorer areas, is a good example of a housing-oriented community investment institution. The ICA Revolving Loan Fund in Massachusetts, financing worker-owned businesses, is a job-creating community investor.

"Conventional wisdom holds that you can't mix money and ethics. Conventional wisdom is wrong. Socially responsible investors have proven it so." [1]

Notes

1. Most commonly, such shareholders are voting on "proxy resolutions"—so named because, rather than attending the annual meeting, most shareholders take the option of authorizing a stated representative to vote as they direct.

2. The CERES Principles are a ten-point code of corporate environmental conduct, which, by the end of 1999, had been adopted by more that fifty corporations, including about ten *Fortune* 500 companies.

Summary of

Corporate Social Responsibility

by George A. Steiner and John F. Steiner

[Published in *Business, Government and Society: A Management Perspective. Text and Cases.* 8th ed. (New York: McGraw-Hill, 1997).]

While the concept and the practice of corporate social responsibility varies widely from nation to nation and from firm to firm, overall it is expanding to respond to stakeholders previously ignored and situations previously unknown. Modern business theory has its roots in classical economic theory. That theory has not evolved to keep pace with an implicit social contract in which business must respond to changing realities, needs, and expectations.

The Case Against Expansive Social Responsibility

Milton Friedman (1962) has been one of the strongest supporters of the classical view, stating that "Few trends could so thoroughly undermine the very foundations of our free society as the acceptance by corporate officials of a social responsibility other than to make as much money for their stockholders as possible. This is a fundamentally subversive doctrine." Friedman's argument begins with the assumption that only the owners (who, in the case of a corporation, are most often shareholders) have a right to determine the firm's goals. It is illegitimate for firms to undertake social actions that either reduce profits for the owners or raise prices for the consumers. Friedman also fears that social action by firms turns them into political as well as economic agents; the fusion of the two kinds of power represents a danger to democracy. Friedman expects that the market mechanism, freed as much as possible from government or civil society interference, will keep economic power fragmented but can still serve as an adequate counterweight to other sources of political power.

Others share Friedman's belief that corporate managers are not the right people to solve social problems. Representatives of the political left believe that such problems should be left to government. Members of the political right would leave all problems in individual hands. Promoters of civil society, such as Peter Drucker, look to the third sector. Yet Drucker (1973) also notes that "a healthy business and a sick society are hardly compatible." This observation is behind the slowly growing belief among top executives that social responsibility is, in fact, in the self-interest of corporations.

A Broader View of Corporate Social Responsibility

Modern views on the corporation in the United States—articulated both in academia and in the corporate community—find a number of reasons to accept more social responsibility. One such reason is that corporations have an ever-greater impact on society as their technological and economic power grows. This observation is supported by the definition of an expanded group of stakeholders who are likely to be affected by corporate actions. Going beyond stockholders, these include customers, employees, governments, and the communities where the firm does business. These groups are increasingly aware of corporate impacts, as, for example, advancing science reveals the presence and the effects of carcinogens in industrial effluents; or better statistics make it harder to ignore racial discrimination in hiring. As these trends change the expectations of the stakeholders, firms must adapt or lose legitimacy.

Corporate owners and managers have seen plenty of cases during the last two centuries in which corporate refusal to address growing social issues has provoked government action. The lesson seems to be sinking in that self-regulation is less painful than new laws. Corporate leaders also recognize that, in the long run, "violent cities, deteriorating schools, pollution, poverty, and other problems are the ingredients of economic stagnation, not corporate welfare." [116] All the same, the trend to social responsibility would be hard to maintain if the more responsible companies were clearly at an economic disadvantage.

Between 50 and 100 scholarly studies have compared the performance of corporations with a high reputation for social responsibility to those who have no such claims. Overall, the results suggest that social responsibility confers neither a great reward nor a great cost. However, results of the individual studies vary widely, partly due to the difficulty of defining social responsibility, or even reputation. There is also a possibility that the results are skewed by the fact that the more profitable companies are those that can best afford to act responsibly.

Of what does responsibility consist?

"Business must be considered predominantly an economic institution with a strong profit motive. Business should not be expected or required to meet noneconomic objectives in a major way without financial incentives. . . . Social responsibility may complement, but cannot replace, the profit motive." [126] With this said, corporations do not have the right to externalize their costs onto others; they should minimize externalities and pay to compensate for those they cannot eliminate.

Public policy may be seen as a guide to how legitimacy is conferred—or withheld—in a particular national context. Going beyond formal regulations, the stakeholder perspective that is being developed in scholarly writings suggests that compliance, alone, is not enough; the benefits as well as the burdens of corporate operations should be distributed fairly among the various stakeholders.

The issues that are especially relevant to any particular firm will depend on firm characteristics such as its size, products, manufacturing processes, marketing techniques, and places of operation. "Thus, a multinational chemical manufacturer has a much different impact on society than a small, local insurance company and its social responsibilities are both different and greater." [126] The perception and acceptance of responsibilities will also vary according to local problems, culture, and expectations.

As an example, the behavior of Japanese firms reflects historical and cultural as well as economic realities. The Emperor Meiji's decision, in the mid-nineteenth century, to modernize Japan, stemmed from a feeling of national humiliation at the hands of industrialized foreign nations. The national purpose that stemmed from Japan's emergence from isolation was jointly carried out by government and business. Business was seen to have a clear goal: "to make the country dominant and ensure preservation of the Japanese race in a hostile world." [122] This rational goal was combined with a Confucian tradition that spells out duties in terms of direct relationships. Thus it has seemed normal for Japanese businesses to take responsibility for their employees by building housing, roads, and other public facilities for them. History and culture give less support to the claims of other stakeholders, such as consumers and the environment; however, there is a slowly growing movement to broaden corporate responsibility in Japan.

In Europe, by contrast, there has been no such identity of purpose between government and business; and labor unions, too, have tended to assume a conflict between their interests and their employers'. Government regulations have focused on labor issues such as wages, working conditions, and employment security. "In France, for example, companies must spend 1 percent of total wages on worker education programs. The French parliament also required in 1977 that large companies draw up an annual social report for the government, focused mainly on employee relations." [122] Other social issues in Europe are left to governments, which levy higher taxes than the United States in order to fund far-reaching social programs. As compared to the United States, "European companies are more likely to believe that they have met their obligations by paying taxes and following regulations." [123]

India is an example of a less-developed country whose history and culture emphasizes strong corporate social responsibility based on Mahatma Gandhi's doctrine of trusteeship. In other less-developed countries it seems evident to many that the primary duty of business is to promote economic growth. Foreign multinationals, with greater resources and experience of greater expectations in other places, are more likely to take on social responsibilities. "However, there is a worldwide movement, now confined mainly to industrialized nations but spreading, to encourage voluntary responsibility." [125]

Summary of

Toward a New Conception of the Environment-Competitiveness Relationship

by Michael E. Porter and Claas van der Linde

[Published in *Journal of Economic Perspectives,* 9, 4 (Fall 1995), 97–118.]

Standard economic doctrine assumes that firms exist in an equilibrium in which they have already optimized the products and services they offer and reduced production costs as much as possible. In such a static model economists do not expect to find additional cost savings waiting to be exploited by firms—any more than they expect to find a $10 bill lying on the ground. If it had been there, it would already have been picked up. This doctrine views regulation always as a burden, imposing some additional cost above the minimum that has theoretically been reached.

In reality, however, this static model of competition has become increasingly obsolete. Today competition is dynamic and based on innovation. In recent years companies have been discovering that, when their attention is focused by properly designed environmental regulations, their innovative responses can improve, or at least do not hurt, their ability to compete with other companies domestically and internationally.

Creative Responses to Regulation

Environmental standards, if well designed, can trigger innovation that may partially or more than fully offset the costs of complying with them. They can improve a firm's competitive position in a number of ways. They may direct the firm's attention to the cost of incomplete utilization of resources and encourage the collection of more information about wastes—for example, by increasing the number of activities that are monitored, or by installing higher-quality systems and devices for monitoring and reporting. When companies improve their measurement and assessment methods to detect environmental costs and benefits, they raise corporate awareness and increase the incentive to encourage and reward innovations that enhance resource productivity.

Such innovations may reduce product cost by eliminating expensive materials, reducing unnecessary packaging, or simplifying design. This was the result, for example, of a 1991 Japanese recycling law, which led firms to emphasize reducing disassembly time. Innovative responses to appropriately designed regulations can also change production processes in the direction of better material utilization or finding valuable uses for production by-products. Discharges, scrap, and emissions should be regarded as clues to opportunities for cost reduction. Until corporations accept this approach, pressure must be applied through regulations.

The most limited type of response to environmental regulation involves "end of the tailpipe" solutions, seeking ways to deal with pollution problems after they have occurred. Innovations aimed solely at this goal may reduce the cost of complying with regulations, but are unlikely to achieve more than that. The responses that are more likely to serve the aims of the company, as well as of society, are those that take two steps beyond such pollution control measures as waste processing and waste disposal.

The first step is *pollution prevention,* for example, using material substitution or closed-loop processes to limit the waste generation. Firms that are sensitized to the need to understand their environmental impact may acquire valuable information about their production processes. "A recent study of process changes in 10 printed circuit board manufacturers, for example, found that 13 of 33 major changes were initiated by pollution control personnel. Of these, 12 resulted in cost reduction, eight in quality improvements, and five in extension of production capabilities." [106]

The second step—important for regulators as well as for firms—is to reframe environmental issues in terms of *resource productivity,* which is "the efficiency and effectiveness with which companies and their customers use resources." [106] When this is the focus, it becomes evident that wastes generated by a firm are symptoms of an avoidable opportunity cost, whether it derives from wasted resources, wasted efforts (e.g., avoidable downtime), or diminished value of the final product.

Better Regulations Will Achieve Better Results

Unfortunately, under the prevailing economic assumption of an inevitable tradeoff between social benefits and private costs, an adversarial relationship between regulators and regulated has often resulted in requirements that imposes higher-than-necessary compliance costs. If competitiveness is to be better aligned with environmental improvement, environmental standards must be designed to foster innovation in products and production technologies. This requires that regulations focus on outcomes, not technologies.

Standard-setting agencies should not try to second-guess what industry might invent; environmental rules need to be phrased as goals that can be met in a variety of ways. Moreover, regulations should be designed to apply to the latest practical stage in the production chain that goes from raw materials and equipment to the producer to the consumer. This will maximize the producer's flexibility to find opportunities for innovation upstream of the point of regulation. Additionally, regulations should stress the use of market incentives.

Environmental regulations should strive for clarity and good coordination.

When it is clear what the regulations are, who must meet them, and how long they will be in effect, industry is more likely to address them through fundamental innovation rather than adopting incremental solutions or trying to delay or relax their implementation. There is also a need for appropriate coordination between industry and regulators, among regulators at different levels and places in government, and among regulators in different countries.

Industry should participate in standards formulation from early on in the process as is common in many European countries. Companies should not need to deal with multiple regulatory bodies posing inconsistent goals and approaches. On the national level, regulatory policies should be consistent with the practices of other countries—and ideally be slightly ahead of them.

> This will eliminate possible competitive disadvantages relative to foreign competitors who are not yet subject to the standard, while at the same time maximizing export potential in the pollution control sector. Standards that lead world developments provide domestic firms with opportunities to create valuable early-mover advantages. However, standards should not be too far ahead of, or too different in character from, those that are likely to apply to foreign countries, for this would lead industry to innovate in the wrong directions. [114]

Governments can play other useful roles in aligning business interests with the social need for environmental protection. They can help to create demand pressure for environmental innovation, for example, by supporting eco-labeling. They should also position themselves as demanding buyers of environmental solutions and environmentally friendly products. They can create forums for settling regulatory issues so as to minimize litigation, for example, through mandatory arbitration. And they can play an important role in collecting and disseminating information about innovative ways for companies to reduce their environmental impact at minimum cost or even to come out ahead in the process.

Response to Critics

Not all environmental damages can be avoided without cost. For example, society cannot tolerate the generation of toxic substances and may have to increase the cost to firms of generating them. It is then up to firms to seek innovations that avoid toxicity while going as far as possible toward offsetting the cost of doing so. While no claim is made that fully offsetting technologies can always be found, this possibility is far greater than economists have tended to project.

Some critics simply address the question of frequency: they say that innovative offsets to the cost of environmental compliance are very rare phenomena. Logically, however, there are reasons to believe in the convergence of social and private costs, at least in the area of pollution prevention and resource productivity. Pollution indicates that resources are being wasted, often requiring a firm to perform non-value-creating activities such as handling, storage, and disposal.

Critics of environmental regulations cite studies finding that compliance with such regulations is costly for firms. These costs have been exaggerated in studies depending upon (often-inflated) estimates of compliance costs furnished by the industry in advance of the regulation, or looking only at the early stage, before the innovation response has emerged. In addition, net compliance costs are often overestimated by assuming away innovation benefits. In opposition to these findings, there are plenty of other studies that show no evidence that environmental regulations hurt industrial competitiveness—in itself a striking result, when one considers that regulations have so often been designed in ways that decreased industry's ability to respond intelligently.

> The notion of an inevitable struggle between ecology and the economy grows out of a static view of environmental regulation, in which technology, products, processes and customer needs are all fixed. In this static world, where firms have already made their cost-minimizing choices, environmental regulation inevitably raises costs. . . . The new paradigm of international competitiveness is a dynamic one, based on innovation. [97]

Summary of

Civil Associations *and* Toward a Global Civil Economy

by Severyn T. Bruyn

[Published in *A Civil Economy: Transforming the Market in the Twenty-First* Century (Ann Arbor: University of Michigan Press, 2000) Chs. 6 and 8.]

> By civil markets, we mean *systems of exchange in which competing actors agree to standards for the common good and are capable of enforcing them.* This means situations in which trade, professional, labor, and community associations set codes of conduct, require certification procedures, and establish neutral observers (monitors) and regulatory systems that are authorized to issue penalties for members who break contracts. [207]

Evolving Roles for Governments, Business, and the Third Sector

It is becoming increasingly evident that the existing regime of inadequate corporate regulation is highly injurious to human and environmental health. Toxic substances that are being released into the air, soil, and water are increasing rates of asthma, cancer, birth defects, learning deficiencies, and species extinctions. Yet we cannot look to government as the sole solution. There are too many different firms and facilities (over 15 million business establishments in the United States, and millions more that reach into this country from the rest of the globe) and too many specialized products and processes.

Governments have neither the resources nor the expertise to keep track of all of them, even though their bureaucracies have grown to stifling sizes and complexities. To give a couple of examples: the number of foods being imported into the United States more than doubled between 1992 and 1997, from 1.1 million to 2.7 million. At the beginning of this period the Environmental Protection Agency (EPA) was able to inspect 8 percent of total imports; by 1997 it was inspecting less than 2 percent. "Of the 70,000 chemicals in commercial use in 1995, only 2 percent had been fully tested for human health effects. At least 1,000 new chemicals are introduced into commercial use each year, largely untested. If all the laboratory capacity currently available in the United States were devoted to testing new chemicals, only 500 could be tested each year." [150] The challenges facing government agencies charged with protecting consumers and the environment are literally overwhelming.

At the same time, a civil and civilized society cannot be sustained by unfettered market competition; that force must be tempered by cooperation among competitors and monitored by stakeholders and nongovernmental organizations (NGOs). Civil society groups such as NGOs have a long record of organizing to protect workers, consumers, and the environment against the self-interested actions of firms. These movements have usually taken the form of efforts to get governments to coerce firms, through laws and regulations, to internalize the costs they have been externalizing. This is not the most effective role for civil society groups. Instead they should encourage government to change its emphasis: to focus more on overseeing the rules according to which businesses cooperatively self-regulate, rather than trying to be the primary regulators, and to assist owners, managers, workers, and consumers to solve problems at the point where they originate.

Significant steps toward this ideal have already been taken in the activities of trade associations such as the American Chemical Society, the Defense Industry Initiative on Business Ethics, or the American Industrial Hygiene Association. Trade associations set standards for the size or quality grading of shoes, lightbulbs, lumber, fireproof clothing, and myriad other products. Even though the firms represented in such associations compete with one another in product markets, they cooperate to pay the cost for experts who help to create and, to a varied extent, enforce the standards that are necessary to maintain the public's trust. "The fact is, however, that self-regulation requires systems of accountability. Without government laws or civil associations to enforce agreements, companies forget the rules." [162]

Examples of Cooperation and Self-regulation

The International Organization for Standardization (ISO) is a worldwide federation of businesses whose members come from more than 110 countries. ISO develops international manufacturing, trade, and communications standards whose adoption by firms is voluntary. However, countries and industries are making use of ISO standards in ways that often make them obligatory for firms that want to stay in business. The most publicly known activities of this organization are the ISO 14000 series of standards for managing the environmental impacts of business. They include basic management systems, auditing, performance evaluation, labeling, and life-cycle assessment, and in places they invite third-party certification.

Environmentalists often complain that ISO 14000 does not go nearly far enough, and may give industry an excuse to resist more stringent standards proposed by groups less closely allied with industry. Despite this criticism, the continued evolution of ISO can be seen as moving toward closing the gap between industry preferences to act any way they please and society's requirement that businesses take responsibility for their impacts.

Toxic Use Reduction (TUR) is an environmental protection program that was established in Massachusetts to promote cleaner and safer production processes. It was founded in response to the Toxic Use Reduction Act passed by the state legislature in 1989. This act was sponsored by both business and environmental interests after long negotiations between the two sides. It requires firms to file an annual report and to pay a fee that supports the TUR Institute at the University of Massachusetts at Lowell, where education and training is provided for professionals and the general public. The Institute also sponsors research and encourages the spread of cleaner and safer technologies. TUR levels the playing field, making it harder for noncomplying firms to gain an unfair cost advantage. It is a model for business self-regulation in that it is self-financing and self-accountable.

A case involving the Federal Trade Commission (FTC) shows how a government monitoring system has worked in the past, and hints as to how it could be improved. Sears, Roebuck and Company repeatedly advertised inexpensive products, then told customers that the advertised product was not available, and suggested a higher-priced alternative. FTC served an injunction against this "bait and switch" practice. The point here is that such practices diminish the value of all advertisements, teaching consumers not to believe them. It is therefore in the interest of industry itself to create and enforce the kind of regulation for which the FTC has been responsible. The new task of the FTC should be to encourage and pressure competitors and stakeholders to solve problems of this sort on their own.

We can, in fact, see this taking place in the new warranty rules that have been created by a coalition including the FTC, nonprofit consumer groups, and automobile makers. Similarly, "[n]onprofit environmental groups and oil companies meet at the Interior Department to formulate air-pollution standards for offshore drilling rigs. Labor unions and manufacturers meet at the Occupational Safety and Health Administration to decide on new factory exposure limits for widely used toxic solvents." [160]

Extending Existing Practices

A variety of kinds of regulatory systems are possible in the private sector. Standards are codified in written contracts and formal agreements. Various kinds of tribunals and final arbiters are established, by the consent of the governed, to judge and define penalties for those who infringe on the standards. A more sharply defined public policy could help to extend such practices further. Key to this is increasing transparency, starting with better accounting and more disclosure. "[G]overnment can require disclosure most effectively through trade, nonprofit and professional associations. . . . Regulatory authority often requires a balance of inside (private sector) and outside (government sector) authority for monitoring, but today the great need is to cultivate more public standards in the private sector." [174]

When we think about the economy along the division "public" versus "private," we preserve a vision from the past that can distort our appreciation of today's reality. The "private" (business) sector is composed of firms that compete; it is equally true that it is composed of associations that cooperate. Markets are as much public as they are private, in their effects on stakeholders, in the ways that stakeholders outside of firms claim a piece of their action, and in the ways that businesses respond to pressures from the rest of society. "When the government requires corporate transparency, it allows non-profit groups to be monitors of a business sector. When this happens, the economy moves further towards being considered public." [175] As all three sectors—government, business, and nonprofit associations—learn to collaborate to deal with the technological and institutional complexity of modern economies, civil rules will reduce social costs.

Summary of

Sustainability and the Accountable Corporation

by Allen L. White

[Published in *Environment*, (October 1999), 30–43.]

Corporations are moving toward more voluntary reporting of social, economic, and environmental information. In the long run, will this trend turn out to be merely a form of greenwashing, or will it be a part of a deeper shift in which corporations genuinely take responsibility for their social, economic, and environmental impacts? This article discusses the development of legal and institutional frameworks for corporate transparency and accountability.

History of Corporate Regulation

In the United States, before the Civil War, the states exercised firm control over corporate behavior. Corporations generally were chartered for a specific public purpose, for instance, to build a road or a canal, and were disbanded when the purpose was accomplished. However, by the 1870s

> major corporate interests had pressed the federal and state governments to treat them in ways that allowed essentially uncontrolled accumulation of wealth with minimal liability for harm to workers or the public at large. Another watershed came in 1886, when the Supreme Court ruled that a corporation was a "natural" person subject to all the protections of the Constitution. This decision effectively reversed hundreds of state laws governing the wages, working conditions, ownership, and tenure of U.S. corporations. It also heralded a period of more than 40 years during which governments and corporations showed little inclination towards transparency. Burgeoning corporate power was accompanied by secrecy; greater accountability would have to wait several decades, one world war, and the collapse of the stock market. [33]

Regulations that emerged in the first quarter of the twentieth century focused on maintaining competition and breaking up monopolies. After the collapse of the U.S. stock market in 1929, the Securities and Exchange Commission (SEC) was established with a broad mandate "to reinstate some kind of social control over corporate behavior through the instrument of public disclosure." [34] However, disclosure was defined primarily as financial information relevant to an investor. The SEC charged the Financial Accounting Standards Board with the development of generally accepted accounting practices to ensure that all companies would report financial information in a standardized format to create consistent and comparable information across companies. When, in the 1970s, the Natural Resources Defense Council tried to enlarge the scope of disclosure required by the SEC, the courts maintained that the requirements were limited to disclosure of only such information as would be directly relevant to the decisions of a prudent investor. The courts in effect reaffirmed that relevance was confined to traditional financial information.

In 1986, after the catastrophic release of air toxics at Union Carbide's plant in Bhopal, India, Congress enacted the Superfund Amendments and Reauthorization Act, which "fundamentally redefined the reporting landscape by creating the Toxics Release Inventory (TRI). Thousands of medium and large facilities were now required to annually report all of their releases to all media—air, water, and land—a provision that would enable interested stakeholders to obtain a complete profile of a facility's performance without having to assemble regulatory compliance information piece by piece." [35]

TRI dramatically raised the level of corporate disclosure. Soon thereafter similar initiatives were established in Canada and selected OECD (Organization for Economic Cooperation and Development) countries. Today, Australia, Ireland, the Netherlands, and the United Kingdom have operating Pollutant Release and Transfer Registries.

Mandatory Versus Voluntary Reporting

TRI is a facility-based disclosure system. However, for many stakeholders, such as securities analysts, investors, and human rights groups, corporate-level performance is equally or more useful. It is extremely difficult to piece together a comprehensive picture of company-wide impacts from information on individual facilities. And even an NGO with the ability to do this would still lack information about a company's environmental management systems, wage equity, gender equality, stakeholder engagement processes, or policies on plant shutdowns and community reinvestment.

Another problem with "compliance reporting" is that it is generally limited to "lagging indicators"—that is, data describing past releases, energy use, water use, and other retrospective information. Judging a corporation's prospects in relation to sustainability requires qualitative and forward-looking information absent from mandatory compliance reporting. Thus far, governments have shown little inclination to mandate such leading indicators, even those with well-developed programs.

In contrast to government mandates, voluntary corporate environmental reports (CERs) began to appear around 1990. A decade later the total number of CERs produced annually probably exceeds 1,000 (including both stand-alone reports and the environmental portions of financial reports); firms from the United States, Canada, Japan, Germany, the United Kingdom, and other OECD countries, as well as a smattering of other nations, are represented in the burgeoning number of such reports.

Even broader than corporate environmental reports is an emerging genre known as sustainability reporting. An example was the 1998 annual report of Freeport-McMoran Copper and Gold, Inc.—a New Orleans–based multinational that operates, among other places, in the fragile social and physical environment of Irian Java, Indonesia. In addition to the usual kinds of financial information, the report included information on the three dimensions of sustainability: the corporation's economic, social, and environmental impacts. Details included taxes, royalties, and dividends paid to Indonesia, as well as information about the corporation's medical and educational facilities, and its impacts on patterns of migration, ethnic conflicts, and alleged human rights abuses. It also described its environmental commitments, its management, and its monitoring and auditing processes.

The small but growing movement toward sustainability reporting is fueled by both external stakeholders and internal management drivers. The former include social investors, NGOs, and human rights groups, as well as the groups that stand to gain from broadened reporting requirements: accountants, auditors, and verifiers. At the same time, companies themselves are interested in stronger management information systems to support internal decision-making as well as comparisons of their own company's practices with others in their industry.

In the future, voluntary reporting will have an especially important role to play in developing nations, where information technology will make voluntary disclosure at least as powerful as governmental regulation as an instrument to advance responsible corporate practices.

Challenges, and a Response

If the progress in corporate accountability is to be achieved in the long term, two challenges will have to be overcome. First is the disjuncture between a social view of the corporate purpose versus the traditional corporate balance sheet, which tracks performance primarily to serve shareholder interests. Revitalizing the early social purpose of corporations will require a gradual process of both legal reform (e.g., revisiting corporate charters laws) as well as broadening disclosure law to embrace social and environmental information.

Second, on the voluntary side, a troubling paradox is emerging. The rapid growth and proliferation of voluntary environmental and sustainability reports "has led to an enormous volume of inconsistent and unverified information. If the information of interest to stakeholders is not presented in a coherent, uniform framework, the resulting confusion and frustration may well stall the momentum toward greater disclosure." [34] Is a generally applicable framework feasible? To be sure, different business sectors should, to some degree, disclose different social, environmental, and economic indicators. At the same time, a generic framework might cover 75 percent of the sustainability information applicable to all companies, while the remainder is tailored to the particular circumstances based on sector, size, and location. But if report users are to make comparisons across nations and companies, it is essential to achieve such a generic framework, analogous to those used in financial reporting.

The leading response to the need for standardized sustainability reporting is the Global Reporting Initiative (GRI) launched by the U.S. nongovernmental organization CERES (Coalition for Environmentally Responsible Economies) and implemented in partnership with the United Nations Environment Programme (UNEP) in collaboration with a wide range of business, accounting, labor, human rights, investor, and environmental organizations. Relative to dozens of reporting initiatives worldwide, GRI has several unique strengths. Its steering committee represents all major stakeholders. Its report framework encompasses traditional environmental health and safety issues within a broader sustainability framework that also includes social and economic aspects of corporate performance. Further, GRI is firmly grounded in the vision of comparable, consistent, and verifiable information in which financial reporting has evolved during the last 60 years.

With the release of Exposure Draft Guidelines in March 1999, GRI is working to develop the reporting guidelines that give equal weight to past and future corporate performance. It is committed to generating information on "the degree to which management is forward-looking, resilient, poised to innovate, and capable of exercising leadership in the face of rising expectations for environmental, social and economic performance. This aspect of company performance is especially valued by securities analysts, who seek barometers of management quality." [39]

The 1999 Exposure Draft Guidelines of the GRI asked pilot test corporations (including some *Fortune* 500 firms such as General Motors, Procter & Gamble, and Bristol-Myers Squibb) to provide social and economic indicators in five areas: "corporate principles (e.g., freedom of association and workforce diversity); local and global community relations (e.g., community involvement and skills transfer); relations with suppliers (e.g., procurement standards); and relations with customers (e.g., labeling and advertising standards)." [40] However, in this early version, GRI did not feel ready to specify quantitative metrics for these aspects of corporate performance.

While corporate environmental reporting has a ten-plus-year history, the measurement—even the definition—of the social and the broader economic impacts of corporations remain elusive. Ongoing, intensive stakeholder consultation coupled with voluminous feedback from the pilot program will help achieve a cleaner articulation of social and economic indicators.

GRI seeks to strike a balance between generic, generally applicable measures of corporate performance and cultural differences that cross nations and regions on such matters as gender equality and minimum wages. Further, GRI must seek to design a reporting framework that generates information that adds value to managers, so that accountability is viewed as a benefit and not merely a cost. Fortunately, the information revolution creates globally well-informed consumers and investors whose good will corporations must cultivate. To the extent that these stakeholders demand to know if and how corporations are contributing to sustainability, social, economic, and environmental information will become as essential to managers as financial data already are.

PART IX

Local and National Strategies

Overview Essay

by Timothy A. Wise

The most visible sustainability debates have taken place in international conferences, yet some of the most important actions are found at the local, regional, and national levels. This section surveys some of the most innovative of these efforts to enhance social and environmental sustainability. This is by no means a comprehensive survey; the range of initiatives is far too broad to cover in a book chapter. Instead, we focus on some of the strategic frameworks that underpin many of the sustainability initiatives around the world while highlighting important analytical approaches to the issue. Although our emphasis is primarily on issues facing developing countries, many of the strategies and frameworks are applicable in developed nations as well. We examine efforts to promote social and economic development as well as initiatives to preserve the environment, recognizing that many of the best practices are able to advance all of these goals simultaneously.

The United Nations Conference on Environment and Development (UNCED), held in Rio de Janeiro, Brazil, in 1992, inspired the development of many local and national sustainability plans. After a brief examination of these efforts, this essay will address the importance of government action in promoting economic development in general and sustainable development in particular. It then examines current strategies for rural development, with a particular emphasis on the impact of globalization and the well-tested strategy of community-based natural resources management. After a brief overview of sustainable communities initiatives, it concludes with an examination of microenterprise finance programs, which have emerged as a popular and market-friendly strategy for fighting poverty.

From Global Discussion to Local Action

Much of the literature on sustainability initiatives brings to mind the wisdom and the limitations of the advice to "think globally and act locally." On the one hand, many of the most inspired efforts to promote sustainable communities and societies seek ways to spur individual, localized action on issues of local and global importance. On the other hand, at the international level we have seen far more thinking and talking than action, and action is needed urgently on problems such as global climate change that can only be resolved through concerted international action. At the global level, the language of sustainability has won wide acceptance—an important achievement—yet we still await meaningful international agreement on many of the most important issues.

Action at all levels—international, national, regional, and local—will be needed to achieve or enhance sustainability. In this regard, it is worth examining briefly one of the efforts to promote sustainable national practices: Agenda 21.

Agenda 21 was the main outcome of the 1992 UNCED Summit (UNCED 1992; Robinson 1993). It set goals for social and environmental sustainability for the twenty-first century and called on national and local governments to develop clear plans to achieve those goals. Advertised as "moving sustainable development from agenda to action," Agenda 21 has certainly produced more agendas than action. Many at the "Rio+5" Conference (convened in 1997 to follow up on UNCED) decried the lack of concrete progress and tepid government commitments to policy change (UNCSD 1997). With few exceptions, implementation has lagged behind sustainability planning. Dernbach (1998) observes that with the notable exception of population growth, for which long-run projections have improved, nearly every negative trend noted at the Earth Summit, held in Rio de Janeiro in 1992, remained unchanged five years later. He goes on to suggest that until countries adopt or modify new laws to bring governance structures in line with sustainable practices, concrete achievements will be limited. A review of the U.S. plan (President's Council on Sustainable Development 1996) found little evidence that Agenda 21 affected U.S. laws or policies (Dernbach et al. 1997). One study of a pre-Agenda 21 National Environmental Policy Plan in the Netherlands demonstrated how difficult it is to maintain the political commitment needed to implement such plans (Lucardie 1997).

At the local level, the mandate for governments to develop local Agenda 21s has given activists a framework within which to advocate for changes in local policies. A study of British municipal government experience suggested a high level of participation, with 70 percent of local councils developing local sustainability plans (Selman and Parker 1999). In Peru, Agenda 21 provided the stimulus for a "Cities for Life" network promoting and developing local environmental action plans and activism (Miranda and Hordijk 1998). It is still too early to assess the full impact of Agenda 21–inspired programs at the national and local levels, but it is an important area for future study.

State Intervention Is Critical

One reason Agenda 21 plans are difficult to implement is that the notion of national planning, which clearly involves the government, conflicts with the dictates of free-market economic policies, which call for a diminished role for the state in economic activities. Indeed, for developing countries still seeking economic growth with social and environmental sustainability, this issue remains at the heart of the dilemma. How can one pursue social and environmental goals when overwhelming economic forces and institutions mandate a limited role for the state?

Alice H. Amsden, in the first article summarized in this section, examines the role of the state as it relates to "late industrialization" in East Asia. Based on an extensive review of the literature and detailed study of several countries, Amsden demonstrates that several East Asian countries were able to industrialize precisely because the state took an activist role in promoting and protecting strategic industries. She further concludes that among the many factors contributing to economically sustainable growth in East Asian countries, low levels of income inequality were among the more important. While East Asian industrialization does not offer a model for environmental or social sustainability, the experience offers important lessons for some of the world's poorer countries, which need not just sustainable practices but also economic growth.

Amsden's findings have been confirmed by many analysts. In an earlier study, Irma Adelman presented a compelling comparative analysis of equity and development, identifying a pattern in cases of successful late industrialization. Prior to the stage of rapid growth, all of the studied countries experienced a state-led redistribution of assets, particularly land, coupled with limits on financial capital. This was followed by heavy state investment in education and training and a focus on labor-intensive industrial development, generally supported with foreign capital (Adelman 1975, 1980). Birdsall et al. (1995) found positive interactions between the rate of growth, investment in education, and relatively equal income distributions in eight East Asian countries. In particular, the authors noted two important "virtuous circles" related to investment in education. First, such investment stimulated economic growth, which further stimulated investment in education. Second, such investment, when focused on primary and secondary education, decreased inequality, which further fueled the demand for education investment.

The need for such investment in "human capital" is now quite widely accepted (as is noted in Part I of this volume). Yet the contradictions remain, with World Bank programs, for example, simultaneously encouraging both a decrease in the role and budget of the state and an expansion of education programs. Such factors make replication of the East Asian model more difficult. Peter Evans (1998) warns that globalization based on rules that benefit those who already control financial capital and intellectual property will make it much more difficult for countries mired in highly unequal economic structures to overcome those limitations. Still, he notes the continued possibility for an activist state committed to industrial development to promote competitive growth based on lessons from the East Asian experience.

Active state involvement is at least as important to environmental sustainability as it is to social and economic sustainability. The state often plays a decisive role in supporting the kind of research and development that can lead to innovation and economic growth. To the extent such innovation furthers environmental sustainability, for example, by providing incentives for research and development (R&D) in renewable energy, the state can provide the stimulus lacking in the unregulated market (de Jongh and Captain 1999). The state also has a critical role to play in promoting infrastructure that favors sustainability. Infrastructure investments determine the direction of development for years to come: "path dependence" that can promote or derail future efforts to achieve sustainability. In many parts of the developing world these decisions are critically important not only to national sustainability but also to global survival; the extent to which China opts for a fossil-fuel-dependent development path, for example, will have great impact on the rate of global climate change (see Byrne et al., summarized in Part VI, and Lenssen 1993).

The state can also use tax policy to encourage sustainable practices. In the second article summarized in this chapter, **M. Jeff Hamond** and his coauthors present the argument for "ecotax reform." They propose reducing taxes on activities society values—labor, innovation, capital formation—and replacing them with taxes on behaviors society wants to discourage—pollution and waste. They argue that such an approach can be revenue-neutral—a tax shift rather than a tax increase—and can be targeted in ways that are not regressive. They also suggest that such policies can produce a "double dividend," enhancing environmental protection while stimulating sustainable forms of economic growth. This kind of "green tax reform" has been implemented in Sweden, Denmark, Norway, and some other European countries. A study by the European Environmental Agency (1996) found some evidence of the double-dividend effect, though the subject remains a topic of some debate in the field of environmental economics (see Bovenberg 1999; Kahn and Farmer 1999).

While the tax shift proposal is far-reaching and has been tried only relatively recently, other forms of environmental taxation have now been implemented successfully in a variety of places (see Table IX.1). The benefits of such policies are by now well-demonstrated as an important market-based instrument for environmental reform, but some researchers note that such policies will succeed only if they are accompanied by other actions that favor sustainability (Durning and Bauman 1998). While more widely practiced in developed countries, ecotaxes have also been advocated for developing countries where taxing resources and pollutants may be a more practicable way to raise government revenues than taxing income, commodities, trade, or luxuries (Sterner 1996).

Table IX.1. Ecotaxes: Selected "Green" Tax and Permit Systems

Environmental Problem	Policy	Country, Year	Description
Overfishing	Fishing permit systems	New Zealand, 1986	Overfishing reduced. Many stocks appear to be rebuilding. Fishing industry, unlike that of most countries, seems stable and profitable despite lack of subsidies.
Excessive water demand	Tradeable water rights	Chile, 1981	Existing users grandfathered. Rights to new suppliers auctioned. Total water use capped.
Solid waste	Toxic waste charge	Germany, 1991	Toxic waste production fell more than 15% in three years.
	Solid waste charge	Denmark, 1986	Recycling rate for demolition waste shot from 12 to 82% over 6 to 8 years.
Water pollution	Fees to cover wastewater treatment costs	Netherlands, 1970	Main factor behind the 86–97% drop in industrial heavy metals discharges and substantial drops in organic emissions.
	Fertilizer sales taxes	Sweden, 1982 and 1984	One charge, 1982–1992, funded agricultural subsidies; the other pays for education programs on fertilize-use reduction. Use of nitrogen dropped 25%; potassium, 60%; phosphorus, 64%.
Acid rain	Nitrogen oxide charge on electricity producers	Sweden, 1992	Refunded as electricity production subsidy. Contributed to 35% emissions reduction in two years.
SO_2 air pollution	Sulfur permit system	USA, 1995	Nearly all permits allocated free to past emitters. Forced total emissions to about half the 1980 level by 2000; cost of compliance far lower than predicted.
Ozone depletion	Ozone-depleting substance tax	USA, 1990	Smoothing and enforcing phase-outs.
	Chlorofluorocarbon permit system	Singapore, 1989	Half of permits auctioned, half allocated to past producers and importers. Smoothing and enforcing phase-out.
Global climate change	Carbon dioxide tax	Norway, 1991	Emissions appear 3–4% lower than they would have been without the tax.
Uncontrolled development	Tradable development rights	USA, New Jersey pinelands, 1982	Land-use pan sets density limits on development in forested, agricultural, and designated growth zones. In growth zones, developers may build beyond density limits if they buy credits from landowners agreeing to develop less than they could. Owners of 5,870 hectares in more protected areas have sold off development rights.

(*continues*)

Table IX.1. *Continued*

Environmental Problem	Policy	Country, Year	Description
General	Linking investment tax credits to environmental and employment records	USA, Louisiana, 1991	Tax credits reduced up to 50% for firms that pollute the most and employ the fewest. Twelve firms agreed to cut toxic emissions enough to lower the state's total by 8.2%. Repealed after one year.

Source: David M. Roodman, *Getting the Signals Right: Tax Reform to Protect the Environment and the Economy,* 1997.

People-Centered Development

It is only relatively recently, with the renewed push for neoliberal economic policies, that the question of state involvement in promoting economic development has even become an issue. Most development debates concern not *whether* the state should intervene but *how* it should intervene. The East Asian countries are considered successful in their promotion of industrial growth, but far less so in their advancement of environmentally and socially sustainable human development. With the replicability of the East Asian model limited, particularly for many of the world's poorer nations, it is worth exploring some of the attempts to "put the poor first" in the development process.

There are many such examples at the local or project level but few to examine at the national level. The model of the centrally planned economy has largely been discredited, though some still defend aspects of central planning in allocating scarce resources to basic needs (Pastor 1998). Among the more developed examples of "people-centered development" is the three-decade-long communist-led rule in Kerala, India, a state larger than many nations. Kerala has attracted widespread attention for its redistributive policies and its heavy investment in public health and education, which have produced human development indicators that compare with those in the developed world. Rates for literacy, infant mortality, life expectancy, and population growth are among the best in India, a remarkable achievement in a state with limited resources.

Patrick Heller's article summarized in this chapter presents some of the background and results from Kerala's experience. Recalling Amsden's observation that prior relative levels of equality favor development, Heller notes in some analytical detail the ways in which Kerala's radical land reform not only equalized rural assets but also dispossessed the traditional elite of the economic base from which it ruled. This paved the way for a peasant-worker alliance to assume political power and advance its reform program.

Heller's analysis, and those of others who cite Kerala as an example (see, for example, Alexander 1994; Franke and Chasin,1994; Dreze and Sen 1989; Kannan 1995), are not without their critics. Tharamangalam (1998) argues that the model has produced economic stagnation, stifled innovation and capital formation, produced a fiscal crisis despite high taxes, and failed to generate adequate employment. Indeed, economic growth rates in Kerala were quite low in the 1980s compared to the rest of India, prompting local government leaders in the early 1990s to acknowledge a crisis that had the potential to undermine the very redistributive policies on which the model is based. Slow growth also raises questions about the extent to which investments in social welfare programs, particularly health and education, can stimulate economic growth over a sustained period.

Rural Development Still Essential

Kerala's experience holds important lessons for many of the world's poorest people. In much of the developing world, large portions of society remain marginal to the market economy. Many have lost their traditional livelihoods to expanding market forces, but unlike the dispossessed peasants of earlier industrialization processes they have found little gainful employment awaiting them in industry.

David Barkin, in the book chapter summarized here, explores this dilemma in some depth. He argues that prevailing economic development models contribute to a dual economy, with increasing concentrations of wealth alongside intractable poverty. He points out that from an economic perspective it is only more efficient for the rural poor to give up local production of food and other agricultural goods to areas with a comparative advantage if both the land and labor used in that production can find productive use elsewhere. In many rural areas, there are neither productive alternatives for the land nor adequate employment for the displaced. Moreover, peasant farmers who remain on the land often serve as stewards of genetic diversity, a service largely unrecognized and unremunerated in the marketplace (Brush 1993, 1998). Displaced peasants, on the other hand, are often forced into environmentally destructive practices (see summary of Boyce in Part VII). Barkin proposes that rural communities de-link from the market economy in strategic ways, developing greater food self-sufficiency and creating other forms of autonomous production.

Other researchers and practitioners offer a range of strategies to sustain rural development as part of a diverse strategy for poverty reduction. Despite the continued march of urbanization, four-fifths of the world's poor live in rural areas (Jazairy et al. 1992). Arguing that macroeconomic stabilization and democratization allow hope for rural development, de Janvry and Sadoulet (1996) conclude that macroeconomic stabilization is "necessary but not sufficient for successful rural development." They further find that contractions in government services caused by structural adjustment represent smallholders' most serious obstacle to growth. Noting that successful agricultural development alone will not solve rural poverty, they advocate strategies that go beyond agriculture, such as microenterprise development to deliver goods and services needed by agriculture.

Linking Poverty Eradication with Conservation

Robert Chambers points out in his summarized article in Part II of this volume that there is an urgent need to address rural poverty by fostering "sustainable livelihoods"—diverse, integrated forms of production and labor consistent with local customs and resources. This approach has translated into local and regional strategies to build development strategies out of the most urgently felt needs of the poorest residents, in the process linking livelihoods with stewardship of natural resources (Singh and Kalala 1995).

A more limited but well-tested strategy to link rural people's needs with environmental concerns is "community-based natural resource management" (CBNRM), in which rural communities are empowered to manage a local natural resource, such as timber, in a manner that allows residents to sustain livelihoods for at least some residents. Offered as an alternative to top-down and exclusive conservation set-asides, CBNRM initiatives have met with some success, even in the world's poorest areas. In an interesting article synthesizing the lessons from twenty-three case studies taken from ten different African countries, **Peter G. Veit** and his coauthors find CBNRM to be far from a panacea. Still, they note that it can be successful if accompanied by adequate government support, market incentives that make conservation profitable, security of land tenure, and access to existing resources and livelihoods, and also when external support complements indigenous knowledge and resources. As many researchers have pointed out, such projects can only succeed if implementing agencies set aside their preconceived notions to allow residents to develop strategies consistent with local practices and needs (Gupta 1995).

A more critical view of CBNRM is offered by **Melissa Leach et al.** in the article summarized in this chapter. Calling the results from many such projects disappointing, the authors suggest there are flaws in the underlying assumptions about "community," "environment," and the ways in which the two interact. On the one hand, the term "community" implies a certain unity of culture, purpose, and interest. This masks important social, class, ethnic, and gender conflicts that often prevent residents of a village from acting in unison. On the other hand, assumptions about the "environment" often rest on static, linear models for ecological systems. Drawing on the work of Amartya Sen, they offer the notion of "environmental entitlements," which recognizes both the evolutionary nature of the environment and the political and social factors that often limit community members' ability to benefit from local resources.

Sustainable Communities

The community-based approach to resource management recognizes that there is no substitute for local knowledge and participation, and that neither is likely to be forthcoming in the absence of community control. This is as true in urban areas and developed countries as it is in rural Africa. While a community (loosely defined) needs supportive international and national policies and institutions to achieve sustainable human development, it is remarkable what a determined group of reformers can achieve at the local level even in the absence of such factors. Moreover, action at the local level builds support for action at the national and international levels.

Sustainable communities initiatives have sprung up in many parts of the United States to respond to issues of sprawl and environmental degradation. Local governments or community groups are usually the instigators, supported by a fast-developing methodology that has evolved from the fields of community development, urban planning, and environmental management (Roseland et al. 1998). "Community indicators projects" use quality-of-life indicators to set goals and to measure progress toward them, while the "ecological footprint" approach has helped local communities—and even nations—assess and reduce their ecological impact (Redefining Progress 2000; Wackernagel and Rees 1996). Such efforts have made measurable progress on many issues, spurring successful recycling programs, curbs on growth, expansions of green space, and other issues that can only be solved locally. Examples are not limited to the developed world; Curitiba, Brazil, is offered by many as a model for visionary local sustainability policies (see Box IV.3 in Part IV of this volume). While many local strategies are not transferable or are easily overwhelmed by the sheer power of market forces, a growing set of "best practices" databases seek to maximize the potential for such successes to be replicated elsewhere.[1]

Some researchers take this analysis further to argue that only a profound "relocalization" can reassert community control in a rapidly globalizing world (see the final section of Mander and Goldsmith 1996, including contributions by Helena Norberg-Hodge, Wendell Berry, David Morris, and others). Acknowledging that this will be a difficult transition to achieve, such theorists argue that only a return to economies based on local production for local consumption can produce sustainable and just societies.

Microcredit: A "Market-Friendly" Alternative?

Relocalization and other strategies that call for large-scale restructuring of the economy receive scant attention in the economics field or among mainstream development agencies. Most development strategies remain based on the premise that the best way to improve the lot of the world's people is to expand their connections to the global market. While the market can play an important role in improving the economic lot of the poor, there are serious flaws in approaches that are blindly market-driven (see Part VII). Among them is the assumption that government actions that limit the functioning of market forces will, in the long run, undermine economic growth. In the development field this bias has led to a litmus test of "market realism," which development initiatives must pass. As a result, many sound strategies are excluded out of hand, because they are considered either too expensive or disruptive of the market.

Microcredit, however, is a strategy that passes the litmus test of market realism and has still gained adherents among advocates for the poor. Microcredit programs vary greatly but commonly are based on revolving loan programs that make small amounts of short-term credit available to individuals or groups generally denied resources by large financial institutions. Pioneered by institutions like Chicago's South Shore Bank in the United States, and the Grameen Bank in Bangladesh, microcredit programs have demonstrated that with the appropriate supports to reduce transactions, costs repayment rates on loans to low-income people can be quite high. This not only makes such programs economically sustainable, it also puts credit into the hands of the poor, stimulating entrepreneurial activity and small-enterprise development, which provide needed employment and income for residents.

As stimulants for small-scale capitalist enterprises, microcredit programs certainly remain consistent with the bias toward market-realism. In developing countries, such programs offer the added attraction that, compared with expensive and often ineffective large-scale development programs, once capitalized they can be relatively self-sustaining and can even gain the support of private-sector financial capital. They also extend market relations to sectors of the developing world otherwise marginalized by development.

The field of microcredit is relatively new (the Grameen Bank began making loans in 1976 but was formally established in 1983), and there has been as much cheerleading as analysis of the long-term benefits of such programs. Some studies, carried out by practitioners in the field, argue that microcredit programs are succeeding in providing access on a massive scale to the poor, and some data back them up (Otero and Rhyne 1994). Some 350 microenterprise programs in the United States reach an estimated 2 million people, while the Grameen Bank since 1976 has lent over $2.6 billion to 2.4 million borrowers, the overwhelming majority of which have been women (Bonavogilia 2000). While it is generally promoted for its impact in reducing poverty, some researchers tout microenterprise as the sustainable alternative to mass industrialization (Mayur and Daviss 1998). Unfortunately, many studies stop at the question of access, failing to ask what the long-term impact is on poverty and well-being. In the summary included in this section, **Linda Mayoux** attempts such an overview, taking a hard look at the impact of microenterprise programs for women. This has been a rapidly growing area of microcredit, lauded by the

World Bank, the U.N. Secretary General, and other influential advocates as a critical tool in fighting poverty among women (Scully and Wysham 1997). Proponents argue that microcredit programs targeting women can not only raise women out of poverty but can also increase women's economic and political power.

Mayoux surveys microcredit programs for women and finds that the majority have failed to have a significant impact on women's incomes over time, have generally benefited relatively better-off women, and have done little to alter women's subordinate role in society. She suggests that the market orientation leads to a bias against the poorest women, who represent higher credit-risks for institutions dedicated first to recovering their loans and only second to alleviating poverty. This is confirmed by other research. One seven-country study found that the income-impact of current microcredit programs diminished greatly for lower-income participants (Mosley and Hulme 1998). Another study of three large Bangladesh programs, including Grameen, found significant impacts on income but saw such poverty-reduction as short-lived unless the programs fostered high-productivity nonfarm activities with strong forward and backward linkages with agriculture, a feature uncommon in lending directed mainly at the informal-sector (Khandker et al. 1998).

Mayoux also finds that women's empowerment is limited because women often do not control the loans they receive or the income generated by the enterprise. This has been confirmed by other studies (Goetz and Gupta 1996). Some researchers have found that gender dynamics in the household and the village actually deteriorated with women's microcredit programs (Gibbons 1995). One anthropological study of a Bangladeshi village with a Grameen Bank women's program found that in many cases male family members enrolled the women, controlled the loans, and used the loans to support activities other than the proposed enterprise. The researcher also reported an observable increase in domestic violence as male family members pressured women participants to find ways of repaying loans so that the flow of funds could be maintained (Rahman 1999).

Mayoux argues that microcredit programs can be an important part of anti-poverty efforts, but only in the right context. She observes that often such programs are promoted in conjunction with neoliberal reforms that remove welfare supports and reduce labor protections. She concludes that for microenterprise programs to succeed in reducing poverty and empowering women they need strong welfare support (health care, child care, education, etc.), improved labor rights, and measures to address gender inequality. The most successful programs are those implemented by women's or poor people's organizations with a broader transformative agenda in which microcredit is but one of many tools for fighting poverty and empowering women.

Conclusion

A common theme that emerges from these studies, whether they are examining local, regional, national or international policies and strategies to promote sustainability, is that there are severe limitations to action at only one level.[2] Local actions are needed to draw on local knowledge, motivate wide participation, and make needed structural and policy changes. National leaders must ensure not only that the macroeconomic environment is stable but also that policies favor sustainable practices. Finally, at the international level, agreement is needed on structural and institutional reforms that can regulate market forces in such a way that the inevitable process of international economic integration proceeds in ways that foster not just economic growth or increased consumption but also human development for all.

Notes

1. See, for example, the UNHCS Best Practices database (http://www.bestpractices.org/); "Local Sustainability," the European Good Practice Information Service found on the Internet at http://www.iclei.org/europractice/index.htm, and the ICLEI Project Summaries found on the Internet at http://www.iclei.org/leicomm/leicases.htm.
2. The question of the coordination of actions at different levels is discussed at length in Goodwin (2000).

Summary of

A Theory of Government Intervention in Late Industrialization

by Alice H. Amsden

[Published in *State and Market in Development: Synergy or Rivalry?* ed. Louis Putterman and Dietrich Rueschemeyer (Boulder: Lynne Rienner Publishers, 1992), Ch. 5, 53–84.]

Developing countries since World War II have been engaged in a process of industrialization that is distinctly different from previous industrialization processes. In the first industrial revolution in England, the market was the catalyst, while in the second in the United States and Germany, the control of pioneering technology was the lever that gave leading enterprises their competitive advantage. Late industrialization in developing countries, however, must proceed on the basis of borrowed technology, not innovation, and their competitive advantage is largely reduced to lower wages. Contrary to neoclassical economic theory, low wages are not a sufficient condition for industrialization. This section explains why higher levels of government intervention, such as that in South Korea and Taiwan, are necessary for industrialization to take place, a strategy that stands in contrast to the market-led prescriptions for reducing state intervention in developing countries. It further examines the conditions that have led to successful state-promoted industrialization and finds that relatively equal income distribution is a critical factor.

Late Industrialization

"As late industrialization has unfolded, it has become clear from observing its 'leading sector'—cotton textiles—that low wages are no match for the higher productivity of more industrialized countries." [53] Even in Korea and Taiwan, where wages were very low and education and infrastructure were relatively developed thanks to foreign aid, leading textile manufacturers still could not compete with Japan's more productive cotton textile industry. "Under these circumstances, and *a fortiori* in industries requiring greater skills and capital investments, governments have to intervene and deliberately distort prices to stimulate investment and trade. Otherwise industrialization won't germinate." [53] This is not a case of "market failure," but necessarily involves state intervention in the process of getting the prices "wrong."

In late industrialization, governments must subsidize key industries for them to achieve international competitiveness, which was far less true in the industrialization of Germany and the United States. There, subsidies were only needed during the time lag between the stages of innovation and commercialization, after which the competitive advantage of the new technology could be expected to ensue. Not so for late industrializers, who suffer from the virtual impossibility of internationally competitive innovation and the near-total dependence on technology transfers. The institutionalization of research and development in multinational enterprises created effective entry barriers to outside innovators, a condition German and U.S. firms did not face. This allowed them to leapfrog their more established British competitors.

In late industrialization, economists argue that low-wage countries should be able to compete internationally in labor-intensive industries using borrowed technology. But the evidence suggests otherwise, with success determined largely by heavy government intervention in the form of subsidies and protection. South Korea's automobile industry received heavy subsidies for at least thirty years, during which time foreign cars were scarcely seen in Korea.

Nor have we seen a closing of the productivity gap between developed countries and late industrializers, which prevailing economic theories would predict. Instead, productivity gaps have widened, further undercutting developing countries' comparative advantage in cheap labor. Also reducing that competitive advantage is the presence of low-wage sectors in high-wage developed countries.

It has been found to be empirically true for advanced capitalist countries that the lower a country's productivity level the faster will be the productivity growth rate. Though the same is assumed to be true for late industrializers, the evidence suggests that it is not. Obviously, this theory of industrialization only works above a certain threshold of development; below it the laws of industrialization change. Under those laws, there is no reward for low productivity, technology transfer cannot close the large productivity gaps, and "lower wages are generally inadequate to overcome the penalty of lateness." [56] This is true even considering the strategy of reducing real wages through foreign exchange rate devaluation. In practice, there is a physiological or political limit below which real wages cannot be reduced through devaluation.

Without low wages as the competitive asset needed to trigger late industrialization, market theory is robbed of an effective mechanism for industrialization. Market theorists, therefore, resort to the tautology of "market failure," arguing that "market distortions" prevent industrialization from taking place. In fact, the theory is flawed. "Late industrializers have faced far easier conditions of technology transfer and far more competitive downward pressures on their wages . . . than in previous industrial revolutions. Yet they have still found it extremely difficult to industrialize—precisely because markets have been working, not failing." [59]

Government Intervention as Catalyst

This opens the door to the argument that government intervention is the necessary catalyst for late industrialization. In assessing the role of such intervention it is important to gauge whether its benefits exceed its costs. For market-oriented economists, the post-independence state in the developing world "is corrupt to the point of aborting economic development," [60] making the costs of government intervention far outweigh the risks in virtually every case. This assumption is not supported by evidence on how states have operated. The evidence suggests that in those countries where industrialization is possible, corruption ("rent-seeking") may persist but not necessarily to the point of derailing industrialization. "This is because in the long run . . . rent-seekers can probably enrich themselves more by sustaining a systematic process of capital accumulation (through savings and investment) than by ransacking the economy in the short run through gross misappropriation of revenues." [60]

The evidence suggests that in many of the poorest countries corruption has indeed undermined effective state intervention in economic development. "On the other hand, the fastest-growing late-industrializing countries—South Korea, Taiwan, Singapore, Thailand, Malaysia, and, to a lesser extent, Japan . . .

—have all had extensive government intervention and highly active industrial policies." [60–61] One of the keys to success in the latter countries is that subsidies have not been giveaways but rather have been allocated in exchange for specific performance standards, including output, exports, product quality, investments in training, and research and development. Performance standards raise productivity, increasing cost-competitiveness and making such subsidies more affordable than they would be otherwise. Performance standards discipline not only private enterprise but also the state, whose bureaucrats can be judged by the same criteria as business.

"In all late-industrializing countries, the state has disciplined labor, driving wages down as far as politically possible. What accounts for differences in rates of growth of industrial output and productivity among late-industrializing countries is not the degree to which the state has disciplined labor but the degree to which it has been willing and able to discipline capital. The discipline of capital constitutes a major factor in the success or failure of state intervention in late industrialization." [61]

In addition to the above, there are other factors to consider in determining the success or failure of late industrialization. First, while there has been some of the expected movement of multinational companies' more labor-intensive production to developing countries, such investment is not generally the catalyst for industrialization. Rather, government intervention and subsidization have transformed such investment into an industrialization process. Generally, direct foreign investment is a relatively small fraction of capital formation in that process, and it often lags behind industrial growth. More often, such growth has been financed by domestic saving, particularly once growth rates began to raise domestic incomes. Foreign capital more often takes advantage of those factors that are already producing rapid growth, further accelerating it.

Second, the labor supply is more "unlimited" than it was in earlier periods of industrialization, making the ratio of wages to productivity lower as well. There is a more abundant labor supply because (1) population growth rates are higher, (2) international migration has decreased dramatically, limiting the relief for excess labor, and (3) trade union organization is more difficult, in part due to political repression and in part due to the absence of dispossessed skilled artisans who were the catalyst for unionization in earlier industrialization periods.

Third, real exchange rates can only be devalued to a limited extent, reducing developing countries' ability to create low-enough wages to compete internationally in labor-intensive industries. This is partly due to the physiological and political limitations on low real wages discussed earlier and partly because devaluation raises the prices of imported goods, which are a more significant portion of the economy than in earlier periods of industrialization.

Finally, it is critical to subsidize credit below the market-determined rate of interest by offering low real interest rates in key sectors. In South Korea and Thailand, for example, real interest rates were negative due to government intervention in credit markets for much of their industrialization processes.

The Conditions for Success

The question, then, is under what conditions is late industrialization likely to succeed or fail? The evidence suggests the following conclusions:

1. *Performance standards for industry must be established in exchange for subsidies*—These were successful in both Taiwan and Thailand, for example. In Taiwan, subsidies to exporters in the 1960s were tied to export targets, with exporters penalized if they did not meet goals. By the 1990s, subsidies were conditioned on firms' research and development spending, training programs, and even environmental protection standards.
2. *A more equal distribution of income raises the probability of successful late industrialization*—This is due to many related factors, including "worker motivation, the expected returns to investments in education, cost-push inflation, the effectiveness of currency devaluations, and other micro- and macro-economic variables. To the extent that more equal income distribution increases the growth of output and productivity, it makes the state more committed to industrialization and *willing* to impose performance standards on business. In addition, a more equal income distribution makes the state more able to impose performance standards on business. . . ." [73] The quality of government intervention tends to go up the more equal the income distribution. In general, where income distribution is particularly skewed, small elites tend to overwhelm the state's ability to discipline business. In addition, particularly low earnings differentials between managers and workers in industry help motivate both to raise productivity levels. The author's preliminary data analysis suggests that "the variable of income distribution had greater explanatory power by far than other plausible influences on the growth rate of manufacturing labor productivity . . . " [77] in several late-industrializing countries.
3. *Government intervention remains important late in the industrialization process*—In fact, where late industrialization has been successful, such as in Taiwan, South Korea, Singapore, and even Hong Kong, government intervention remained extensive even as the country entered high-technology industries. "Thus, as late industrializes march closer to the world technological frontier, the relevant question is not when all subsidies should be withdrawn but when particular ones should commence and others cease." [79]

Summary of

An Idea Whose Time Has Come

by M. Jeff Hamond, Stephen J. DeCanio, Peggy Duxbury, Alan H. Sanstad, and Christopher H. Stinson

[Published in *Tax Waste, Not Work* (Washington, D.C.: Redefining Progress, 1997), Chs. 1 and 3, 1–7, 25–44.]

There has been a great deal of discussion regarding the use of fiscal policies to encourage environmental conservation, and many such tax policies have been implemented at the local level in the United States and in some European countries. While such proposals take many forms, the premise of most is that government should tax more heavily those activities it most wants to discourage and tax less heavily the behaviors it wants to promote. These two chapters argue that to those ends government should reduce taxes on labor, innovation, and capital formation and replace those revenues with new taxes or fees on pollution and waste. Such a change could be called a "resource-based tax shift."

The Tax-Shift Concept

The United States faces many long-term problems that a resource-based tax shift could help solve. Among them are high payroll taxes, expanding entitlement programs, global climate change, and limited economic opportunity in inner cities. A revenue-neutral tax shift can be designed that is also neutral in its impact on income distribution and that encourages environmentally sound practices. Such a tax shift could involve new taxes or auctioned emission permits, but the impact would be similar. This approach could gain support across the political spectrum "and potentially create incentives for more investment in both human and physical capital—an economic stimulus package with no revenue cost." [2]

"The current tax system sends the wrong signals to virtually everyone. It discourages work, enterprise, and capital formation while it encourages sprawl, pollution, waste, and the inefficient use of resources." [2] Through its taxation policies, a government chooses what to tax and what not to tax, or to tax at a significantly reduced rate. When government wants to promote a social goal, such as investment in inner cities, one of the tools it currently uses is reducing taxes, usually in the form of tax breaks, on such activities. Such principles can also work as they relate to the environment. Often, the tax debate comes down to a conflict over how much of the tax burden is borne by labor and how much by capital, with the tax burden often shifting from one to the other. Higher taxes on capital can discourage savings and investment, while raising taxes on labor discourages work. But in a society where we value both labor and entrepreneurship, this trade-off is not productive.

"Why not develop a socially useful tax system that would tax those things the country needs less of, and untax those things of which society wants more?" [3] It is reasonable to suggest that government could reduce taxes on payroll, individual income, and corporate profits while raising taxes on environmentally destructive practices. We could tax carbon dioxide emissions, air and water pollution, or overconsumption of virgin materials. "While such a shift from taxing 'goods'—the creation of wealth through labor and investment—to 'bads'—the depletion of wealth through pollution and environmental degradation—cannot be a magic bullet for every economic and environmental ill, it does offer a promising chance for promoting work and investment while moving toward market-based policies that would be an improvement over the current regulatory structure." [3]

Such proposals have foundered in the United States for a number of reasons. They have often been offered as tax increases, rather than revenue-neutral proposals. The business community has often opposed such taxes for this reason. Some proposals have been criticized for being regressive, increasing the relative tax burden on the poor. Critics have also argued that such policies are risky, as they haven't been tried before. This proposal addresses all such concerns because it can be both revenue-neutral and distributionally neutral, and it is now based on the successful experiences in many countries with such "green taxes." (See Box IX.1.)

Economic and Environmental Impact

There are several compelling rationales for such a tax structure. First, it restores legitimacy to public finance, imposing the logic that people should keep more of what they earn but should pay more heavily for costs they impose on others. This gives a coherent rationale to our beleaguered tax structure, restoring the notion that future generations should not bear the costs of today's actions. It allows the public to benefit from revenues collected on publicly owned resources. And it empowers people to reduce their own taxes by engaging in environment-friendly activities—buying energy-efficient vehicles, homes, and equipment, for example.

The economic and fiscal rationales for a tax shift are also compelling. It could reduce inefficiencies in the tax system, stimulating growth. It could incorporate the now-uncounted externalities of social and environmental costs into prices, making the economy as a whole more efficient. It would promote greater efficiency, because any pollution or waste is an example of an input purchased that is then not used to create a product or a service. And lowering taxes on individuals to stimulate work, savings, and investment is an attractive goal.

Finally, the environmental rationale is that it could "provide a least-cost approach to reducing pollution, waste, and the long-term threat of climate change." [5] There are three principal benefits of such a tax shift. The first is averting long-term environmental damage from climate change and pollution. With estimated annual global costs of global warming of between $270 billion and $316 billion, it is imperative that we find ways to reduce carbon emissions.[1] Carbon taxes or tradable permits could raise significant revenues while encouraging the switch to cleaner sources of energy. Other forms of pollution also have long-term costs; acid rain from air pollution reduces productivity and increases public health costs.

Second, an efficient tax on common property resources such as the atmosphere will help conserve such resources by raising the costs of polluting. A resource tax can reduce or eliminate low-value uses of this resource, promoting conservation while providing tax revenues to further the goals of society as a whole.

Third, there are many cases where relatively inefficient government pollution-control regulations could be eliminated in favor of resource taxes, a market-based approach to environmental control. Higher gas prices, for example, would be a more efficient way to achieve fuel efficiency than the enforcement of federal fuel efficiency standards. An increasing number of countries are experimenting with such approaches, including Germany, Chile, and several Scandinavian countries. The United States should follow their lead.

Notes

1. Bruce, James P., Hoesung Lee, and Eric F. Haites. *Climate Change 1995: Economic and Social Dimensions of Climate Change*. Contribution of Working Group III to the Second Assessment Report of the Intergovernmental Panel on Climate Change (Cambridge: Cambridge University Press, 1996).

Summary of

From Class Struggle to Class Compromise: Redistribution and Growth in a South Indian State

by Patrick Heller

[Published in *Journal of Development Studies,* 31, 5 (June 1995), 645–672.]

"People-centered development" is often put forward as the alternative to many of the top-down, business-led development strategies so widely criticized by those advocating social and environmental sustainability. Such a strategy, featuring a redistributive-welfarist state investing heavily in public health and education, has been followed in the Indian state of Kerala. This article describes the Kerala model and examines the conditions that contributed to its success, and to its limitations. (While the article focuses a great deal on the political dynamics surrounding state power, this summary focuses more heavily on the part of the article that describes the model, its achievements, and its challenges.)

The Fruits of the Basic Needs Approach

"Under the impetus of a broad-based working-class movement organized by a communist party, successive governments in Kerala have pursued what is arguably the most successful strategy of redistributive development outside the socialist world." [645] Since 1957, when the communist party was first elected to power, the features of the Kerala model have included a far-reaching land reform; entitlement programs that have given the overall population more equitable access to education, health, and subsidized food than any other Indian state; and labor market interventions that have raised both urban and rural wages.

As a result, Kerala has achieved impressive gains in all measures of the physical quality of life. Literacy in 1991 was 91 percent (87 percent for women) compared to the national average of 52 percent (39 percent for women). Infant mortality figures show similar gains (17 per 1,000 live births, compared to 91 for India), and life expectancy is high (68 for men, 72 for women). In just two decades, Kerala's decadal population growth rate fell from 26.3 percent to 14 percent, compared to the national rate of 23.5 percent. Kerala is also considered to have the most developed human capital in the country, with the highest ratio of science and technology personnel of any state in the country (5.9 per 1,000, compared to 2.4 for India as a whole).

What is unique about the Kerala experience is that it made the transition from a semifeudal, rural state to a capitalist economy based primarily on the mobilization of the lower classes. A strong worker-peasant-tenant alliance, under communist leadership, gained power through parliamentary means backed by strikes and land takeovers. Its most important early act was a sweeping land reform in 1970 that "spelled the end of feudal relations of production by abolishing all forms of rent and transferring property rights to tenants, thus eviscerating both the economic and political power of the landed oligarchy." [649] This was accompanied by measures to formalize labor relations, regulate wages, and enhance the collective bargaining power of agricultural laborers. Because land reform was carried out under the direction of a parliamentary party and was backed by organized workers and peasants, the transition to a capitalist economy also resulted in a strengthening of democratic institutions.

There was also a significant increase in cooperative production and in the public sector. The latter was associated with a dramatic expansion of the welfare state, reflecting an attempt to socialize basic consumption with India's most comprehensive set of entitlement programs. By 1991, Kerala's education expenditures as a percentage of Net Domestic Product were twice the national average and the most in any state. Overall, "social and development services" account for over half of total government spending. Not surprisingly, Kerala's public sector is the largest in India, consuming 59 percent of all government receipts.

The Limits of Labor Militancy and the Welfare State

While these welfare measures have "made Kerala a model of 'social development,' two decades of what are even by Indian standards low levels of economic growth threaten to unravel those redistributive gains. The case of Kerala would thus appear to substantiate the oft-asserted thesis that, in the long run, high levels of labor militancy produce negative-sum economic outcomes. There is indeed little doubt that high levels of social consumption and regulated labor markets have adversely affected capital investment." [647]

Growth rates have been sluggish, with agriculture showing a negative growth rate of 0.4 percent and the industrial sector growing only 3.5 percent per year from 1961 to 1989. The unemployment rate has been the highest in India. Kerala also has one of the highest tax burdens in the country, which has contributed to a serious fiscal crisis. Even communist leaders have acknowledged the need to expand the production base if redistributive policies are to be sustainable.

Kerala's development has been hampered by a weak industrial base, limited natural resources, and isolation from national markets. While few Indian states can claim strong records of economic growth during the same period, Kerala's redistributive policies have inhibited growth in two ways. First, the state has a poor track record in promoting capital accumulation. Second, labor militancy has inhibited private investment despite relatively high labor productivity and little evidence that industrial workers earn higher wages. Though strike data for the 1980s show Kerala with fewer lost days than India's four main industrial states, it is clear that Kerala suffers from the perception created by its redistributive policies and its history of labor militancy.

Stagnation has made clear that the state needs to transform the Kerala model into a pro-growth development model that can retain its social character. To that end, the state has embarked on an interesting corporatist experiment, building on its history and experience in brokering class conflicts. "While there are many usages of the term corporatism, it is used here in the sense in which it has been applied to the processes of organized intermediation between state, labor and capital characteristic of social democratic countries." [658] What makes Kerala's corporatism unique is that the class in control of the negotiations is labor, not the elite, as is usually the case. In addition, "because of the close integration of the state and working class organizations . . . agreements are negotiated as part of a larger social pact in which growth is tied to the expansion of the social wage." [658]

"The political logic of class struggle in Kerala has exhausted itself. The transformative capacities of a redistributive and welfarist state have been extended to their limits. The state's institutions and political practices, which evolved over the past three decades in response to lower class mobilization, are now grappling with the challenge of negotiating the transition to a growth-led strategy of development." [659] To that end, the communist leadership and its union federation have become the primary advocates for peaceful labor relations and productivity agreements with business. "In agriculture as well as industry, the party's new position is that the 'Kerala Model' can no longer be sustained without increases in output." [662] This has interesting parallels with the evolution of the European social democratic parties, where "the traditional Marxist project of collectivizing the means of production was abandoned in favor of a class compromise that took the form of accepting the prerogatives of private property while socializing wages and control rights." [663]

The impact of this strategic shift is still difficult to gauge. Its unique character is clear, however, as a hegemonic working class seeks to promote growth on the strength of the welfare state, not at its expense. It is a strategy that "takes into account the social and political costs of blindly unleashing market forces. As such, it has made the possibility of the transition to a high-growth capitalist economy that much more viable." [666]

Summary of

New Strategies for Rural Sustainable Development: Popular Participation, Food Self-sufficiency, and Environmental Regeneration

by David Barkin

[Published in *Wealth, Poverty and Sustainable Development* (Mexico City: Editorial Jus, 1998), Ch. 4, 49–69.]

Today's global economy, based on the international expansion of capital, is integrating resources and people into a dual economy in which great wealth is generated alongside great poverty and despoliation. Such polarization imposes a burden on society, wasting human and natural resources. To defend their cultures and livelihoods, and to save the environments on which they depend, the rural poor need to consciously de-link from the global economy in critical areas, creating autonomous productive systems they can control and defend.

New Strategies for Sustainability

"While the trickle-down approaches to economic progress enrich a few and stimulate growth in 'modern' economies and 'modern' sectors within traditional societies, they do not address most people's needs; moreover, they have contributed to depleting the world's store of natural wealth and to a deterioration in the quality of our natural environment. . . . In the ultimate analysis, we rediscover that in present conditions *the very accumulation of wealth creates poverty.*" [51]

"The search for sustainability involves a dual strategy: on the one hand, it requires releasing the bonds that restrain people from strengthening their own organizations, or creating new ones, in order to use their relatively meager resources to search for an alternative and autonomous resolutions to their problems. On the other hand, a sustainable development strategy must contribute to the forging of a new social pact, cemented in the recognition that the eradication of poverty and the democratic incorporation of the disenfranchised into a more diverse productive structure are essential." [51]

Several conclusions emerge from a review of the literature on sustainable development. First, "sustainability is a process rather than a set of well-specified goals. It involves modifying economic and social processes so that nature can better adjust to more modest demands from humanity." [52] Second, sustainability must focus on local participation and control over the way in which people live and work. As such, sustainability goes beyond issues of environment, economic justice, and development. It is about diversity in all its dimensions, not just flora and fauna but also human communities and cultures. Achieving sustainable development "requires challenging not only the self-interest of the wealthy minority, but also the consumption package which is defining our quality of life." [52]

Food self-sufficiency is the first issue to deal with in developing a strategy for rural development. It is a controversial objective, raising questions of autonomy in contrast to the assumed process of increased integration into the global market, with specialization based on monocropping. From an economic point of view, it is generally argued that it is inefficient for a society to favor local production of basic goods when such goods can be produced more efficiently elsewhere. But this is true if and only if both the land and labor involved in such production could find productive use elsewhere. Where land or labor is rendered unproductive through the substitution of tradable goods, a strategy of self-sufficiency can be more efficient.

On the other hand, "in the context of today's societies, in which inequality is the rule and forces discriminating against the rural poor legion, a greater degree of autonomy in the provision of the material basis for an adequate standard of living is likely to be an important part of any program of regional sustainability." [54–55] At the least, local food production contributes to higher nutritional health standards. Food self-sufficiency must be one part of a larger strategy of productive diversification based on principles of greater self-reliance. Rural communities have always been characterized by a diversity of productive activities. This diversity, much of which has been lost to the mistaken importation of large-scale commercial agriculture, must be reintroduced.

Sustainability is also about direct participation by the rural poor in the real power structures that control development. While official development practitioners recognize that such empowerment is a prerequisite for sustainability, few programs actually empower local communities. State agencies and development agencies, following the logic of the neoliberal economic paradigm, generally end up promoting participation that fails to give meaningful power to local communities.

Sustainability must also deal with poverty. "Economic progress itself will depend on involving the grassroots groups to help the affluent find ways to control their consumption and in the organization of development programs that offer material progress for the poor and better stewardship of the planet's resources." [58]

Because international economic integration does not affect all peoples equally, different strategies for sustainable development will apply in different regions. In the regions that have been largely left behind, people have "the unique opportunity to take advantage of their marginal status." [61] Many are of indigenous origin and retain knowledge of ethnobotany, ethnobiology, as possess other cultural knowledge that can contribute to increases in productivity de-linked from the global economy. Further research into technological advances compatible with traditional knowledge, combined with increased transfers of knowledge among cultures, can improve productivity and reduce labor-time in production.

In such regions, redeveloping the peasant economy is essential. "It is not a question of 'reinventing' the peasant economy, but rather of joining with their own organizations to carve out political spaces that will allow them to exercise their autonomy, to define ways in which they will guide production for themselves and for commerce with the rest of the society." [61] These regions have many opportunities to diversify their productive base, develop new uses of renewable energy, and seek new ways of adding value to traditional productive processes.

Local control is especially important in areas now valued internationally for their biodiversity. These are battlegrounds, as local residents struggle to not lose control of their land and knowledge to scientific and environmental elites seeking to identify and protect such biodiversity. Although "biosphere reserves" are one way to protect the earth's dwindling genetic diversity, they have led "to conflicts between local populations which have traditionally coexisted with these species, exploiting them in sustainable ways, until the powerful forces of the market led to increased kill rates that threatened their very survival." [62] Alternative approaches include the "peasant reserve of the biosphere" or "neighborhood restoration clubs" in which local residents are given responsibility for managing resources sustainably, while the international community agrees to guarantee an acceptable quality of life for residents.

Autonomous Development

Given the stark juxtaposition of winners and losers in the dual economy, we need a new strategy that recognizes that the vast majority of rural producers cannot compete with commercial agriculture. Farmers in richer nations simply have too much technological and financial advantage, and overproduction in their home countries creates the political imperative to export to the Third World, often at prices below local production costs. Yet marginal rural producers can support themselves and contribute to society. "The approach suggested by the search for sustainability and popular participation is to create mechanisms whereby peasants and indigenous communities find support to continue cultivating in their own regions. Even by the strictest criteria of neoclassical economics, this approach should not be dismissed as inefficient protectionism, since most of the resources involved in this process would have little or no opportunity cost for society as a whole." [64–65]

This would formalize an autonomous productive system, recognizing the permanence of a stratified society. This can benefit marginalized rural peoples and those developing links with the global economy, making productive diversification possible. "More importantly, such a strategy will offer an opportunity for the society to actively confront the challenges of environmental management and conservation in a meaningful way, with a group of people uniquely qualified for such activities." [65]

Summary of

African Development That Works

by Peter G. Veit, Adolfo Mascarenhas, and Okyeante Ampadu-Agyei

[Published in *Africa's Valuable Assets: A Reader in Natural Resource Management,* ed. Peter Veit (Washington, D.C.: World Resources Institute, 1998), Ch. 8, 223–267.]

Despite the widely recognized failure of many traditional development programs in sub-Saharan Africa, many community-based initiatives have met with success. Between 1988 and 1991, the World Resources Institute (WRI), Clark University, and a number of African development institutions prepared twenty-three case studies of effective community-based natural resource management (CBNRM) from ten countries as part of their joint "From the Ground Up" program. This article, which summarizes a longer WRI report titled "Lessons From the Ground Up: African Development That Works," draws on those studies to identify seven key factors associated with effective CBNRM. Based on those lessons, six policy recommendations are offered.

Seven Key Factors in Effective Community-based Resource Management

1. *Reducing risks to existing sources of livelihood:* The goal of rural Africans in resource management is not conservation but rather sustainable use: it is designed to raise agricultural and livestock productivity to satisfy social and economic needs. Threats to rural livelihoods often increase the pressure on local resources. Those threats come from both within and outside the community, can be either natural or caused by humans, and may be gradual (population increases) or sudden (natural disasters, civil unrest). Whatever the source or severity, such threats cause communities to jeopardize long-term sustainability by overexploiting local resources.

   On the other hand, when livelihoods are relatively free from danger, communities will practice CBNRM. Where household economies are diversified, such as with multiple crops or off-farm activities like trading or hired labor, rural livelihoods are more secure. Where the environment too is diverse and rich, economic opportunities are varied and farmers need not depend exclusively on one natural resource.

2. *Market incentives for sound CBNRM:* If environmental conservation is profitable, effective CBNRM will be promoted and will replace practices that degrade resources. People will engage in environmentally sound farming and animal husbandry only if they see such activities as profitable and if they believe such incentives will remain in place long enough for them to realize a profit. For example, people will engage in agroforestry only if they

believe the market for future timber sales will be sustained until the trees mature and if they can support themselves until they do.

Sometimes, market incentives encourage CBNRM, for example, in communities dependent on nuts or fruit that require them to protect forests in a more sustainable manner. Sometimes, they do not. One Tanzanian community, responding to government incentives to increase maize production, brought so much land under cultivation that the local irrigation system collapsed under the strain.

3. *Cultural practices that promote CBNRM:* "Resource management is most likely to be sustainable when a culture—a shared system of values, beliefs, and attitudes, grounded and governed by traditional norms—encourages it." [229] Given the difficulty of reshaping ingrained cultural practices, CBNRM tends to be practiced when local cultural practices encourage it and is difficult to implant when it does not.

   Religious practices can encourage sound resource management. For example, in Liberia some villagers consider the Gbelaya River the home of the gods of fertility and rain, so all fishing is prohibited.

   Traditional divisions of labor limit CBNRM by restricting the work of certain groups due to gender, age, or other reasons. For example, land management is primarily women's responsibility in most African societies. Due to household and child-raising responsibilities, women rarely have the time to fully practice CBNRM, such as in building terraces or improving irrigation systems. Similarly, traditional authority structures also influence resource management because they empower certain people, generally male elders, to control resources they do not manage.

4. *Security of access to land and other productive resources:* Security of land tenure is critical to sound resource management. Where people believe they have secure access to land and other resources, they make long-term investments that promote sustainable land use. The more secure they are, the greater will be their investment.

   In most African states, most rural land is government-owned, a practice dating back to colonialism. Most land use is by leasehold. Even where such arrangements allow customary tenure, the farmer has no real legal protection, so no real security. Many government policies and practices discourage effective resource use. Often, to gain title inhabitants must clear the land, which promotes soil erosion.

   Because many African governments cannot effectively manage much of the land they claim, most land remains in customary tenure, which often results in more sound resource management. Such systems are relatively adaptable to changing conditions.

5. *Organizational development and management skill in the community:* Resources are managed effectively where communities are well organized and

have the skills to manage their resources. CBNRM also takes hold most often when resource users are able to coordinate their actions. Many factors contribute to such group cohesion: extended kinship ties; ethnic identity; similar and interdependent socioeconomic activities; shared interests; and mutual perceptions. Such factors inhibit competition and conflict while increasing cooperation.

Many of the most effective CBNRM practitioners derive from traditional organizational forms. They often rely on village rules and practices, customary users' rights, and local leaders. Organization can take many forms: village development committees in The Gambia, mobilization squads in Ghana, resistance councils in Uganda, village councils in Tanzania.

6. *Access to appropriate technology, materials, and resources:* The most effective CBNRM practices rely on local land, natural resources, household labor, capital, and indigenous knowledge. Given the growing scarcity of all of these elements, external inputs are often critical to the success of local CBNRM initiatives.

   "Poverty and environmental degradation feed on each other in myriad ways, particularly in resource-based household economies. Most significantly, poverty keeps small-scale farmers from gaining access to critical inputs, thus limiting their management options and constraining sustainable development. Without the resources to participate and invest effectively, local people are often better off not even trying new management practices." [238]

   Government representatives, development agencies, and local nongovernmental organizations (NGOs) can provide needed ideas, knowledge, information, technologies, skills, capital, and material, but most assistance does not arrive when farmers most need it.

7. *Central government support:* CBNRM practices have the best chance of success where central governments support and legitimize such efforts. Though such support has the greatest impact when it takes the form of concrete project assistance that reaches local communities, even without such assistance governments can encourage sound resource management through policies and statements. These encourage local initiatives and help shift community behaviors and practices.

Recommendations for Effective Community-Based Natural Resource Management

Based on these findings, the "From the Ground Up" researchers developed six policy recommendations, each based on concrete initiatives currently under way in at least one African nation.

1. *Policies and legislation that support sustainable development:* National policies, with supporting legislation, need to link socioeconomic development with environmental management, clearly stating the country's goals in the areas of development and the environment.
2. *Market incentives for natural resource management:* Market forces can be more effective than command-and-control regulations in promoting effective resource use. Governments should adopt a three-pronged approach. First, governments should encourage the development of resource-based economic opportunities in rural areas. Second, economic policies need to accurately account for resource-depletion in calculating the relative yields of production practices and ensure that sustainable production is more profitable than unsustainable practices. Third, the government must help put valuable resource-dependent economic activities in the hands of local people, particularly the poor and the women.
3. *Security in land tenure and access to productive resources:* Farmers' access to land and resources must be protected, and policies and legislation should support sound land use and resource management. Governments should also recognize and build on the effective and equitable aspects of customary tenure rights.
4. *Decentralization:* In general, local governments, NGOs, and grassroots organizations are better positioned to respond to local needs than is the central government.
5. *Incorporation of independent input into government decision-making:* Grassroots organizations, NGOs, and other civil groups should play a critical role in ensuring that government policies are environmentally sound. Business leaders are increasingly being incorporated into economic policies, but institutions driven by profit motives often do not make the best decisions when it comes to resource use.
6. *Direct support for rural farmers:* Governments need to increase support for local farmers, including technical assistance, training, and financial assistance.

"The 23 'From the Ground Up' case studies testify to the wealth of unheralded local knowledge and capability in rural Africa, and to the interest and desire of millions of rural resource-users to manage their own resources. CBNRM will grow as more responsibility, authority, and capacity are handed over to the resource users." [256]

Summary of

Challenges to Community-Based Sustainable Development

by Melissa Leach, Robin Mearns, and Ian Scoones

[Published in *IDS Bulletin,* 28, 4 (1997), 4–14.]

Community-based approaches to both environmental and development issues have been highly touted in recent years, yet results have often been disappointing to the implementing agencies and, most important, to segments of the communities involved. This article suggests that there are shortcomings in the underlying assumptions about "community" and "environment" and the relationship between the two. An alternative is offered, based on the notion of "environmental entitlements," which incorporates the politics of resource access and control among diverse social actors.

Flawed Assumptions about Community-Environment Linkages

Community-based approaches to resource management are quite diverse, yet most approaches rest on a set of common assumptions about community, environment, and their relationship. One is that a distinct community exists. "[C]ommunities are seen as relatively homogenous, with members' shared characteristics distinguishing them from 'outsiders.' Equally fundamental is the assumption of a distinct, and relatively stable, local environment which may have succumbed to degradation or deterioration, but has the potential to be restored and managed sustainably." [4] The community, so defined, is deemed to be capable of acting collectively to restore and manage local resources, usually while satisfying its livelihood needs. The community is further assumed to have the goal of achieving harmony between livelihoods and resources, an equilibrium generally assumed to have existed previously until disrupted by other factors—population growth, modernization, the breakdown of traditional authority, inappropriate state policies, and so forth.

While such assumptions offer a useful critique of more harmful practices, they reflect outdated social theory. "The assumptions about community and environment which they rest on are basically flawed, as is the resulting image of functional, harmonious equilibrium between them." [5] First, communities are not homogenous or even geographically bounded. They are diverse, with gender, caste, wealth, age, and belief systems often producing conflicting values and resource priorities. Second, those differences express themselves through power relations and institutions. Powerful actors do not necessarily act in the collective good and often reproduce and reinforce the unequal status of marginalized groups, such as women or poor people.

Similarly, the assumptions about the environment are often flawed, resting on static, linear and equilibrium models of ecological systems. For example, succession theory has guided management of rangelands and forests with its assumptions of linear vegetation development resulting in a stable and natural climax vegetation for any given ecosystem. Since the 1970s, the emerging field of "new ecology" has challenged this approach by examining variability in both space and time. The dynamic interaction of various factors "is thus less the outcome of a predictable pattern of linear succession, but more due to combinations of contingent factors, conditioned by human intervention, sometimes the active outcome of management, often the result of unintended consequences." [6]

Such approaches to social and ecological processes produce different sets of questions:

- Which social actors see which elements of their changing ecologies as resources at different times?
- How do groups with different modes of livelihood or different roles within the division of labor use resources and view their value?
- How do different groups gain access to and control over local resources?

"Seen in this way, the environment both provides a setting for social action and is clearly also a product of such action." [7]

Environmental Entitlements

"Whereas Malthusian perspectives, and conventional approaches to community-based sustainable development, tend to frame problems in terms of an imbalance between overall society/community needs and overall resource availability, an emphasis on social and environmental differentiation suggests that there may be many different, possible problems for different people. In mediating these differentiated relationships, questions of access to and control over resources are key. Hence, the perspective shifts to focus on the command which particular people have over the environmental resources and services which they value, and the problems they may experience should such command fail." [7]

Economist Amartya Sen developed the concept of entitlements to explain that scarcity refers to people not *having* enough, rather than there not *being* enough. The same concept can be applied to the environment, recognizing that the lack of resources is only one of a number of possible reasons for people to lack secure access to the resources they need to sustain livelihoods. Adapting this framework, we can define key terms as follows:

- *Endowments* refer to the rights and resources—such as land, labor, and skills—people have.
- *Environmental entitlements* "refer to alternative sets of benefits derived from environmental goods and services over which people have legitimate effective command and which are instrumental in achieving well-being." [9] These can include direct uses of food, water, fuel, and so forth as commodities; the market value of, or rights to, resources; and the benefits of ecological processes, such as the hydrological cycle or pollution sinks.
- *Capabilities* are what people can do or be with their entitlements. For example, control over fuel allows someone to cook and remain warm.
- *Legitimate effective command* refers to what Sen calls entitlement mapping and is crucial to community-based sustainable development. This implies recognition that resource claims are often contested and that some people will not be able to make effective use of their endowments. "Legitimacy" refers both to statutory rights and to customary rights to resources.

"Through processes of 'mapping,' environmental goods and services become endowments for particular social actors; i.e., they acquire rights over them. Endowments may, in turn, be transformed into environmental entitlements, or legitimate effective command over resources. In making use of their entitlements, people may acquire capabilities, or a sense of well-being." [9] The importance of such an approach is in understanding the mapping process itself, because this is the multistaged process that structures access to and management of resources.

Such an approach quickly leads to a focus on institutions, understood in their complexity. First, institutions are not organizations. They are the rules of the game in society, with organizations acting as the main players in that game. Some institutions, such as laws, have an organizational manifestation, in this case government. Others do not, such as markets, marriage, money, and the like. Understanding institutions in this way allows us to recognize that the law operates within a social context dominated by power relations, which can define access to resources.

Second, transaction costs in the institutional sphere can play an important role in resource use and management. If the transaction costs of state regulation of forest or grazing land are high, for example, those resources will be poorly protected; other institutional arrangements may be more effective.

Third, in contrast to prevailing community-based project models that focus exclusively on local institutions, "it is clear that people's resource access and control, or the 'mapping' processes by which endowments and entitlements are gained, are shaped by many, interacting institutions." [11–12] Some are formal and exogenous, such as the rule of state law. Others are informal and subject to relations of power, authority, and trust among social actors. This approach recognizes that each actor's perception of the "collective good" is shaped by his or her position in society, producing competing visions. This is in contrast to the assumptions of benign complementarity underlying most community-based resource projects.

Fourth, institutions at various levels shape resource use, including at the international level. International trade relations and donor agency policies affect local resource options, as do domestic macroeconomic policies and national laws, such as land reform. "[I]t is frequently the interactions between institutions which lead to conflicts over natural resources, or to competing bases for claims." [12] This contrasts with the overemphasis on local-level institutional development and action in most community-based resource management projects.

"The relationships among institutions, and between scale levels, is of central importance in influencing which social actors—both those within the community and those at some remove from it—gain access to and control over local resources. . . . [T]his perspective uses the insights of landscape history, and of historical approaches to ecology, to see how different peoples' uses of the environment in this context act, and interact with others' uses, to shape landscapes progressively over time." [12]

Summary of

What Is Micro-Enterprise Development for Women? Widening the Agenda

by Linda Mayoux*

[Published in *From Vicious To Virtuous Circles? Gender and Micro-Enterprise Development* (Geneva: United Nations Research Institute for Social Development, 1995), Part 3, 50–58.]

Since the early 1990s there has been a rapid increase in interest in and funding for micro-enterprise programs for women in developing countries. Multilateral and bilateral donor agencies like the World Bank and USAID have supported programs as part of their anti-poverty initiatives, while other development

*Since the publication of this paper the author has been involved in an international project on women's empowerment and micro-finance and has also written a paper for ILO on Enabling Environments for Women's Enterprise. For further information, please contact the author at e-mail address: LMayoux@dial.pipex.com.

agencies adopted micro-enterprise as part of a new "market realism," hoping to increase efficiency in women's income-generation projects. With Bangladesh's Grameen Bank as a model, many have argued that women-focused micro-finance programs are effective not only in reducing poverty but also in empowering women. In this paper, prepared for the United Nations Research Institute for Social Development in preparation for the 1995 Fourth World Conference on Women, the author surveys the evidence and finds that despite occasional successes the majority of programs fail to have a lasting impact on women's incomes.

A Limited Record of Success

While there is some evidence that in certain contexts it is possible to expand small-scale enterprises and increase women's incomes through micro-enterprise programs, "the majority of micro-enterprise interventions to date have failed." [50] Specifically:

- "First, the majority of programs fail to make any significant impact on women's incomes over a sustained period.
- "Second, programs have on the whole mainly benefitted better-off women. Poorer women have either not been reached at all by the programs, or where they have been successfully targeted, have had lesser levels of success.
- "Third, in most cases gender inequalities continue to seriously constrain women's entrepreneurship activities. Even where women's income has increased this has not necessarily radically altered other aspects of their subordination.
- "Finally, what little evidence exists indicates that although female entrepreneurs are more likely to employ women, they frequently employ unpaid family labour and do not necessarily pay higher wages to employees than men." [51]

This raises questions about whether current programs can be changed to increase their likelihood of success, and whether such programs in isolation can ever achieve the many goals claimed in development agency rhetoric.

"The diversity of the small-scale sector on the one hand, and the complexity of constraints posed by poverty and inequality on the other, make the likelihood of any easy 'blueprint' for successful women's micro-enterprise development extremely slim." [51] There are inherent problems in both the "market approach" to microenterprise, which emphasizes assistance to individual women, and the "empowerment approach," which stresses group activities to increase not only incomes but also bargaining power and solidarity.

The Market Approach

Within the market approach, gender lobbies have worked to broaden the agenda to include women's concerns, but they have not questioned the agenda itself. "There are some fundamental inherent contradictions in any attempt to integrate gender concerns into the market framework. In relation to micro-enterprise development it is clear that attention to purely economic factors, particularly rigid definitions of 'efficiency' and 'cost-effectiveness,' are unlikely to enable significant numbers of poor women to become entrepreneurs." [51–52]

In the first place, the economic emphasis fails to address the varied needs of women entrepreneurs. For many poor women, increasing income may not be as important as gaining greater income security, reducing overall work time, or improving control over income. Second, the economic and technical biases in the approach limit consideration of power relations or gender inequality. Even though gender inequalities are clearly identified as obstacles to female entrepreneurship, policies give them scant attention, rarely addressing macro-level welfare policies or the unequal burdens between women and men. Finally, the emphasis on "cost-effectiveness" and "efficiency" requires a focus on an extremely small number of women who already possess the skills, resources, and experience to make use of microenterprise opportunities and who work in industries and regions with growth potential. "This begs the question of what is to be done for the vast majority of women who do not fit into this very small category." [52]

The Empowerment Approach

"The empowerment approach has attempted to address some of the shortcomings of the market approach—attention to social as well as economic issues, greater recognition of the importance of structural inequality—and has an explicit commitment to poor women per se. In practice, however, the evidence suggests that the more conventional types of 'participatory' project and program have generally failed to make significant impact." [53] While group-loan programs are more relevant to the needs of poor women and generally reach larger numbers of poor women, they are still limited by the circumstances in which women find themselves. Often the income gains are small. Many times the men in participants' households control the use of the funds and retain the profits, contributing little to women's well-being or empowerment. Some "small participatory groups of women which include at least some women with higher levels of skill and greater access to resources, who are prepared to challenge norms of gender subordination, have been relatively successful." [53] These successes, however, are the exception.

The more populist and instrumentalist forms of empowerment programs, hitherto the dominant forms in practice, suffer from a number of inherent tensions. First, poor women generally want and need increases in income and have little time for activities that do not produce some immediate benefit. The incorporation of social goals that are difficult to measure raises problems of evaluation and accountability. Second, participation has its own costs. Studies have shown that the women who participate are those with few family responsibilities—unmarried girls and older women past childbearing age—and those better-off women less subjected to norms of gender subordination. Most of the poorer women lack the time for such programs. Third, women's needs are diverse and complex and may require very different types of micro-enterprise assistance. "The more conventional types of projects have often failed because, in their response to women's immediate needs, they have failed to address longer term underlying constraints." [54]

On the other hand, feminist attempts to incorporate measures that address those constraints have often failed because microenterprise programs in isolation are too limited, and because models imposed from outside, no matter how well intentioned, offer poor women few answers to their needs. "Ultimately, there is a need to combine more participatory strategies at the local level with sectoral strategies at the macro-level." [54]

Widening the Agenda

The conclusions from this study are not all negative. Attention to gender within mainstream programs is a necessary corrective, both to gender biases in many microenterprise programs and to programs that saw women only in terms of their reproductive role. Also, some of the most innovative initiatives are recent and still undocumented.

However, the tensions within each of these approaches are not necessarily resolved by current moves toward a middle ground of more participatory market-led programs. "The enthusiasm for programs like collective credit schemes is still mainly in terms of cost-effectiveness. . . . Within the empowerment approach, moves towards 'market realism' can only increase the problems associated with addressing the needs of poor women." [54] The ultimate logic of "efficiency" and "cost-effectiveness" in both the market and empowerment approach is to exclude poor women.

In any move toward the center ground any microenterprise strategy for women will somehow have to reconcile competing tensions between cost-effectiveness, participation, and wider-impact—each of which in turn has its own inherent problematic. "Arguably, a commitment to grassroots participation only makes sense within a broader political commitment to equity. Although this has been recognized in general terms in relation to class and caste, many development agencies are still unwilling to accept this in relation to gender. At the same time, it is unlikely that micro-enterprises will succeed in addressing women's aims unless they also link to wider movements for change." [55]

"What is disturbing about much of the recent enthusiasm for micro-enterprise development for women is its promotion in the wider context of neo-liberal market reform, particularly 'rolling back the state,' the removal of welfare provision and the dismantling of all forms of labor protection. It is also widely seen as a viable and less socially and politically disruptive alternative to more focused feminist organizational strategies. All the evidence indicates that there are likely to be serious limitations on any micro-enterprise strategy for poor women in isolation." [56]

First, inadequate welfare provisions constrain microenterprise success for poor women. Lack of childcare, health care, decent housing, and basic infrastructure like accessible, safe water increase the time women spend on unpaid domestic work. Limited educational opportunities hamper the success of training programs.

"Second, for many poor women, improving labor legislation and labor rights is likely to be more important than micro-enterprise provision. . . . Many poor women would actually prefer stable employment rather than insecure entrepreneurship." [56]

"Finally . . . lack of resources and lack of power are crucial constraints on women's entrepreneurship and the effectiveness of micro-enterprise programs to date. . . . Without measure to address gender inequality, micro-enterprise programs may merely increase women's workload and responsibilities without increasing their control over income." [56–57]

"It is unlikely that micro-enterprise development will prove to be the rosy 'all-win' solution assumed in much of the promotional literature. Even in terms of narrow aims of increasing beneficiary incomes, micro-enterprise development is unlikely to succeed for the vast majority of poor women (rather than the small number of better-off women) unless it is part of a transformed wider agenda. There are particularly serious implications for any reliance on micro-enterprise programs as the main focus of any wider strategy for poverty alleviation and change in gender inequality." [57]

PART X
Reforming Global Institutions

Overview Essay

by Kevin P. Gallagher

> Markets are sustainable only insofar as they are embedded in social and political institutions. . . . It is trite but true to say that none of these institutions exists at the global level.
> —Dani Rodrik, "The Global Fix" (1998)

The growing mismatch between prevailing economic practice and sustainable development imperatives has been a major theme throughout this volume. Reliance on market forces as the major solution to this mismatch has proved not only to be ineffective, but also to exacerbate existing problems. For a transition to a sustainable society to be successful, markets have to rest on a framework that enables their energies to flourish and to be used for socially and ecologically sustainable development.

This essay presents a growing number of writers who are envisioning a new era of global policymaking for sustainable development. Basing their positions on critiques of the current global structure, these authors advocate reforming the international financial architecture, revitalizing and strengthening existing global institutions, and replacing existing institutions with new ones. Virtually all of these authors stress the need to shape such institutions to reflect the new realities of the global arena. Chief among these new realities is the entrance of civil society, in all its forms, into the world economy.

Reforming the International Financial Architecture

There are as many sweeping proposals to reform the international financial architecture as there are journals, magazines, televisions, and web sites devoted to the world economy. A recent article in the *Journal of Economic Perspectives* (JEP) notes:

> Many of these ideas are not new, but they are being vented more forcefully, and taken more seriously, than at any time since Harry Dexter White and John

> Maynard Keynes masterminded the creation of the World Bank and the International Monetary Fund at the Bretton Woods conference at the end of World War II. (Rogoff 1999, 21)

The starting point for discussion of the global financial architecture is with the institutions founded in Bretton Woods after World War II: the World Bank and the International Monetary Fund (IMF), in addition to the international framework in which modern multilateral trade negotiations took place, the General Agreement on Tariffs and Trade (GATT), and its successor, the World Trade Organization (WTO).

International Monetary Fund

Robert Browne notes that while the initial purpose of the International Monetary Fund (IMF) was to maintain stable exchange rates in the developed countries, it altered and extended its functions following the debt crises of the early 1980s. During the 1980s, the IMF stepped in to issue a number of short-term loans to countries in need. However, these loans became linked to a number of "conditionalities." While the aim of such conditionalities is to secure fiscal and monetary balance, they have also involved extensive social and environmental effects. For example, the IMF has recently forced Brazil to cut environmental spending by two-thirds. Devesh Kapur, in summing up these effects, comments that "despite the IMF's intentions, the realities of local politics often resulted in outcomes that were socially regressive, economically myopic, and only modestly able to put countries on a sustainable growth path" (Kapur 1998).

There are many proposals for reforming the IMF. An exhaustive, recent review of such proposals is offered by Rogoff (1999). From across the political spectrum, these proposals can be grouped into four categories: tinkering with the organization to help it fulfill its current goals; returning the IMF to the role of international lender of last resort; revamping its mandate to respond to contemporary realities; abolishing the IMF altogether (Fischer 1999; Gianini 1999).

The World Bank

Also founded at Bretton Woods, the World Bank was to redistribute the credit and resources supplied by developed country members in the form of loans to emerging developing countries. The bulk of this support was for large-infrastructure projects and the implementation of Structural Adjustment Programs (discussed at length in Part VI of this volume). To its credit, the World Bank has been far more open than the IMF to the idea of sustainable development. However, World Bank economists have recently acknowledged that "economy wide reforms have been implemented without adequate consideration of their environmental and social, as well as economic impacts" (Munasinghe 1998).

While poverty reduction had been debated in World Bank circles for quite some time, before the 1980s sustainable development had no consistent place in the World Bank's development strategy or lending activities. Environmental damage resulting from project implementation was relegated to the realm of negative externalities considered to be an unavoidable trade-off of the development process (Reed 1997). This policy resulted in a highly organized and effective public outcry in the form of nongovernmental organization (NGO) advocacy campaigns on a global scale.

This outcry led the World Bank to make a number of institutional, procedural, and policy reforms. A Vice Presidency for Environmentally Sustainable Development has been created with authority over other departments, and environmental technical units have been formed at the regional level. Procedurally, the World Bank has expanded the number and scope of project-level environmental assessments and now allows for the public release of project information documents. In addition, the public can now submit complaints to review panels. Finally, the World Bank has also revised the composition of its lending portfolio and encourages governments to prepare National Environmental Action Plans (NEAPs) as the basis for lending operations to all economic sectors (Reed 1997; World Bank 1995).

Carlos Heredia acknowledges that changes have been made on the social front as well. However, Heredia argues that the World Bank's changes have not lead to many concrete results because these issues have been treated as "add-ons" rather than as a central objective of World Bank operations. In addition, he and others criticize the World Bank for focusing on the symptoms of unsustainable development rather than on their structural roots (Seymour and Dubasch 1999).

A major study conducted by **Jonathan Fox and David Brown** highlight the limitations of World Bank policy reforms in both environment and poverty areas. While advocacy campaigns over past projects did secure promises of policy changes that would affect future projects, Fox asserts that many of these promises did not become reality. Efforts at reform have had some effect: as a result of social and environmental "trip wires" set by earlier NGO campaigns, the number of especially damaging projects has slightly decreased. However, **Hilary French** notes that "a review of fifty recent loans found that few paid much heed to environment and social matters. In addition, whereas in 1993 it was found that some 60 percent of adjustment loans included environmental goals, a recent study concluded that this share has now decreased to 20 percent."

Fox concludes that if the World Bank is to meet minimum sustainable development goals, the following three reforms must occur: at the international level, policies must be supported and controlled by committed reform elements within the Bank; at the government level, they must be designed to target support specifically to agencies already controlled by reformist, pro-accountability elements within the state; within civil society, they must include informed participation by representative social organizations from the beginning of the design process.

World Trade Organization

The World Trade Organization (WTO), originally proposed at Bretton Woods, was established in 1994 as an outgrowth of global trade talks under the GATT. The GATT had two principal functions. First, it served as the home for a number of agreements on nondiscriminatory reductions in tariffs, quotas, and other restrictions on trade in goods and services. Second, it managed dispute settlements arising under these agreements. In 1995, GATT members voted to increase the role, expand the scope, and change the name of the organization to the World Trade Organization. French notes that while many cheered the birth of the WTO as the beacon of a new era of global prosperity, others criticize the global trade regime for elevating the rights of global corporations over national governments, local communities, and the environment (see, e.g., Mander and Goldsmith 1998). Some go so far as to refer to the global trading system as a "corporate managed trade" regime (Wallach 2000).

Like the World Bank and the IMF, the WTO has recently come under intense scrutiny. Indeed, the November 1999 Seattle Ministerial Conference of the WTO was marked by large-scale demonstrations and sharp deadlock between the negotiators themselves. Proposals for changes can be grouped into three broad categories: those that call for working within the WTO's current framework; those that demand a complete overhaul of the WTO mandate, organization, and rules; and those that argue that the WTO should be abandoned for new, more accountable institutions. The first two are discussed here. Calls for new institutions are dealt with in the last part of this essay.

Dan Esty represents those wishing to work within the system. He has suggested a number of environmental reforms for the WTO. Among others, he stresses that there are "win-win" issues where both trade and environmental objectives can be pursued at the same time. The most notable of these is the opportunity to expand market access by attacking government policies such as environmentally harmful subsidies that are also trade barriers. In addition, Esty asserts that environmental policies should be pursued whenever possible without resorting to trade measures. Where Multilateral Environmental Agreements (MEAs) represent an attempt to internalize negative environmental externalities, efficient operation of world markets demands reinforcement of these efforts (Esty 1996). He has also called for increased transparency and participation by NGOs so that civil society can act as a watchdog of the WTOs policies (Esty 1998).

A far more dramatic overhaul is proposed by Martin Khor, a leader in the Third World Network. He argues that agreements on international property rights (TRIPS) should be amended to allow countries the right to choose patenting life forms; that food produced for domestic consumption and the products of small farmers should be exempted from the Agriculture Agreement's disciplines on import liberalization, domestic support, and subsidies; and that investment provisions of the WTO (TRIMS) should be reformed to allow developing countries the right to have "local content" policy (i.e., to require firms or projects to use a certain minimum amount of local materials) so to provide domestic industrial development.

A further set of proposals is aimed at making the WTO more democratic, transparent, and accountable. Specifically, Khor and others argue that any changes to rules or proposals for new agreements should be made known in their draft form to the public at least six months before decisions are taken so that civil society in each country can study them and influence their parliaments and governments; all WTO members must be allowed to be present and participate in discussions and negotiations (including in informal groups and meetings where many key decisions are made). The practice of small informal groups making decisions for all members should be discontinued; parliaments should be constantly informed of proposals and developments at WTO and should have the right to make policy choices regarding proposals in WTO; and civil society should be given genuine opportunities to know issues being discussed and to express their views and influence the outcome of policies (Khor 1999).

Revitalizing and Strengthening Existing International Institutions for Sustainable Development

In contrast to the international financial architecture, there exists a set of international bodies that *are* mandated to work on social and environmental issues—mostly established at world meetings in and around the United Nations. Just a few of them are featured in this essay: United Nations Development Programme (UNDP); United Nations Environment Program (UNEP); United Nations Commission on Sustainable Development (UNCSD); and the many outgrowths of Multilateral Environmental Agreements. Such efforts are quite small relative to the attention and funding enjoyed by the international financial architecture. It is therefore not surprising that the countless international conferences and meetings on sustainability have not resulted in significant change.

Revitalizing and strengthening these institutions will be essential for a transition to a sustainable society.

UNDP was established by the U.N. General Assembly in 1965 and merged two existing U.N. entities: the Expanded Programme of Technical Assistance and the Special Fund. UNDP's mission is "to help countries in their efforts to achieve sustainable human development by assisting them to build their capacity to design and carry out development programmes in poverty eradication, employment creation and sustainable livelihoods, the empowerment of women and the protection and regeneration of the environment, giving first priority to poverty eradication" (UNDP 1999).

UNEP was launched by the U.N. Conference on the Human Environment held in Stockholm in 1972. Its mandate is to catalyze and coordinate activities to increase scientific understanding of environmental change and to develop environmental management tools. In the form of environmental management tools, UNEP's efforts have led to conventions to protect stratospheric ozone, to control the transboundary movement of hazardous wastes, and to protect the planet's biological diversity (UNEP 1999).

A watershed event that merged the development and environment agendas occurred in 1992. More than 100 heads of state met in Rio de Janeiro, Brazil, for the United Nations Conference on Environment and Development (UNCED). The Earth Summit, as it is often referred to, was convened to address environmental protection *and* socioeconomic development. The assembled leaders signed the Framework Convention on Climate Change and the Convention on Biological Diversity; endorsed the Rio Declaration and the Forest Principles; and adopted Agenda 21, a 300-page plan for achieving sustainable development in the twenty-first century. The Commission on Sustainable Development was created in December 1992 to ensure effective follow-up of UNCED and to monitor and report on implementation of the Earth Summit agreements at the local, national, regional, and international levels (CSD 1997).

UNCED formally charged UNDP, UNEP, and UNCSD with the goal of elaborating strategies and measures to halt and reverse the effects of environmental degradation while continuing to raise the standards of living in all countries. As **Alexandrea Timoshenko and Mark Berman** note, for UNDP the challenge has been to integrate an environment component into its operational activities, while for UNEP the challenge has been to add a human development component to its environmental efforts. The UNCSD has distinguished itself from these organizations by being the primary forum for the review of the implementation of Agenda 21 (Mensah 1996).

One of the successes of these institutions, especially of UNEP, has been in providing a forum for international treaty negotiations. There are now more than 230 international environmental treaties, more than three-fourths of them having been negotiated since UNEP was established. These accords cover such issues as atmospheric pollution, ocean despoliation, endangered species protection, hazardous waste trade, and the preservation of Antarctica (Susskind 1994). However, many of these treaties address a single threat and are not thought of in broad sustainability terms.

Since UNCED there has been an effort to broaden the scope of global treaty making. Nowhere is this more evident than in the global climate change regime—a pair of treaties that attempts to address the environmental and social ramifications of the world's entire industrial system. The United Nations Framework Convention on Climate Change (UNFCCC) and the subsequent Kyoto Protocol develop an apparatus that has the goal of significantly altering current patterns of global energy consumption to reduce greenhouse gas emissions. Both in process and product, these treaties have expanded the set of traditional actors and tools usually called upon in the global process (Moomaw et al. 1999). Innovative elements of the global climate regime include allowing civil society—from corporations to community groups—to play a larger role in the process; allowing developing nations to have "common but differentiated" responsibilities in the name of fairness and equity; setting up a billion-dollar fund (the Global Environment Facility [GEF]) to help with implementation; and proposing to use a set of proactive economic instruments to give nations an incentive to comply with treaty requirements.

UNCED has also created the space for international institutions to work together to implement such treaties. Timoshenko and Berman outline how UNDP and UNEP have joined forces with the United Nations Industrial Development Organization (UNIDO) and the World Bank to work on program implementation under the Multilateral Fund of the Montreal Protocol. Collaboration between UNEP and UNDP, along with the World Bank, has also occurred in the context of the Global Environment Facility under the global climate change and biodiversity regimes.

Not only was UNCED a watershed event in forging the links between development and environment, and in creating space for institutions to work together, but it also began a new era of inclusion. Over 20,000 NGO representatives attended UNCED, outstripping official representatives by more than two to one. This trend has carried over into the many negotiations for international treaties and conferences that have occurred since then. Over 4,000 participated in the Cairo Conference on Population and Development in 1994. Many also attended the World Summit for Social Development in Copenhagen the following year. While these trends are positive, NGOs still face many obstacles. At this writing, there are still "no formal provisions for public review and comment on international treaties, nor are there mechanisms for bringing citizen suits to the World Court" (French 1995).

While there are some success stories, there is a growing consensus that these treaties have so far failed to overcome the barriers to sustainable development. The ambitious goals of the global climate-change treaties have proved difficult to translate into practice, as differences among nations and negotiating deadlocks persist. A number of studies have evaluated whether these treaties have been able to change the behavior of states, solve environmental problems, and do so efficiently and equitably. A sweeping review of these studies concludes that most have not (Bernauer 1995). French argues that the main reasons for these shortcomings are that member governments have generally permitted only vague commitments, lax enforcement, and weak funding for treaty implementation. To revitalize and strengthen these institutions and treaties, proposals include further increasing the transparency and degree of civil society participation; imposing sanctions on violators; creating economic incentives for compliance; lowering transaction costs; fostering social learning; and increasing the funding and therefore the legitimacy and authoritativeness of the existing institutions (Young 1999).

Vandana Shiva presents a number of deeper criticisms. She argues that the existing global institutions and treaties are veiled attempts to conserve economic and political power rather than to work toward sustainable development. Using the Global Environment Facility as a case in point, she asserts that the GEF operates on a "pay-the-polluter rather than on a polluter-pays principle." The GEF helps channel funds to developing countries for carbon sequestration to meet climate change commitments. In Shiva's view, such a strategy promotes nonsustainable forestry in the South as a way to offset emissions from new power plants in the United States. Thus an institution whose declared purpose is global environmental protection serves to promote the interests of wealthier nations in practice. This critique leads Shiva and other critics of existing international institutions to look to new actors and the possibility of new institutions to fill the gap.

New Actors, New Institutions, New Roles

Those who propose a more fundamental departure from existing institutions are worried about the rising power of multinational corporations (MNCs). In 1970 there were some 7,000 MNCs. Today there are at least 53,607 with at least 448,917 subsidiaries (Worldwatch 1998). One indication of rising corporate power, the North American Free Trade Agreement (NAFTA) marks the first time in history where corporations have been granted the ability to overturn domestic laws of their host governments. NAFTA's Chapter 11 permits private investors to initiate direct action lawsuits to challenge domestic law. A number of such cases have been brought before NAFTA, many targeting environmental, health, and safety regulations (Mann and von Moltke 1999). Some have resulted in the compulsory payment of millions of dollars in compensation to corporations by host governments.

A similar move to expand corporate rights, was attempted on a global scale in 1995, when twenty-nine nations of the Organization for Economic Cooperation and Development (OECD) began negotiations for a Multilateral Agreement on Investment (MAI). Under the MAI, the world's corporations were to receive legal privileges that would allow them to sue governments for monetary damages—an agreement that would have forever changed the balance of power between governments and citizens on one hand and MNCs on the other (Stumberg 1998). After extensive pressure from NGOs and developing country negotiators, the MAI has been set aside, but it could reappear in another form.

While concern has risen over the increasing power of MNCs, a countervailing trend can be seen in the expanding influence of international NGOs. The number of NGOs working across international borders has soared, climbing from 176 in 1909 to over 23,000 in 1998. Environmental groups are rising steadily as a proportion of the total, rising from 2 percent in 1953 to 14 percent in 1993 (French 1999). In addition to networking and organizing in such innovative arenas as e-mail and the Internet, NGOs have also educated millions of people about sustainable development issues, and then harnessed the power of a knowledgeable citizenry to effect change. Such groups can take credit for raising sustainable development concerns at the World Bank and in negotiations for NAFTA, and for bringing international attention to the flaws of the Multilateral Agreement on Investment and the WTO.

Proposals for new global institutions are often written off as grand schemes that are politically unrealistic. Yet such criticisms fail to recognize the many global institutions created just in the last decade. Indeed, the reunification of Germany, the unification of Europe, the establishment of the Russian Federation, the WTO, the creation of the new South Africa, and many others are all recent institutional accomplishments of great magnitude. Building on this momentum, many writers hope that a new set of institutions will evolve into a regime more compatible with sustainable development.

Proposals for alternative global institutions date as far back as the early 1970s when the stability of world currency markets was in question and developing nations called for a New International Economic Order. Such proposals, far too many in number to review in this essay, can be grouped into three categories: financial institutions, institutions for equity and development, and institutions for the environment. Most of the proposals for alternative financial institutions are mainly concerned with replacing existing financial institutions so as to better meet their original mandates (for an in-depth survey, see Rogoff 1999). Some proposals go beyond this narrower goal to encompass the goals of sustainable development.

A leading advocate of new institutions to foster global equity and development is the economist Paul Streeten. In addition to advocating a strong peacekeeping apparatus, Streeten defines three essential functions of a new institutional structure: a center that generates balance-of-payments surpluses for the benefit of the developing world; mechanisms to convert such surpluses into long-term loans or equity investments; and the industrial and technological capacity to produce the capital goods or intermediate products required for industrialization. Streeten argues that a number of institutions would have to buttress these functions, ranging from world central bank and international debt facilities, to global health, technology, and environmental agencies (Streeten 1995).

David Vogel has shown how new institutions can lead to a strengthening of consumer and environmental protection regimes. His book documents how in the establishment of the European Union, European nations were able to design an entire parliament and regulatory system to counter and steer European economic integration to better ends (Vogel 1996).

On the environmental front, the establishment of a Global Environmental Organization (GEO) that would provide a counterpart (or counterweight) to the WTO has been proposed. Rather than surrendering national sovereignty, a GEO would allow countries to regain their control over their environment by giving them an instrument by which to coordinate with other countries on measures aimed at transboundary pollution and other threats to the global commons (Esty 1996; see also Runge 1994). Indeed, such ideas have even been proposed by leaders of the WTO. In March of 1999 the outgoing secretary general of the WTO remarked:

> With the WTO we are poised to create something truly revolutionary—a universal trading system bringing together developed, developing, and least-developed countries under one set of international rules, with a binding dispute settlement mechanism. I would suggest that we need a similar multilateral rules-based system for the environment—a World Environment Organization to also be the institutional and legal counterpart to the WTO. (Ruggiero 1999)

The increasing globalization of both economic and environmental issues requires a global institutional response. The international organizations of the twenty-first century will need to combine the financial and social functions proposed by Streeten with the environmental mandate urged by Esty, Runge, and Ruggiero. Certainly, the scope and importance of the issues surrounding sustainable development will dominate the debate over the practices of existing institutions, and proposals for new institutions, during the decades to come. Only when a stable international structure has been established that can respond to global social and environmental issues on a proactive rather than reactive basis will sustainable development on a global scale become a reality.

Summary of

Alternatives to the International Monetary Fund

by Robert S. Browne

[Published in *Beyond Bretton Woods: Alternatives to the Global Economic Order*, ed. John Cavanagh et al. (London: Pluto Press, 1994), 57–73).]

The International Monetary Fund (IMF) has been widely criticized for ignoring the political, social, and environmental realities of the developing countries it has operated in. Much of the criticism of the IMF arises from the fact that its original mandate is no longer suitable for today's complex global economy. This chapter offers a reconceptualization of the type of global monetary institution most appropriate for today's world.

The First Fifty Years of the IMF

Founded as one of the Bretton Woods institutions at the conclusion of World War II, the IMF was charged with maintaining stable exchange rates in the global economy. Exchange rate stability was seen as key to expanding international trade, which, in turn was viewed as a prerequisite for a prosperous international economy. All currencies were set at an agreed-upon rate of exchange with the U.S. dollar. The IMF was also charged with expanding the supply of money (liquidity). During the IMF's first thirty years, the IMF was most directly involved with activity in the developed countries.

In the early 1970s, however, the United States de-linked the dollar from gold, which undercut the IMF's role in the maintenance of international currency values. In the years that followed, the IMF gradually shifted the focus of its activities from the developed to the developing countries. The emergence of the debt crisis in 1982 created a new opportunity for the IMF to play a major role in shaping international financial affairs. It became a major lender to the growing number of poor countries besieged by unmanageable international debt obligations. It chose to link these loans to a number of "conditionalities."

It soon became evident that, often, the indebted countries were suffering not from temporary payments imbalances but from longer-term structural problems that were bringing the countries close to insolvency. Country after country began failing to meet IMF loan criteria with the result that IMF loans were canceled. Simultaneously, angry publics began taking to the streets to protest against the negative social effects that the loans were inflicting. Until then the IMF had an official position of being ideologically neutral, but as its involvement with the poor, indebted, developing countries expanded, the IMF's bias toward the market system became ever more evident. By the 1990s, countries' access to IMF resources had become highly conditioned on the extent of their adherence to market policies.

An International Monetary System for the Twenty-First Century

The massive expansion of the global economy, the quadrupling of the number of independent countries, the rise of a plethora of new global economic actors such as multinational corporations, new developments in communications, and the pressure from enormous amounts of Third World debt are a few of the factors whose impact on the global economy warrant a reconceptualization of the kind of international monetary fund necessary for today's world.

Criticisms of the IMF can be grouped into those that object to the conceptual outdatedness of the original mandate given to the IMF and those that are concerned with structural and operational flaws in the ways the IMF carries out its mandate. The conceptual inadequacies include:

- Absence of an independent reserve currency
- Absence of a mechanism to stabilize exchange rates
- Absence of a mechanism to provide macroeconomic direction to the global economy
- Absence of a lender of last resort.

The structural and operational deficiencies that prevent the IMF from fulfilling its current limited role as the monetary overseer for the world economy include:

- Undemocratic voting structure
- Absence of public participation and excessive confidentiality
- Medium-term loans used for long-term needs
- Duplication of efforts at the World Bank

- Freeze on the issuance of Special Drawing Rights (SDRs)
- Irrational allocation of SDRs.

These inadequacies and deficiencies can only be addressed through the replacement of the IMF with a new institution.

A Visionary Alternative

John Maynard Keynes foresaw that the attempt to use the currency of a particular country as the reserve for the world would not be viable for the long term, and argued that a more far-reaching institution than the IMF was needed, one that would have the ability to use its own currency and to function similar to a global central bank.

Replacing the dollar with a neutral reserve currency would introduce a greater measure of equity in relations between the poorer and richer nations. Sticking with the dollar links the fate of virtually all countries to U.S. economic policy. Of course, the very thought of so powerful institution may seem daunting, but it must be remembered that the G-7 is implicitly acting in such a manner at present. Allocating this responsibility to the G-7 undermines the concept of a global democratic order. However, it would have to be ensured that a new global central bank did not mimic the IMF and become a tool of rich nations.

A Pragmatic Alternative

Unfortunately, a global central bank such as the one Keynes had envisioned is little more than a hope. However, the need for an international monetary institution has not disappeared. Despite the prevalence of floating exchange rates, countries still need short-term payment funds to carry them through disequilibria. The need for a more effective vehicle for providing international monetary stewardship remains pressing. But the current IMF arrangements, in which the vast majority of countries are denied any real influence, must be altered. The asymmetry in the treatment meted out to the poor and the rich countries should be eliminated.

Serious attention should be given to the creation of a country-neutral reserve currency in the form of an expanded SDR. Whereas SDRs were initially distributed on the basis of IMF quotas, they might better be distributed on a needs-determined basis, as the developing countries have urged. One might wish to revisit the proposal for the IMF to make a one-time issuance of "special purpose" SDRs, allocated to the poorest countries on a needs-determined basis, with the proviso that they could only be used for repayment of official debt.

Conclusions

The current global economic system favors rich over poor nations. While the monetary system cannot solve this problem itself, it can assist in the solution. A logical next step would be for the IMF to test the feasibility of an entirely new type of international currency that would maintain a constant purchasing power based on a basket of commodities. Democratizing currency so to speak, should be coupled with democratic voting arrangements. However, a new mindset could bring on a visionary alternative of an entire new (and more democratic) institution possibly funded by a tax on international transactions or from revenues yielded from the sustainable use of some of the international commons.

Summary of

The World Bank and Poverty

by Carlos A. Heredia

[Published in *Lending Credibility: New Mandates and Partnerships for the World Bank*. ed. Peter Bosshard et al. (Washington, D.C.: World Wildlife Fund, 1996), 229–242.]

> Sustainable poverty reduction is the World Bank's fundamental objective. It is the benchmark by which our performance as a development institution should be judged.
>
> —Lewis Preston, World Bank president, 1993.

While the statement above shows an acknowledgment of the seminal role that poverty reduction should play in all aspects of the World Bank's development efforts, thus far its rhetorical commitments have not been translated into results on the ground. This paper critically examines the World Bank's poverty reduction efforts and describes five challenges that the Bank faces in realizing its poverty reduction goals.

Critique of the Bank's Role

Perhaps the principal reason why poverty reduction is not being achieved by the World Bank is that, contrary to its stated goals, the Bank has adopted a "compensatory" approach to poverty reduction—one that is an "add-on" rather than a central objective of its operations. In addition, existing World Bank policies toward poverty reduction focus on the symptoms of poverty rather than on poverty's structural roots. The structural roots of poverty can be grouped into seven categories:

- Lack of democracy
- Lack of access to means of production and resources by the majority of a population
- Lack of adequate mechanisms for savings and distribution
- Orienting national economies to foreign markets rather than meeting local needs
- Erosion of the role of government's role in administering social services
- Overexploitation of natural resources and contamination of ecosystems
- Policies that generate greater monopolization and therefore polarization.

The World Bank's macroeconomic policy prescriptions, even when coupled with compensatory programs for the poor, have sometimes exacerbated the structural roots of poverty and inequality. The Bank's promotion of structural adjustment programs (SAPs), its failure to address the issue of debt reduction, and its close relationships with economic and political elites, all compromise its ability to serve as an agent of poverty-reducing development.

The Bank has argued that SAPs are necessary to get countries in economic crisis back on a growth trajectory that will lead to long-run increases in income and employment. But many of these programs have immediate, disproportionate impacts on the poorest sectors of society. UNICEF reports have shown how SAPs have adversely affected human and ecosystem health, nutrition, education status, and labor interests. In Mexico, SAPs aimed at deregulating the rural sector reduced Mexico's output of basic grains and pushed many peasants to migrate to cities and to the United States. This deterioration of "social capital" is not captured in macroeconomic performance indicators.

A related issue is the burden of foreign debt. From 1982 to 1990, developing countries received $927 billion in total resource flows from the developed countries while remitting $1,345 billion in debt service. In 1995, the debt burden of developing countries was $1.9 trillion, of which $304 billion (roughly 17 percent) was owed to the World Bank and the IMF. By diverting funds away from productive investments, debt presents a major obstacle to eradicating poverty. Politically, debt forces governments to be more accountable to external donors than to their own citizens. NGOs have long asserted that only serious debt reduction or cancellation, along with a commitment to maintain aid levels, will solve the problem.

World Bank and IMF projects are often biased toward international commercial and political interests at the expense of domestic capital markets and political stability. This is manifest in the Bank's unwillingness to allow market intervention for poverty reduction while making large loans to bail out private banks and investors. The World Bank has also sent wrong signals on democracy and accountability, presenting countries such as Chile (under Pinochet) and Indonesia as models of successful economic management.

The World Bank's Poverty Strategy

The World Bank has stated a three-pronged strategy for poverty reduction:

- Promote broad-based growth that makes efficient use of the poor's main asset: labor
- Provide the poor with access to social services
- Recommend that safety nets be established to protect the most vulnerable in societies.

To implement these strategies, the World Bank has promoted labor-intensive growth, social sector lending, poverty-focused adjustment, a program of targeted interventions, social investment funds (SIFs), and consultative groups to assist the poorest. These programs have had varying degrees of success. Programs designed to stimulate labor-intensive growth have sometimes had perverse consequences, as in the case of Mexico. Lending in the energy sector appears to favor capital-intensive investment, and efforts to deregulate labor markets have undermined minimum wages and health and safety standards. Although the World Bank has significantly increased social sector lending, dependence on external loan funds requires countries to accept the Bank's guidelines on social policy.

Targeted Interventions and Social Investment Funds are intended to reach those sectors of the population who are least likely to benefit from economic growth. These efforts serve to alleviate some of the symptoms of poverty but fail to reach its structural roots. For example, the Zapatista rebellion started in the Mexican state of Chiapas, where the Mexican SIF had the highest per capita social expenditure. The Consultative Group to Assist the Poorest (CGAP), launched in 1995, is based on the Grameen Bank's microlending model; it is too early to predict how successful this program will be.

Challenges

Current World Bank president James Wolfensohn has reaffirmed previous World Bank president Lewis Preston's commitment to poverty reduction as a main goal of the Bank. The World Bank faces many challenges in realizing this goal. The overarching issue is the need to redirect the Bank's poverty strategy away from reliance on compensatory strategies to one that focuses on the structural roots of poverty. Five dimensions of this challenge are

1. **Putting Poverty on the International and National Agendas.** While many nations have shifted their language from poverty reduction to poverty eradication, altering policies accordingly will require a great deal of political will on the part of governments and multilateral institutions. The World Bank could play a key role by showing willingness to work with other institutions such as the United Nations and by employing leverage to encourage its borrowers to focus on poverty.
2. **Equity.** The World Bank needs to incorporate social equity in its approach to sustainable development. The Latin American and Southeast Asian experiences show that countries with a less-skewed income distribution are more likely to be able to reduce poverty. Moreover, it has been shown that unequal societies tend to be politically and socially unstable, and that this is reflected in lower rates of investment and growth.
3. **Forging Consensus on Structural Adjustment and Debt.** Recent constructive dialogue on reforming structural adjustment policies needs to continue. The World Bank has shown openness to changing its SAP criteria to include poverty-reduction objectives. It has also begun to consider the mobilization of multilateral resources for debt reduction. NGOs critics, who once fundamentally opposed SAPs, are now focusing on analyzing the impacts of particular elements of SAPs. However, they continue to stress that compensatory programs are insufficient to address the structural roots of poverty.
4. **Measurement.** Conventional macroeconomic indicators such as Gross Domestic Product (GDP) often mask poverty and inequality. The World Bank needs to place greater reliance on measures such as the Human Development Index (HDI) and the Gender-related Development Index (GDI). The Bank's recent "Wealth of Nations" analysis is a step toward a better measure of sustainable development, incorporating measures of human, natural, and social capital, but it does not deal with distributional issues.
5. **Participation.** The final challenge to the World Bank is to include the views of the poor in decision-making regarding the opportunities and constraints that face them. Important country-level policy planning processes that need more broad-based input include the following:
    - Participatory Poverty Assessments (PPAs) that examine the extent

and nature of poverty in borrowing countries and develop policy recommendations for reducing poverty.

- Country Assistance Strategies (CASs) that put forth the priorities for World Bank lending in a borrower country over a three- to five-year period.
- Public Expenditure Reviews (PERs) that examine government expenditure patterns across sectors.

The World Bank's current policy of treating these assessments as confidential must be altered to provide for effective contribution by NGOs and other elements of civil society. Reorienting the Bank's approach to poverty reduction will require support from stakeholders in both shareholder and borrower countries, as well as systematic participation by the poor.

Summary of

Assessing the Impact of NGO Advocacy Campaigns on World Bank Projects and Policies

by Jonathan Fox and David Brown

[Published in *The Struggle for Accountability: The World Bank, NGOs, and Grassroots Movements.* ed. Jonathan Fox and David Brown (Cambridge: MIT Press, 1998), 485–539.]

NGO advocacy campaigns have greatly influenced the World Bank's decision to put in place a number of policies related to sustainable development. Unfortunately, the implementation of these new policies has been very limited in practice. By reviewing studies of compliance with reform policies at the World Bank, this chapter evaluates the impact of such reforms on development practice. The chapter concludes by proposing conditions under which sustainable development reforms can actually be effective.

Assessing Institutional Change

While World Bank officials stress that the Bank has indeed changed its ways, many of its critics argue that the institution does not comply with its own reforms. Assessing the degrees of change at the Bank can be viewed from three perspectives: portfolio trends, impact on projects, and impact on policies.

Portfolio Trends

If the World Bank's changes in policies and discourse managed to influence its overall lending patterns, the proportion of "good, bad, and ugly" loans within its portfolio would change. "Good" projects are good from a sustainable development point of view because they offer support for such goals as expanded educational access for girls, primary health care, reproductive choice, safe drinking water, biodiversity protection, pollution control, and so on. "Bad" projects are loans that contribute to ongoing environmental degradation or social inequity, or that are largely wasted through corruption, patronage, or support for local elites. These "bad" projects reinforce existing negative trends, while "ugly" projects are more actively destructive. "Ugly" projects directly immiserate large numbers of low-income people, endanger fragile indigenous cultures, encourage the dangerous spread of toxics, promote irreversible biodiversity loss, and prop up dictatorships that might have otherwise been toppled. Have nongovernmental organizations (NGOs) affected the relative mix of good, bad, and ugly projects?

Impact on Projects

The impact of NGOs on World Bank projects varies greatly over time, from country to country, across regime types, and through the project cycle. The studies reviewed here have found no direct link between the intensity of grassroots mobilization and impact on projects, and they suggest that additional factors are required to explain the outcome of most cases. However, NGO campaigns do matter. The chapter outlines twenty-four cases that involved some NGO impact. These impacts can be grouped into four categories:

- *Blocking, canceling, or temporarily suspending a project*—Perhaps the most well-known example of a loan cancelled due to social and environmental campaigning is the Narmada Dam Project in India.
- *Social and environmental impact mitigated, which is the most common*—Such mitigation can take many forms, such as redesigning a project to reduce its impact, as in the case of Thai Dam project; creating channels for negotiating solutions with NGOs, such as with Brazil's Planafloro project; and the recognition of indigenous organizations, as occurred in Ecuador.
- *New projects in response to past protests*—A good example is Brazil's Itaparica Resettlement and Irrigation project. Rather than funding a huge hydroelectric dam itself, the World Bank supported alternative livelihoods for those who were relocated.
- *Spillover effects on World Bank policies*—Several NGO campaigns have set precedents for future World Bank policy. For example, the international

campaign against Indonesia's transmigration program led the World Bank to ease funding settler-colonization programs in rainforests in general.

In these and other cases, NGO activism appears to have had some tangible impact on the World Bank, either in mitigating impacts or influencing subsequent policy. But, to what degree do reform policies in turn influence actual projects?

Policies versus Projects

The World Bank's sustainable development policies are now quite varied and require Bank staff to carry out environmental impact assessments (EIAs), consider alternative investments, minimize involuntary resettlement, prevent immiseration of those resettled, buffer the impact of projects on indigenous peoples, and encourage NGO collaboration in project design and implementation. Three types of studies have reviewed the implementation of these policies: independent studies conducted outside the World Bank; internal studies conducted by the World Bank but with relative autonomy from the operational staff responsible for projects; and internal World Bank studies limited to a "desk review" based on official Bank documentation and supplementary interviews with project managers.

Independent, external reviews for energy policy, indigenous peoples, and water resources have been largely negative. Following a 1992 World Bank energy policy that encouraged increased attention to integrated resource planning and energy efficiency, an Environmental Defense Fund (EDF) and Natural Resources Defense Fund (NRDC) report assessed all power loans under consideration the following year (amounting to $7 billion). The study found that except for the area of pricing, the Bank "failed to incorporate its own policies into the loan preparation process." [519] Of forty-six loans covered, only two complied with World Bank policy. While indigenous peoples policy was one of the World Bank's first reforms, it is still one of the weakest. Violation of peoples' rights remains one of the most frequent causes of conflict over Bank projects. A study of water resources did find some responsiveness to water policies in projects, although alternative style projects involving watershed protection and smaller-scale initiatives received little funding.

Internal reviews that were conducted with relative autonomy for involuntary resettlement and gender offer more mixed results. The involuntary settlement review of 1994 indicates that evictions were part of 192 projects active between 1986 and 1993 and displaced roughly 2.5 million people. Moreover, the report said that "projects appear often not to have succeeded in reestablishing resettlers at a better or equal living standard and that unsatisfactory performance still persists on a wide scale." [520] The World Bank has a more positive record on gender as of late. One study estimates that 28 percent of World Bank operations had some gender provisions (often small components). The majority of such provisions are education and health, and a small but growing number of micro-credit initiatives.

Internal "desk reviews" have been conducted for poverty-targeted lending, forest policy, agricultural pest management, and EIAs. Since 1990, the World Bank has set up an indicator to track its targeted anti-poverty activities called the Program of Targeted Interventions (PTI). PTI projects now account for a growing and significant number of World Bank lending, reaching 54 percent of the investments to the poorest countries in 1995. Such assessments have been criticized, however, because even though only a small fraction of a loan may be allocated toward anti-poverty measures, the entire loan is counted as poverty-targeted.

In 1991, the World Bank issued a comprehensive forestry policy. An internal World Bank review that compared the periods between 1984–1991 and 1991–1994 found that lending for forest protection and restoration increased from 7 to 27 percent, alternative livelihood support rose from 1 to 14 percent, plantations fell from 32 to 23 percent, and road construction fell from 10 to 0.4 percent. The report acknowledges that social assessment and community participation were limited, and issues of land tenure and agrarian reform were not considered.

In 1985 the World Bank issued pesticide guidelines favoring integrated pest management (IPM). While an independent review that assessed the policy from its inception to 1988 found that the policy was ignored, an internal World Bank review covering 1988 to 1995 found improvement. Of the ninety-five projects that involved pest management, forty-two involved pesticide purchases, and forty-eight claimed an IPM component, but only twenty-two actually planned to implement an IPM approach. Finally, an internal review of environmental impact assessment policies found significant progress in "end of pipe" issues with much weaker performance in areas such as seriously considering alternatives, broader sectoral and regional EIAs, as well as public transparency and participation.

Preventing Problems

If the World Bank's policies actually worked as written, they would not continue to be so controversial. However, there are three institutional mechanisms within the World Bank that may influence the degree to which "ugly" projects are vetoed, perhaps even before external critics begin to mobilize. The first is the possibility that the World Bank's new cadre of environmental staff may serve as a "tripwire" that alerts the Bank to potential disasters before they happen. Such capacity is limited, however, because the Bank's project task managers are rewarded for moving money quickly through the system. Another avenue is for World Bank staff to resort to "back channels," such as leaking documents into concerned local and international NGOs (many early NGO campaigns were informed by such insider tip-offs). The third institutional mechanism is the Inspection Panel that serves as an in-house watchdog. The mandate of the panel is to investigate when parties directly affected by projects submit complaints that official World Bank policies were not followed. The Arun III dam in Nepal is a case where the Inspection Panel was instrumental in overturning a project. However, the existence of a review panel has led to a staff backlash in favor of watering down the World Bank's social and environmental policy standards.

Conclusion and Epilogue: Sustainable Development and Accountability

In sum, NGOs have not been very effective in changing projects that are already under way—once launched, projects have too much momentum. However, advocacy campaigns over past projects did secure promises of policy changes that would affect future projects. Because many of these promises did not become reality on the project level, NGO impact has been limited overall. However, because of social and environmental tripwires set by earlier NGO campaigns, the number of "ugly" projects may decrease.

The studies reviewed here indicate that if they are to meet minimum sustainable development goals, World Bank project loans must include the following three elements:

- At the international level, be directly controlled by committed reform elements within the Bank
- At the government level, be designed to target support specifically to agencies already controlled by reformist, pro-accountability elements within the state
- Within civil society, include informed participation by representative social organizations from the beginning of the design process.

If any one of these three conditions is missing, results will likely fall short of the World Bank's minimum sustainable development criteria.

Summary of

Coping with Ecological Globalization

by Hilary French

[Published in *State of the World 2000,* ed. Lester Brown et al. (New York: Norton 2000), 184–202.]

Ecological globalization in its many forms poses enormous challenges to traditional governance structures. In this chapter, globalization is taken to mean a "broad process of societal transformation in which numerous interwoven forces are making national borders more permeable than ever before, including growth in trade, investment, travel, and computer networking. Ecological globalization is used here to refer to the collective impact that these diverse processes have on the health of the planet's natural systems." [185] While this chapter provides an extensive analysis of the effects of globalization on the planet's ecology in addition to discussing how existing and new institutions can cope with these problems, this summary primarily focuses on the latter.

Trading on Nature

While globalization is increasingly playing a major role in the degradation of the earth's environment, the emerging rules of the global economy are largely failing to recognize the importance of reversing these ecological perils. This mismatch between the prevailing economic practice and ecological imperatives urgently needs to be addressed in order to halt the unraveling of critical ecological systems in the early decades of the next century.

Five key elements that are essential to the maintenance of the earth's ecology are rapidly becoming deeply integrated into the world economy: the world's forests, food supply, fisheries, wildlife, and "exotic" species. The lure of international markets is becoming an inducement for countries to deplete their stocks of these resources far faster than would be required to meet their domestic needs alone. The industrial countries dominate the consumption of these resources and are causing rapid depletion of forest, fish, and wildlife stocks in the developing world. Such degradation threatens the economic and food security of the developing world, as well as damages human health and the health of ecosystems across the entire globe.

The WTO Meets the Environment

The World Trade Organization (WTO) is a leading example of the way in which transnational governance structures are giving short shrift to the urgent need to halt global environmental degradation. The WTO was established in 1994 as an outgrowth of global trade talks under General Agreement on Tariffs and Trade (GATT). While many cheered the WTO's creation as the dawn of a new era of global prosperity, others criticized the WTO for elevating the rights of global corporations over national governments, local communities, and the environment. The passions on both sides of this debate were exacerbated in November 1999 when trade ministers from around the world gathered in Seattle to launch a new round of global trade talks. Tens of thousands of activists were also in Seattle, many of them protesting what they saw as the WTO's environmental blindness.

While the preamble to the WTO agreement includes environmental protection and sustainable development among the organization's goals, critics argue that the WTO has done little to halt global environmental decline. Indeed, there have been a number of cases where the WTO has ruled that the domestic environmental laws of member countries constitute barriers to trade. These acts expose some glaring inconsistencies between the rules of the WTO and emerging international environmental principles and practices.

The much discussed tuna-dolphin decision, where the WTO ruled that an embargo against Mexican tuna under the U.S. Marine Mammal Protection Act violated the GATT, is a case in point. The United States had imposed the embargo because Mexico's fishing practices were ensnaring dolphins in the pursuit of catching tuna. A more recent decision was made by the WTO in 1998 to strike down another U.S. measure aimed at reducing unintended sea turtle mortality as a by-product of shrimp trawling. As environmental policy is moving increasingly toward focusing on the environmental impacts of products through their life-cycle—including production, distribution, use, and disposal—the WTO is moving in the opposite direction. Furthermore, it is feared that these inconsistencies are only the tip of the iceberg, that the WTO will challenge other domestic laws, such as those concerning genetically modified organisms (GMOs), and possible aspects of other international environmental agreements.

As opposition to the WTO continues to mount, some governments are beginning to acknowledge the need for significant reforms in WTO principles and procedures. The following measures could be undertaken at the WTO to help the world community cope with ecological globalization:

- Reduce environmentally harmful subsidies in the agricultural, energy, and forestry sectors
- Incorporate the precautionary principle into WTO rules
- Protect consumers' right to know about the health and environmental impact of products they purchase by safeguarding labeling schemes

- Recognize the legitimacy of distinguishing among products based on how they are produced
- Provide deference to multilateral environmental agreements in cases where they conflict with WTO rules
- Ensure the right of countries to use trade measures to protect the global commons
- Open the WTO to meaningful public participation.

Greening the International Financial Architecture

If the protests in Seattle were a wake-up call for the trade community to recognize the risks of globalization, the Asian financial crisis in 1997 was a similar wake-up call to the investment community. The collapse of many Asian economies at that time has launched a critical international dialogue about how to reform the international financial architecture to meet the demands of the twenty-first century. Such reform is needed in public and private financial institutions alike.

The logical place to start discussion regarding new financial architecture is where it exists—chiefly with the International Monetary Fund (IMF) and the World Bank. Both of these institutions pay insufficient attention to the profound effects of their policies on the ecological health and social fabric of recipient countries. One important aspect of their policies is to encourage recipients to increase foreign exports to generate necessary foreign exchange with which to pay back debts. This pressure to export, as mentioned earlier can lead countries to liquidate natural assets such as fisheries and forests, therefore undermining longer-term economic prospects. Simultaneously, these institutions often require that recipient countries slash spending on the environment. For example, Brazil was recently forced by the IMF to cut environmental spending by two-thirds. However, there have also been cases where the IMF has suspended loans that caused environmental degradation. In the case of Cambodia, loans were suspended after the government gave concessions to foreign firms that threatened to open up the nation's entire remaining forest for exploitation.

On paper, the more development-oriented World Bank has been far more open than the IMF to the idea of incorporating environmental concerns into the development process. The World Bank now has a policy that requires the environmental impact of its adjustment lending needs to be considered. However, a review of fifty recent loans found that few paid much heed to environment and social matters. In addition, whereas in 1993 it was found that some 60 percent of adjustment loans included environmental goals, a recent study concluded that this share has now decreased to 20 percent.

With a renewed commitment to the environment, the IMF and the World Bank could

- Promote environmental beneficial fiscal reforms, such as reduction in environmentally harmful subsidies or the imposition of pollution taxes
- Help promote improvements in environmental accounting, such as incorporating the depletion of natural resources into national income figures
- The IMF could include environmental issues in its assessments of the economic prospects of its member countries by tracking environmental spending levels and legislation that effects the environment.

Private investors are increasingly turning to bilateral export agencies to fund support for projects that are not fundable by the World Bank. Export credit rose from $24 billion in 1988 to $105 billion in 1996. Such projects often are environmentally disruptive, including those of mines, pipelines, and hydroelectric dams. The U.S. government has had environmental policies in place at its export promotion institutions for some time, but in a world economy such policies can easily be undermined by laggards abroad. This became clear when the United States refused to extend credit to a U.S. firm on environmental grounds for a project in China, and the export promotion agencies of other countries stepped into the breach.

Although efforts to harmonize export promotion credit guidelines are under way, private capital markets can still be tapped for environmentally sensitive projects. The United Nations Environment Programme (UNEP) launched an initiative in 1992 that encourages commercial banks to incorporate environmental concerns into their lending programs. So far, 162 banks from 43 countries have signed on to the initiative. Environmental liabilities could also be incorporated into the way stock markets are regulated. Companies in the United States are already required to disclose large environmental risks on the forms they file. Developing countries are well placed to write such rules as they establish their own financial markets. For these efforts to be a success, transparent information about corporate environmental performance is essential. The Boston-based Coalition for Environmentally Responsible Economies (CERES) has started a Global Reporting Initiative where corporations, NGOs, professional accounting firms, and UNEP are working together to produce a global set of guidelines for corporate sustainability reporting.

Innovations in Global Environmental Governance

Creating a global society fit for the new century will not only require reform of economic institutions, but also a strengthening of international environmental institutions and the creation of new institutions to provide an ecological counterweight to today's growing economic powerhouses.

There are more than 230 existing environmental treaties, and more than three-fourths of them have been negotiated since the first U.N. conference on the environment in 1972. These accords cover such issues as atmospheric pollution, ocean despoliation, endangered species protection, hazardous waste trade, and the preservation of Antarctica. While there are some success stories, most treaties have so far failed to turn around deteriorating environmental trends. The main reason for this is that the governments that created these treaties have generally permitted vague commitments and lax enforcement. In addition, governments have supplied weak funding for treaty implementation.

For these reasons, many are looking to new actors and the possibility of new institutions to fill the gap. An idea now gaining momentum is to create a World Environmental Organization (WEO) that would serve as an umbrella organization for the current scattered collection of treaty bodies, just as domestic agencies oversee national laws. The private sector and civil society are also being looked to for support. The private sector has begun its own environmental standard-setting process through the Geneva-based International Organization for Standardization (ISO). While the environmental guidelines that the ISO has set up are useful tools, they do not mandate performance, and their credibility has suffered from the fact that it is widely perceived to be industry dominated.

An encouraging recent development is the growth in international environmental NGOs flourishing at both national and grassroots levels. The number of NGOs working across international borders has soared, climbing from 176 in 1909 to over 23,000 in 1998. Environmental groups have risen steadily as a proportion of the total, from 2 percent in 1953 to 14 percent in 1993. In addition to networking and organizing using such innovative new tools as e-mail and the Internet, NGOs have also educated millions of people about environmental issues and harnessed the power of a knowledgeable citizenry to effect change.

Summary of

The United Nations Environment Programme and the United Nations Development Programme

by Alexandre Timoshenko and Mark Berman

[Published in *Greening International Institutions,* ed. Jacob Werksman (London: Earthscan, 1996), 38–54.]

The United Nations Environment Programme (UNEP) and the United Nations Development Programme (UNDP) are the principal agencies in the United Nations family of international institutions to address sustainable development. Indeed, the United Nations Conference on Environment and Development (UNCED) formally charged each of these bodies with the goal of elaborating strategies and measures to halt and reverse the effects of environmental degradation while continuing to raise the standards of living in all countries. The early report card on these efforts is positive.

UNEP

The process of making UNEP and UNDP into instruments for sustainable development translates differently into their mandates. For UNDP the challenge has been to integrate an environment component into its operational activities, while for UNEP the challenge has been to more fully integrate a human development component into its environmental efforts.

Actually, UNEP has incorporated a development component in its mandate since its inception, and UNCED has freshly invigorated those efforts. As early as 1973, UNEP had launched the concept of "ecodevelopment" and was an early pioneer of examining the "environmental costs and benefits of development, including social costs." In addition, a number of the earlier international treaties that have been adopted under UNEP auspices have been designed to integrate environment and development. A group of regional seas conventions adopted in the 1980s to address marine pollution have each included provisions for ensuring "comprehensive development without environmental damage." [40] Perhaps the quintessential example of integrating environment and development is UNEP's involvement in the 1985 Vienna Convention for the Protection of the Ozone Layer and the subsequent 1987 Montreal Protocol. This regime steers economic production to become more sustainable by substituting with alternatives to ozone-depleting substances and also uses economic instruments such as trade sanctions to enforce its mandate.

UNCED reinvigorates these previous efforts at UNEP but also legitimizes a special emphasis on the development side of the equation, and therefore gives UNEP a new, expanded, and strengthened mandate for the post-UNCED world. Agenda 21 provides a number of areas where UNEP will now concentrate, including catalyzing action, promoting international cooperation, conducting monitoring and assessment, further developing international environmental law, research, and raising awareness. Prior to UNCED, UNEP focused primarily on three levels of work: environmental assessment, environmental policy development and coordination, and environmental management. In addition to the specific mandates outlined above, since UNCED, UNEP now has a more integrated approach to its program.

Finally, UNEP now has a clearer understanding of its niche among the myriad other international organizations, and global society as a whole. While UNEP will put effort into the development process, it will continue to serve in the inter-institutional process as the primary environmental resource through the provision of scientific, technical, and policy information and advice. UNCED, however, now mandates that such efforts are seen and acted upon through the more comprehensive lens of sustainable development.

UNDP

UNDP was established in 1965 to help develop the human and natural resources required to meet basic needs: economic growth and human development. Specifically, UNDP supports projects that are designed to help governments to attract development capital to train personnel and to apply modern technologies that are needed for social and economic improvement. To a certain extent, UNDP worked on environmental issues before UNCED, but UNCED has acted as a catalyst for environmental efforts at UNDP just as it has acted as a catalyst for development efforts at UNEP.

Both UNDP and UNEP have been given significant roles in implementing Agenda 21. While UNEP is the lead agency for policy guidance and coordination on the environment, UNDP is responsible for mobilizing donor assistance and organizing efforts across the U.N. system to build expertise for sustainable development. The new challenge for UNDP is to integrate sustainable and human development into all of its operations: "sustainable human development which not only generates economic growth but distributes its benefits equitably; that regenerates the environment rather than destroying it; that empowers people rather than marginalizing them." [48] UNDP's key comparative advantage for these efforts is its network of 132-country offices that allow it to offer development alternatives that are responsive to local needs and priorities.

To spearhead its new mandate, in 1993 UNDP launched Capacity 21—an initiative to assist countries in developing sustainable development strategies.

This is intended as a catalytic fund to facilitate the integration of environmental considerations in all aspects of a developing country's programs. Capacity 21 also institutionalizes the involvement of all stakeholders into development planning and environmental management.

UNEP/UNDP Collaboration

As UNEP and UNDP continue to maximize their comparative advantages in the manner described above and to collaborate with each other, positive steps toward sustainable development will soon be realized. Through such collaboration, these institutions will be further "greened." While it is still too early for a full assessment, specific examples of cooperative program development indicate that the necessary groundwork has been laid for expanded collaboration.

The two organizations, along with the United Nations Industrial Development Organization (UNIDO) and the World Bank, work on program implementation under the Multilateral Fund of the Montreal Protocol—the financial mechanism to assist countries to comply with the Montreal Protocol. Here, UNEP provides the organizational umbrella, serves as the Fund's treasurer, and provides technical and institutional advice to governments in establishing their national ozone policy frameworks. UNDP assists countries in the planning, preparation, and implementation of country projects and institutional strengthening.

Collaboration between UNEP and UNDP, along with the World Bank, has also occurred in the context of the Global Environment Facility (GEF). UNDP assists the GEF with technical assistance, capacity building, and small grants administration, and UNEP integrates environmental concerns into projects. Following UNCED, the two institutions plan to coordinate their efforts by

- The development of national frameworks for sustainable development
- Assistance to governments in the implementation of post-Rio conventions
- Mobilizing UNDP's country-based strength for dissemination of environmental information.

There are obviously many opportunities for cooperation between these two agencies, and the record shows that there has been much precedent for UNEP to incorporate development concerns in its mandate and for UNDP to integrate environmental concerns into its mandate.

Summary of

Conflicts of Global Ecology: Environmental Activism in a Period of Global Reach

by Vandana Shiva

[Published in *Alternatives* 19 (1994), 195–207]

The origins of the environmental movement were in local efforts to resist ecological degradation and toxic pollution, but there is now a recognition that many environmental threats are caused by multinational corporations (MNCs) and multilateral development banks (MDBs). These institutions have a "global reach" that touches every city, village, field, and forest through their worldwide operations. The emphasis has shifted from local to global environmental problems, and it is widely assumed that the solutions must also be global. Using the Global Environment Facility (GEF) as a case in point, this paper suggests that this interpretation is false. Indeed, it argues that defining these issues as "global" is a veiled attempt to conserve economic and political power rather than to conserve the environment.

The "Global" as a Parochial Interest

The expression "global" is not an expression of universal humanism or of planetary consciousness. Instead, the "global" is the political space in which dominant local interests in a select group of countries seek to control international decision-making and free themselves from local, national, and global accountability. The Group of Seven may dictate global affairs, but they are parochial in the interests that guide them.

Throughout history, when this global reach has been threatened by resistance, the language of resistance has been co-opted, redefined, and used to legitimate further control. This has recently been the case with the World Bank, the IMF, and many multinational corporations who claim they are in the business of "sustainable development." In addition, the notion is created that since such global interests need to be protected, local interests sometimes must be sacrificed. The way that global environmental problems are constructed hides the role and responsibility of dominant interests in the destruction of the environment and the subjugation of local peoples. In so doing, the blame for environmental problems is shifted onto the communities without a global reach.

Biodiversity is just one example where control has shifted from the South to the North through its identification as a global problem. Erosion of biodiversity has taken place because MDBs are financing dams, mines, and highways in ecologically diverse areas, replacing natural ecosystems with systems of monoculture. The most important first step for biodiversity conservation is to reverse the World Bank's planned destruction of biodiversity, which is occurring because, through the GEF and the Biodiversity Convention, the North is demanding free access to the South's biodiversity.

Excluding the Environment

The purely financial logic of the GEF excludes environmental concerns. It assumes that finance-capital-rich regions are "donors" but fails to recognize that the biodiversity and environmental-capital-rich regions are "donors" in a more fundamental sense. In fact, this "donor" contribution in environmental terms is more significant for global environmental management than the traditional "donor" contribution in terms of finances. Also, while $50 billion financial flows move from North to South, financial flows moving from South to North are about $500 billion. And many of the North's financial investments have wreaked havoc on the environment in the South.

The GEF encourages the persistence of the externalization of environmental costs. This is evident from the fact that close to 80 percent of GEF projects are components of, or directly associated with, regular World Bank lending programs. While the GEF provides institutional support for the notion that environmental costs are external, it rewards the World Bank and its loan recipients for environmental destruction. For example, a $242 million World Bank loan to a private hotel industry in Egypt to develop a coastal resort was coupled with a $4.75 million GEF grant for conservation activity in the same region. Another example is in Ecuador, where $4 million of International Finance Corporation credit was given to Ecuador's largest logging company, along with a $2.5 million GEF grant to the same company to purchase forests and local lands from local communities. The GEF grant allowed logging on 8,382 hectares and set up a privately owned reserve on 610 hectares.

Incremental Costs and the Internalization of Costs

The only way to avoid environmental destruction is to make the actors who are responsible for destruction bear the full social and environmental costs involved. The GEF does the opposite. Through the GEF's version of the concept of "incremental" costs, the GEF is working on a pay-the-polluter rather than on a polluter-pays principle.

When the South agreed to incremental costing during UNCED, we thought that such a term meant new and additional resources to compensate the Third World for the inequities in the global economic system and to help the South build capacity to take on new responsibilities under a new set of global treaties.

At the GEF, "incremental" means that national and global benefits are not congruent. For example, a GEF research report identified the United States as the nation with the most carbon emissions and the highest carbon emitted per capita. Rather than identifying strategies for reducing levels in the United States, the GEF externalizes these costs on the South.

First, this is unsound from an ecological perspective. The real sources of carbon dioxide in the United States are not addressed, and by treating trees as carbon sinks, the strategy promotes nonsustainable forestry in the South as a way of reducing carbon. Because trees only absorb carbon when they are growing, the GEF encourages short-rotation forest plantations. However, such plantations should be considered unsustainable because of their lack of hydrological balance, nutrient balance, and biodiversity. Moreover, the costs of carbon dioxide pollution by the United States are socially externalized by failing to treat carbon dioxide in social and environmental terms rather than only in financial terms. Because it is much cheaper to plant carbon-absorptive trees in developing countries, tree-planting projects are suggested to offset emissions from new power plants in the United States.

> Just because Third World peoples and their resources have been devalued by the global economic system, there is no reason to imply that the social costs of diverting land and forests from local needs to serve as timber mines and carbon sinks for northern economic interests are negligible. [203]

Global Action at the Grass Roots

Even though we are witnessing a process that conserves the economic and political power of the status quo rather than conserving the environment, global environmental activism is continuing to grow and to cover new areas of concern. These include the threat of genetic erosion and the patenting of life forms. There are now a number of activists attempting to put pressure on world trade, forestry, and property rights rule-making bodies.

One example is that of the Seed Satyagraha movement launched by Indian farmers to protect the rights of farmers to produce, modify, and sell seeds. The corporate demand to change the farmer's common heritage into a commodity leads not only to ethical and cultural erosion, but also jeopardizes the economic stability of the farmers. Indeed, this new corporate dominance is seen as a further wave of colonization in India. Thanks in part to Seed Satyagraha's efforts, the seed is rapidly becoming a symbol of India's freedom from recolonization.

Bibliography

Ackerman, Frank. *Why Do We Recycle? Markets, Values, and Public Policy* (Washington, D.C.: Island Press, 1997).

Ackerman, Frank. "Globalization and Labor." In *The Changing Nature of Work,* eds. Ackerman et al. Frontier Issues in Economic Thought, Volume IV. (Washington, D.C.: Island Press, 1998).

Ackerman, Frank. "Still Dead After All These Years: Interpreting the Failure of General Equilibrium Theory." Working Paper 00-01, Global Development and Environment Institute, Tufts University, 2000a.

Ackerman, Frank. "Waste Management and Climate Change." *Local Environment* (June 2000b).

Ackerman, Frank, Bruce Biewald, David White, Tim Woolf, and William Moomaw. "Grandfathering and Coal Plant Emissions: The Cost of Cleaning Up the Clean Air Act." *Energy Policy* 27, 15 (1999), 929–940.

Ackerman, Frank, et al., eds. *Human Well-Being and Economic Goals.* Frontier Issues in Economic Thought, Volume III. (Washington, D.C.: Island Press, 1997).

Ackerman, Frank, et al., eds. *The Political Economy of Inequality.* Frontier Issues in Economic Thought, Volume V. (Washington, D.C.: Island Press, 2000).

Adams, Walter, and James W. Brock. "Bigness and Social Efficiency: A Case Study of the U.S. Auto Industry." In *Corporations and Society,* eds. Samuel, Warren J. and Arthur S. Miller (Westport Conn.: Greenwood Press, 1987).

Adelman, Irma. "Development Economics—A Reassessment of Goals." In American Economic Review 65, 2 (May 1975), 302–309.

Adelman, Irma. "Economic Development and Political Change in Developing Countries." *Social Research* 47, 2 (Summer 1980), 213–234.

Agarwal, Anil, Gunita Narain, and Anju Sharma, eds. *Green Politics: Global Environmental Negotiations* (Delhi: Center for Science and the Environment, 1999).

Agarwal, Bina. "The Gender and Environment Debate: Lessons from India." *Feminist Studies* 18, 1 (1992).

Agcaoli, Mercedita, and Mark W. Rosegrant. "Global and Regional Food Supply, Demand, and Trade Prospects to 2010." In *Population and Food in the Early Twenty-First Century: Meeting Future Food Demand of an Increasing Population,* ed. Nurul Islam (Washington, D.C.: International Food Policy Research Institute, 1995).

Ahmed, Iftkhar, and Jacobus A. Doeleman, eds. *Beyond Rio: The Environmental Crisis and Sustainable Livelihoods in the Third World* (New York: St. Martin's Press, 1995).

Alexander, William M. "Exceptional Kerala: Efficient Use of Resources and Life Quality in a Non-Affluent Society." *Gaia (Ecological Perspectives in Science, Humanities and Economics)* 3, 4 (1994).

Alinsky, Saul. *Rules for Radicals: A Practical Primer for Realistic Radicals* (New York: Vintage Books, 1972).

Anand, Sudhir, and Amartya K. Sen. *Sustainable Human Development: Concepts and Priorities.* (United Nations Development Programme, Office of Development Studies Discussion Paper Series, 1996).

Anderson, Molly D., and John T. Cook. "Does Food Security Require Local Food Systems?" In *Rethinking Sustainability: Power, Knowledge, and Institutions,* ed. Jonathan M. Harris, (Ann Arbor: University of Michigan Press, 2000).

Annis, Sheldon, ed. *Poverty, Natural Resources, and Public Policy in Central America* (New Brunswick, N.J.: Transaction Publishers for the Overseas Development Council, 1992).

Arrow, Kenneth. "The Rate of Discount for Long-Term Public Investment." In *Energy and the Environment: A Risk-Benefit Approach,* eds. H. Ashley et al. (New York: Pergamon Press, 1976).

Arrow, Kenneth, et al. "Economic Growth, Carrying Capacity, and the Environment." *Science* 268 (1995), 520–521.

Ayres, Robert U., and Leslie W. Ayres. *Industrial Ecology: Closing the Materials Cycle* (Cheltenham, UK: Edward Elgar, 1996).

Ayres, Robert U., and Leslie W. Ayres. *Accounting for Resources, 1: Economy-wide Applications of Mass-balance Principles to Materials and Waste* (Cheltenham, UK: Edward Elgar, 1998).

Ayres, Robert U., and Paul M. Weaver, eds. *Eco-Restructuring: Implications for Sustainable Development* (New York: United Nations University Press, 1998).

Backhouse, Roger. *A History of Modern Economic Analysis* (Oxford, UK: Blackwell, 1991).

Barkin, David. "Wealth, Poverty, and Sustainable Development." In *Rethinking Sustainability: Power, Knowledge, and Institutions,* ed. Jonathan M. Harris (Ann Arbor: University of Michigan Press, 2000).

Bates, Robin W. "The Impact of Economic Policy on Energy and the Environment in Developing Countries." *Annual Review of Energy and Environment* 18 (1993), 479–506.

Baudot, Barbara, and William Moomaw, eds. *People and Their Planet: Searching for Balance.* (New York: St. Martin's Press, 1999).

Bawa, Kamaljit S., and Madhav Gadgil. "Ecosystem Services in Subsistence Economies and Conservation of Biodiversity." In *Nature's Services: Societal Dependence on Natural Ecosystems,* ed. Gretchen C. Daily (Washington, D.C.: Island Press, 1997).

Beckerman, Wilfred. "Economic Growth and the Environment: Whose Growth? Whose Environment?" *World Development* 20, 4 (1992), 481–496.

Beckerman, Wilfred. "Sustainable Development: Is it a Useful Concept?" *Environmental Values* 3 (1994), 191–209.

Beckerman, Wilfred. "How Would you Like your 'Sustainability,' Sir? Weak or Strong? A Reply to My Critics." *Environmental Values* 4 (1995), 169–179.

Belman, Dale, and Thea Lee. "International Trade and the Performance of U.S. Labor Markets." In *U.S. Trade Policy and Global Growth: New Directions in the International Economy,* ed. Robert A. Blecker (Armonk, NY: M.E. Sharpe, 1992), 61–107.

Bernauer, Thomas. "The Effectiveness of International Environmental Institutions: How Might We Learn More?" *International Organization* 49, 2 (Spring 1995), 351–377.

Bernow, Stephen, and Max Duckworth. "An Evaluation of Integrated Climate Protection Policies for the U.S." *Energy Policy* 26, 5 (1998), 357–374.

Bhalla, A.S. *Uneven Development in the Third World: A Study of China and India* (New York: St. Martin's Press, 1995).

Birdsall, Nancy. "Economic Analyses of Rapid Population Growth." *The World Bank Research Observer* 4, 1 (1989).

Birdsall, Nancy, David Ross, and Richard Sabot. "Inequality and Growth Reconsidered: Lessons from East Asia." *The World Bank Economic Review* 9 (1995), 477–508.

Bishop, R.C. "Endangered Species and Uncertainty: The Economics of a Safe Minimum Standard." *American Journal of Agricultural Economics* 60 (1978), 10–18.

Blum, Lauren, Richard A. Denison, and John F. Ruston. "A Life-Cycle Approach to Purchasing and Using Environmentally Preferable Paper: A Summary of the Paper Task Force Report." *Journal of Industrial Ecology* 1, 3 (1997), 15–44.

Blumberg, Louis, and Robert Gottlieb. *War on Waste: Can America Win its Battle With Garbage?* (Washington, D.C., Island Press, 1989).

Bonavogilia, Angela. "Women's Work: In the U.S., Microenterprise Fuels a Boom in Self-employment." *Ford Foundation Report* 31, 1 (Winter 2000), 10–13.

Boserup, Ester. *The Conditions of Agricultural Growth* (Chicago: Aldine, 1965).

Boserup, Ester. *Population and Technological Change* (Chicago: University of Chicago Press, 1981).

Bouvier, Leon F., and Lindsey Grant. *How Many Americans?* (San Francisco: Sierra Club Books, 1995).

Bovenberg, A.L. "Green Tax Reforms and the Double Dividend: An Updated Reader's Guide." *International Tax and Public Finance* 6, 3 (August 1999), 421–443.

Bowles, Samuel, and Herbert Gintis. *Democracy and Capitalism: Property, Community, and the Contradictions of Modern Social Thought* (New York: Basic Books, 1986).

Braidotti, Rosi, Ewa Charkiewicz, Sabine Hausler, and Saskia Wieringa. *Women, the Environment, and Sustainable Development: Towards a Theoretical Synthesis* (London: Zed Books, 1994).

Broad, Robin, "The Poor and the Environment: Friends or Foes?" *World Development,* 22, 6 (June 1994), 811–822.

Brower, Michael. *Cool Energy: Renewable Solutions to Environmental Problems* (Cambridge Mass: MIT Press, 1992).

Brown, Lester R. "Feeding Nine Billion." Chapter 7 in *State of the World 1999,* eds. Brown et al. Worldwatch Institute Report (New York: W.W. Norton, 1999).

Brown, Lester R. "Challenges of the New Century." *State of the World 2000,* eds. Brown et al. Worldwatch Institute Report (New York: W.W. Norton, 2000).

Bruce, James P., ed. *Climate Change 1995: Economic and Social Dimensions of Climate Change,* Contribution of Working Group III to the Second Assessment Report of the Intergovernmental Panel on Climate Change (Cambridge: Cambridge University Press, 1996).

Brush, Stephen B. "Indigenous Knowledge of Biological Resources and Intellectual Property Rights: The Role of Anthropology." *American Anthropologist* 95, 3 (1993), 653–686.

Brush, Stephen B. "Bio-cooperation and the Benefits of Crop Genetic Resources: The Case of Mexican Maize." *World Development* 26, 5 (1998), 755–766.

Bruton, Henry J. "A Reconsideration of Import Substitution." *Journal of Economic Literature* 36 (June 1998), 903–936.

Buvinic, Myra, Catherine Gwin, and Lisa M. Bates. *Investing in Women: Progress and Prospects for the World Bank* (Baltimore: The Johns Hopkins University Press, 1996).

Camacho, Luis. "Consumption as a Topic for the North-South Dialogue." Chapter 27 in *Ethics of Consumption: The Good Life, Justice, and Global Stewardship,* eds. David A. Crocker and Toby Linden (Oxford: Rowman & Littlefield Publishers, 1998), 552–559.

Campbell, Dave. "Community-Controlled Economic Development as a Strategic Vision for the Sustainable Agriculture Movement." *Alternative Agriculture* 12, 1 (1997), 37–44.

Ciriancy-Wantrup, S.V. *Resource Conservation* (Berkeley: University of California Press, 1952).

Ciriacy-Wantrup, S.V. *Resource Conservation: Economics and Policies,* 3rd ed. (Berkeley: University of California Press, 1968)

Cleveland, Cutler J., and Matthias Ruth. "When, Where, and by How Much do Biophysical Limits Constrain the Economic Process? A Survey of Georgescu-Roegen's Contribution to Ecological Economics." *Ecological Economics* 22, 3 (1997), 203–223.

Cleveland, Cutler J., and Matthias Ruth. "Indicators of Dematerialization and the Materials Intensity of Use," *Journal of Industrial Ecology* 2, 3 (1999), 15–50.

Cline, William R. *The Economics of Global Warming* (Washington, D.C.: Institute for International Economics, 1992).

Cohen, Joel E. *How Many People Can the Earth Support?* (New York: W.W. Norton, 1995).

Common, Mick. "Beckerman and his Critics on Strong and Weak Sustainability: Confusing Concepts and Conditions." *Environmental Values* 5 (1996), 83–88.

Common, Mick, and Charles Perrings. "Towards an Ecological Economics of Sustainability." *Ecological Economics* 6, 1 (1992), 7–34.

Conway, G.R. "The Properties of Agroecosystems." *Agricultural Systems* 24 (1987), 95–117.

Costanza, Robert, et al. "The Valuation and Management of Wetland Ecosystems." *Ecological Economics* 1 (1989), 335–362.

Costanza, Robert, et al. "The Value of the World's Ecosystem Services and Natural Capital." *Ecological Economics* 25, 1 (1998), 3–15.

Costanza, Robert, and Carl Folke. "Valuing Ecosystem Services with Efficiency, Fairness, and Sustainability as Goals." Chapter 4 in *Nature's Services: Societal Dependence on Natural Ecosystems,* ed. Gretchen C. Daily (Washington, D.C.: Island Press, 1997).

Crosson, Pierre. "Future Supplies of Land and Water for World Agriculture." In *Population and Food in the Early Twenty-First Century: Meeting Future Food Demand of an Increasing Population,* ed. Nurul Islam (Washington, D.C.: International Food Policy Research Institute, 1995).

Daily, Gretchen C. *Nature's Services: Societal Dependence on Natural Ecosystems* (Washington, D.C.: Island Press, 1997).

Daly, Herman E. "Elements of Environmental Macroeconomics." Chapter 3 in *Ecological Economics: The Science and Management of Sustainability,* ed. Robert Costanza (New York: Columbia University Press, 1991).

Daly, Herman E. "Operationalizing Sustainable Development by Investing in Natural Capital." In *Investing in Natural Capital: The Ecological Economics Approach to Sustainability,* eds. AnnMari Jansson et al. (Washington, D.C.: Island Press, 1994).

Daly, Herman E. "On Wilfred Beckerman's Critique of Sustainable Development." *Environmental Values* 4 (1995), 49–55.

Daly, Herman E. *Beyond Growth: The Economics of Sustainable Development* (Boston: Beacon Press, 1996).

Daly, Herman E. "Globalization versus Internationalization: Some Implications." *Ecological Economics* 31, 1 (1999), 31–37.

Daly, Herman E., and John B. Cobb Jr. *For the Common Good: Redirecting the Economy Toward Community, the Environment, and a Sustainable Future* (Boston: Beacon Press, 1994).

Daly, Herman E., and Robert Goodland. *Environmentally Sustainable Economic Development: Building on Brundtland.* (World Bank Environment Department Working Paper No. 46, 1991).

de Janvry, Alain, and E. Sadoulet. "Seven Theses in Support of Successful Rural Development." *Land Reform Bulletin* (1996). Food and Agriculture Organization web site: http://www.fao.org/sd/ltdirect/lr96/dejan.htm.

de Jongh, Paul E., and Sean Captain. *Our Common Journey: A Pioneering Approach to Cooperative Environmental Management* (London: Zed Books, 1999).

den Hond, F., P. Groenewegen, and W.T. Vorley. "Globalization of Pesticide Technology and Meeting the Needs of Low-Input Sustainable Agriculture." *Alternative Agriculture* 14, 2 (1999), 50–58.

Derber, Charles. *Corporation Nation: How Corporations Are Taking Over Our Lives and What We Can Do About It* (New York: St. Martin's Press, 1998).

Dernbach, John. "Sustainable Development as a Framework for National Governance." In *Case Western Reserve Law Review* 49, 1 (Fall 1998).

Dernbach, John, and the Widener University Law School Seminar on Law and Sustainability. "U.S. Adherence to Its Agenda 21 Commitments: A Five-Year Review." 27 *Environmental Law Reporter (ELI)* 27 (1997), 10,504–10,525.

Deutch, Lisa, et al. "The 'Ecological Footprint': Communicating Human Dependence on Nature's Work." *Ecological Economics* 32, 3 (2000), 351–355.

Diamond, Peter A., and Jerry A. Hausman. "Contingent Valuation: Is Some Number Better Than No Number?" *Journal of Economic Perspectives* 8, 4 (Fall 1994), 45-64.

Ditz, Daryl, Janet Ranganathan, and R. Darryl Banks. "Environmental Accounting: An Overview." In *Green Ledgers: Case Studies in Corporate Environmental Accounting,* eds., Daryl Ditz, Janet Ranganathan, and R. Darryl Banks (Washington, D.C.: World Resources Institute, 1995).

Donaldson, Graham. "Government-Sponsored Rural Development: Experience of the World Bank." In *Agriculture and the State,* ed. C. Peter Timmer (Ithaca: Cornell University Press, 1991).

Dreze, Jean, and Amartya Sen. *Hunger and Public Action* (Oxford: Clarendon Press, 1989).

Drucker, Peter F. *Management Tasks , Responsibilities, Practices* (New York: Harper & Row, 1973), 341.

Duchin, Faye, and Glenn-Marie Lange. *The Future of the Environment: Ecological Economics and Technological Change* (New York: Oxford University Press, 1994).

Dugger, William M. "An Institutional Analysis of Corporate Power." *Journal of Economic Issues,* XXII, 1 (March 1988).

Durning, Alan. *How Much is Enough? The Consumer Society and the Future of Earth.* Worldwatch Environmental Alert Series, ed. Linda Starke (New York: W.W. Norton, 1992).

Durning, Alan Thein. "The Conundrum of Consumption." In *Beyond the Numbers,* ed. Laurie Mazur (Washington, D.C.: Island Press, 1994), 40–47.

Durning, Alan Thein, and Yoram Bauman. *Tax Shift* (Seattle: Northwest Environment Watch, 1998).

Edwards, Clive A., et al., eds. *Sustainable Agricultural Systems* (Delray Beach, Fla.: St. Lucie Press, 1990).

Ehrlich, Paul R. "Ecological Economics and the Carrying Capacity of the Earth." In *Investing in Natural Capital: The Ecological Economics Approach to Sustainability,* eds. AnnMari Jansson et al. (Washington, D.C.: Island Press, 1994).

Ekins, Paul, Carl Folke, and Robert Costanza. "Trade, Environment, and Development: The Issues in Perspective." In *Ecological Economics* 9 (January 1994), 1–12.

Ekins et al., eds. *The Gaia Atlas of Green Economics* (New York: Anchor Books, Doubleday, 1992) Part II, 48–61.

Elkington, John. *Cannibals with Forks: The Triple Bottom Line of 21st Century Business* (Stony Creek, Conn: New Society Publishers, 1998).

El Serafy, Salah. "The Environment as Capital." In *Toward Improved Accounting for the Environment: An UNSTAT–World Bank Symposium,* ed. Ernst Lutz (Washington, D.C: The World Bank, 1993).

El Serafy, Salah. "In Defence of Weak Sustainability: a Response to Beckerman." *Environmental Values* 5 (1996), 75–81.

Engelman, Robert. "Population as a Scale Factor: Impacts on Environment and Development." Baudot, Barbara and William Moomaw, eds. *People and Their Planet: Searching for Balance* (New York: St. Martin's Press, 1999).

England, Richard. "Alternatives to Gross National Product: A Critical Survey." In *Human Well-Being and Economic Goals,* eds. Frank Ackerman et al. (Washington, D.C.: Island Press, 1997), 373–402.

England, Richard W., and Jonathan M. Harris. *Alternatives to Gross National Product: A Critical Survey.* Global Development and Environment Institute Discussion Paper no. 5 (1997), available at http://ase.tufts.edu/gdae.

Esty, Daniel. "Greening World Trade." In *The World Trading System: Challenges Ahead,* ed. Jeffrey Scott (Washington, D.C.: Institute for International Economics, 1996), 69–84.

Esty, Daniel. "Non-Governmental Organizations at the World Trade Organization: Cooperation, Competition, or Exclusion." *Journal of International Economic Law* VI, 1 (March 1998), 123–147.

Eswaran, Hari, Fred Beinroth, and Paul Reich. "Global Land Resources and Population Supporting Capacity." *Alternative Agriculture* 14, 3 (1999), 129–136.

European Environmental Agency. *Environmental Taxes: Implementation and Environmental Effectiveness* (Copenhagen: European Environmental Agency, 1996).

Evans, Peter. "Transferable Lessons? Re-examining the Institutional Prerequisites of East Asian Economic Policies." *The Journal of Development Studies* 34, 6 (August 1998), 66–86.

Faria, V. *Social Exclusion in Latin America: An Annotated Bibliography* (Geneva: IILS Discussion Paper Series No. 70, 1994).

Fields, Gary. "Income Distribution in Developing Economies: Conceptual, Data, and Policy Issues in Broad Based Growth." In *Critical Issues in Asian Development,* ed. M.G. Quibria (Hong Kong: Oxford University Press, 1995) 75–93.

Fine, Ben. "The Developmental State Is Dead—Long Live Social Capital?" *Development and Change* Vol. 30 (1999), 1–19.

Fischer, Stanley. "On the Need for an International Lender of Last Resort." *Journal of Economic Perspectives* 13, 4 (Fall 1999), 85–104.

Flavin, Christopher, and Nicholas Lenssen. *Power Surge: Guide to the Coming Energy Revolution* (New York: W. W. Norton, 1994).
Food and Agriculture Organization of the United Nations (FAO). *The Sixth World Food Survey* (Rome, Italy: FAO, 1996).
Franke, Richard W., and Barbara H. Chasin. *Kerala: Radical Reform as Development in an Indian State* (Oakland, Calif.: Institute for Food and Development Policy, 1994).
French, Hilary. "Forging a New Global Partnership." *State of the World 1995,* eds. Lester Brown et al. (New York: Norton, 1995), 170–189.
French, Hilary. "Coping With Ecological Globalization." Chapter 10 in *State of the World 2000,* eds. Lester Brown et al. (New York: Norton, 2000).
Friedman, Milton. *Capitalism and Freedom* (Chicago: University of Chicago Press, 1962).
Friends of the Earth. *Green Scissors 2000: Cutting Wasteful and Environmentally Harmful Spending* (Washington, D.C.: Friends of the Earth, 2000).
Funtowicz, S.O., and J.R. Ravetz. "The Worth of a Songbird: Ecological Economics as a Post- Normal Science." *Ecological Economics* 10 (1994), 197–207.
Galeano, Eduardo, and Miguel Bonaso. "The Last Chance Café." *South Visions of Tomorrow* (United Kingdom: Channel 4 television program, April 12, 1993).
Gallagher, Kevin. "World Income Inequality and the Poverty of Nations." *The Political Economy of Inequality,* eds. Frank Ackerman et al. (Washington, D.C.: Island Press, 2000).
Gallarotti, Giulio M. "It Pays to be Green: The Managerial Incentive Structure and Environmentally Sound Strategies." *The Columbia Journal of World Business* (Winter 1995), 38–51.
Gardner, Gary, and Payal Sampat. *Mind Over Matter: Recasting the Role of Materials in Our Lives,* Worldwatch Paper 144. (Washington, D.C.: Worldwatch Institute, 1998).
Gentry, Bradford. *Private Investment and the Environment* (New York: United Nations Development Programme, 1996).
Gentry, Bradford, et al., eds. *Private Capital Flows and the Environment: Lessons from Latin America* (Yale Center for Environmental Law and Policy: 1996).
George, Susan. *The Debt Boomerang: How Third World Debt Harms Us All* (London: Pluto Press, 1992).
George, Susan. *The Lugano Report: On Preserving Capitalism in the Twenty-First Century* (London: Pluto, 1999).
Georgescu-Roegen, Nicholas. *The Entropy Law and the Economic Process* (Cambridge, Mass.: Harvard University Press, 1971).
Georgescu-Roegen, Nicholas. "Selections from 'Energy and Economic Myths.'" Chapter 4 in *Valuing the Earth: Economics, Ecology, Ethics,* eds. Herman E. Daly and Kenneth N. Townsend (Cambridge, Mass., MIT Press, 1993).
Giannini, Curzio. "'Enemy of None, But a Common Friend of All'? An International Perspective on the Lender-of-Last-Resort Function." *Essays in International Finance* 214 (Princeton, N.J.: Princeton University, Department of Economics, International Finance Section, 1999).
Gibbons, Peter. *Structural Adjustment and the Working Poor in Zimbabwe: Studies on Labour, Women Informal Sector Workers and Health* (Uppsala, Sweden: Nordiska Afrikainstitutet, 1995).
Gleick, Peter H. *The World's Water 2000–2001* (Washington, D.C.: Island Press, 2000).
Goetz, Anne Marie, and Rina Sen Gupta. "Who Takes the Credit: Gender, Power, and

Control Over Loan Use in Rural Credit Programs in Bangladesh." *World Development* 24, 1 (January 1996), 45–63.

Goodland, Robert, Herman E. Daly, and Salah El Serafy, eds. *Population, Technology, and Lifestyle: The Transition to Sustainability* (Washington, D.C.: Island Press, 1992.).

Goodland, Robert, Herman E. Daly, and Salah El Serafy. *The Urgent Need for a Rapid Transition to Global Environmental Sustainability* (World Bank Environment Department Divisional Paper No. 45, 1994).

Goodwin, Neva. "Development Connections: The Hedgerow Model." In *Rethinking Sustainability: Power, Knowledge, and Institutions,* ed. Jonathan M. Harris (Ann Arbor: University of Michigan Press, 2000).

Goodwin, Neva R., ed. *As if the Future Mattered: Translating Social and Economic Theory into Human Behavior* (Ann Arbor: University of Michigan Press, 1996).

Goodwin, Neva R., et al., eds. *The Consumer Society.* Frontier Issues in Economic Thought, Volume II (Washington, D.C.: Island Press, 1997).

Gouldner, Lawrence H., and Donald Kennedy 1997. "Valuing Ecosystem Services: Philosophical Bases and Empirical Methods." Chapter 3 in *Nature's Services: Societal Dependence on Natural Ecosystems,* ed. Gretchen C. Daily (Washington, D.C.: Island Press, 1997).

Gowdy, John, and Sabine O'Hara. "Weak Sustainability and Viable Technologies." *Ecological Economics* 22 (1997), 239–247.

Greider, William. *One World, Ready or Not: The Manic Logic of Global Capitalism* (New York: Simon and Schuster, 1997).

Grieg-Gran, Maryanne, et al. "Towards a Sustainable Paper Cycle: A Summary." *Journal of Industrial Ecology* 1, 3 (1997), 47–68.

Grossman, G.M., and A.B. Krueger. "Environmental Impacts of a North American Free Trade Agreement." National Bureau of Economic Research Working Paper No. 3914 (Cambridge, Mass., 1991).

Grossman, G.M., and A.B. Krueger. "Economic Growth and the Environment." *Quarterly Journal of Economics* 112 (1995), 353–378.

Guerard, John B. Jr. "Additional Evidence on the Cost of Being Socially Responsible in Investing." *The Journal of Investing* (Winter 1997), 31–53.

Gupta, Anil K. "Sustainable Institutions for Natural Resource Management: How Do We Participate in People's Plans?" Chapter 15 in *People's Initiatives for Sustainable Development: Lessons of Experience,* eds. Syed Abdus Samad, Tatsuya Watanabe, and Seuing-Jin Kim (Kuala Lumpur: Asian and Pacific Development Centre, 1995), 341–373.

Haberl, H. "Human Appropriation of Net Primary Product as an Environmental Indicator." *Ambio* 26, 3 (1997), 143–146.

Haddad, Lawrence, Lynn Brown, Andrea Richter, and Lisa Smith. "The Gender Dimensions of Economic Adjustment Policies: Potential Interactions and Evidence to Date." *World Development* 23, 6 (1995), 881–896.

Hammond, Allen L. "Natural Resource Consumption: North and South." Chapter 23 in *Ethics of Consumption: The Good Life, Justice, and Global Stewardship,* eds. David A. Crocker and Toby Linden (Oxford: Rowman & Littlefield Publishers, 1998), 437–475.

Han, Xiaoli, and Lata Chatterjee. "Impacts of Growth and Structural Change on CO_2 Emissions of Developing Countries." *World Development* 25, 3 (1997), 395–407.

Hanemann, W. Michael. "Valuing One Environment Through Contingent Valuation." *Journal of Economic Perspectives* 8, 4 (Fall 1994), 19–43.

Hanson, James C., Erik Lichtenberg, and Steven E. Peters. "Organic Versus Conventional Grain Production in the Mid-Atlantic: An Economic and Farming System Overview." *Alternative Agriculture* 12, 1 (1997), 2–9.

Harris, Jonathan M. "Global Institutions and Ecological Crisis." *World Development* 19, 1 (January 1991), 111–122.

Harris, Jonathan M. "World Agricultural Futures: Regional Sustainability and Ecological Limits." *Ecological Economics* 17, 2 (1996), 95–115.

Harris, Jonathan M., ed. *Rethinking Sustainability: Power, Knowledge, and Institutions* (Ann Arbor: University of Michigan Press, 2000).

Harris, Jonathan M., and Scott Kennedy. "Carrying Capacity in Agriculture: Global and Regional Issues." *Ecological Economics* 29, 3 (1999), 443–461.

Harrison, Ann, and Gordon Hanson. "Who Gains from Trade Reform?" *Journal of Development Economics* 59 (1999), 125–154.

Hartwick, J.M. "Intergenerational Equity and the Investing of Rents from Exhaustible Resources." *American Economic Review* 66 (1977), 972–974.

Harvard Business Review. "The Challenge of Going Green." (July-August 1994).

Haskel, Thomas L. "Capitalism and the Origins of the Humanitarian Sensibility." *American Historical Review* 90, 2 and 3 (April and June 1985).

Heiman, Michael. "Community Attempts at Sustainable Development Through Corporate Accountability." *Journal of Environmental Planning and Management* 40, 5, 631–644.

Herendeen, Robert A. "Monetary-Costing Environmental Services: Nothing Is Lost, Something Is Gained." *Ecological Economics* 25 (1998), 29–30.

Herendeen, Robert A. "Should Sustainability Analyses Include Biophysical Assessments?" *Ecological Economics* 29 (1999), 17–18.

Hickman, Kent A., Walter Teets, and John J. Kohls. "Social Investing and Modern Portfolio Theory." *American Business Review* (January 1999), 72–78.

Hicks, Sir John. *Value and Capital 2nd ed.* (Oxford: Oxford University Press, 1946).

Hillel, Daniel. *Out of the Earth: Civilization and the Life of the Soil* (New York: The Free Press/Macmillan, 1991).

Hinterberger, Friedrich, and Eberhard K. Seifert. "Reducing Material Throughput: A Contribution to the Measurement of Dematerialization and Sustainable Human Development." In *Environment, Technology and Economic Growth: The Challenge to Sustainable Development,* eds. Andrew Tylecote and Jan van der Straaten (Cheltenham, U.K.: Edward Elgar, 1997).

Ho, Mae-Wan. *Genetic Engineering: Dream or Nightmare? The Brave New World of Bad Science and Big Business* (Bath, U.K.: Gateway Books, 1998).

Hoffman, Andrew J. *From Heresy to Dogma: An Institutional History of Corporate Environmentalism* (San Francisco: New Lexington Press, 1997).

Holdren, John. "Energy in Transition." *Scientific American* (September 1990), 157–163.

Holmberg, Johan, ed. *Making Development Sustainable: Redefining Institutions, Policy, and Economics* (Washington, D.C.: Island Press, 1992).

Hopkins, Raymond F. "Notes on Agriculture and the State." In *Agriculture and the State,* ed. C. Peter Timmer (Ithaca: Cornell University Press, 1991).

Howarth, Richard B., and Richard B. Norgaard. "Intergenerational Transfers and the Social Discount Rate." *Environmental and Resource Economics* 3 (August 1993), 337–358.

Illich, Ivan. *Shadow Work: Vernacular Values Examined* (London: Marion Boyars, 1981).

Intergovernmental Panel on Climate Change (IPCC). *Climate Change 1995: The Science of Climate Change* (Cambridge: Cambridge University Press, 1996).

International Food Policy Research Institute (IFPRI). *A 2020 Vision for Food, Agriculture, and the Environment: The Vision, Challenge, and Recommended Action* (Washington, D.C: International Food Policy Research Institute, 1995).

International Labor Organization (ILO). "Making a World of Difference for Working People" (Washington, D.C.: International Labor Organization, 1998). Mimeographed.

Islam, Nurul, ed. *Population and Food in the Early Twenty-First Century: Meeting Future Food Demand of an Increasing Population* (Washington, D.C.: International Food Policy Research Institute, 1995).

Jaffe, Adam, Steven Peterson, Paul Portney, and Robert Stavins. "Environmental Regulation and the Competitiveness of U.S. Manufacturing: What Does the Evidence Tell Us?" *Journal of Economic Literature* 32 (March 1995), 132–163.

Jazairy, Idriss, Mohiuddin Alamgir, and Theresa Panuccio. *The State of World Rural Poverty* (Southampton, England: International Fund for Agricultural Development, 1992).

Jubilee 2000/USA. "The Debt Burden on Impoverished Countries: An Overview." On Jubilee 2000/USA web site: http://www.j2000usa.org/debt/edpac/debt.html, February, 1998.

Julka, Abnash C. "Economic Growth, Poverty and Sustainable Development: Window-Shopping in the Marketland." Chapter 2 in *Poverty, Population and Sustainable Development,* ed. S.R. Mehta (New Delhi, India: Rawat Publications, 1997), 38–49.

Kabeer, Naila. *Reversed Realities: Gender Hierarchies in Development Thought* (New York and London: Verso, 1994).

Kahn, J.R., and A. Farmer. "The Double Dividend, Second-Best Worlds, and Real-World Environmental Policy." *Ecological Economics* 30, 3 (September 1999), 433–439.

Kannan, K.P. "Public Intervention and Poverty Alleviation: A Study of the Declining Incidence of Rural Poverty in Kerala, India." *Development and Change* 26 (1995), 701–727.

Kapur, Devesh. "The IMF: A Cure or a Curse?" *Foreign Policy* (Summer 1998), 114–129.

Karshenas, Massoud. "Environment, Technology and Employment: Towards a New Definition of Sustainable Development." *Development and Change* 25 (1994), 723–756.

Kelly, Gavin, et al., eds. *Stakeholder Capitalism* (New York: St. Martin's Press, 1997).

Keyfitz, Nathan. "Consumption and Population." In *Ethics of Consumption: The Good Life, Justice, and Global Stewardship,* eds. David A Crocker and Toby Linden (Oxford: Rowman and Littlefield Publishers, 1998), 476–500.

Khandker, Shahidur R., Hussain A. Samad, and Zahed H. Khan. "Income and Employment Effects of Micro-credit Programmes: Village-Level Evidence from Bangladesh." *Journal of Development Studies* 35, 2 (December 1998), 96–124.

Khor, Martin. "A Checklist of Issues and Positions for the WTO Process, 1999." In *Third World Resurgence,* 108/109 (Aug.-Sept. 1999).

Klingebiel, Stephan. *Effectiveness and Reform of the United Nations Development Programme* (London: Cass, 1999).

Kolodner, Eric. "Transnational Corporations: Impediments or Catalysts of Social Development?" Occasional paper No. 5, prepared November 1994 for the World Summit for Social Development, held in Copenhagen in March, 1995. Commissioned by the United Nations Research Institute for Social Development; available at http://www.unrisd.org/engindex/publ/list/op/op5/op05-02.htm.

Korten, David C. *When Corporations Rule the World* (West Hartford, Conn.: Kumarian Press; and San Francisco, Calif.: Berrett-Koehler Publishers, 1995).

Korten, David C. *The Post-Corporate World* (West Hartford, Conn.: Kumarian Press; and San Francisco, Calif.: Berrett-Koehler Publishers, 1999).

Kozloff, Keith. "Power to Choose: Sustainability in the Evolving Electricity Industry," *Frontiers of Sustainability: Environmentally Sound Agriculture, Forestry, Transportation, and Power Production,* eds. Roger Dower et al. (Washington, D.C.: Island Press, 1997).

Krimsky, Sheldon. *Biotechnics and Society: The Rise of Industrial Genetics* (New York: Praeger, 1991).

Krimsky, Sheldon, and Roger Wrubel. *Agricultural Biotechnology and the Environment: Science, Policy, and Social Issues* (Chicago: University of Illinois Press, 1996).

Krishnan, Rajaram, et al., eds. *A Survey of Ecological Economics.* Frontier Issues in Economic Thought, Volume I (Washington, D.C.: Island Press, 1995).

Krumsiek, Barbara J. "The Emergence of a New Era in Mutual Fund Investing: Socially Responsible Investing Comes of Age." *The Journal of Investing* 6, 4 (Winter 1997), 25–30.

Kuznets, Simon. "Economic Growth and Income Inequality." *American Economic Review* 45 (1955), 1–28.

Lal, Deepak. "Participation, Markets and Democracy." Chapter 13 in *New Directions in Development Economics: Growth, Environmental Concerns and Government in the 1990s,* eds. Mats Lundahl and Benno J. Ndulu (New York: Routledge, 1996), 299–322.

Lampkin, Nicholas H., and Susanne Padel, eds. *The Economics of Organic Farming: An International Perspective* (Wallingford, U.K.: CAB International, 1994).

Lange, Glenn-Marie, and Faye Duchin. *Integrated Environmental-Economic Accounting: Natural Resource Accounts, and Natural Resource Management in Africa* (Winrock International Institute Technical Report No. 13, 1993; also summarized in Krishnan et al. eds., 1995, 262–265).

Lappé, Marc, and Britt Bailey. *Against the Grain: Biotechnology and the Corporate Takeover of Your Food* (Monroe, Maine: Common Courage Press, 1998).

Lenssen, Nicholas. "All the Coal in China." *World Watch* 6, 2 (March/April 1993), 22–29.

"Let Them Eat Pollution." *The Economist* 8 (February 1992).

Lockeretz, W., G. Shearer, and D.H. Kohl. "Organic Farming in the Corn Belt." *Science* 211 (1981), 540–547.

Lotka, Alfred J. *Elements of Mathematical Biology* (New York: Dover Publications, 1956). Original publication 1925.

Lucardie, Paul. "Greening and Un-greening the Netherlands." *Greening the Millennium? The New Politics of the Environment,* ed. Michael Jacobs (Oxford: Blackwell Publishers, 1997), 183–191.

MacEwan, Arthur. *Neo-Liberalism or Democracy?* (New York: St. Martin's Press, 2000).

Mander, Jerry, and Edward Goldsmith, eds. *The Case Against the Global Economy* (San Francisco: Sierra Club Books, 1996).

Mani, Muthukumara, and David Wheeler. "In Search of Pollution Havens? Dirty Industry in the World Economy, 1960–1995." *Journal of Environment and Development* 7, 3 (September 1998), 215–247.

Mann, Howard, and Konrad von Moltke. *NAFTA's Chapter 11* (Ottawa: International Institute for Sustainable Development, 1999).

Markandya, Anil, and Julie Richardson, eds. *Environmental Economics: A Reader,* Part III: Instruments for Environmental Control and Applications (New York: St. Martin's Press, 1993).

Marquette, Catherine, and Richard Bilsborrow. "Population and Environment Relationships in Developing Countries: Recent Approaches and Methods." In *People and their Planet: Searching for Balance,* eds. Barbara Baudot and William Moomaw (New York: St. Martin's Press, 1999).

Martinot, Eric. "Energy Efficiency and Renewable Energy in Russia: Transaction Barriers, Market Intermediation, and Capacity Building." *Energy Policy* 26, 11 (1998), 905–915.

Martinot, Eric, Jonathan E. Sinton, and Brent M. Haddad. "International Technology Transfer for Climate Change Mitigation and the Cases of Russia and China." *Annual Review of Energy and Environment* 22 (1997), 357–401.

Massie, Robert K. *Loosing the Bonds: The United States and South Africa in the Apartheid Years* (New York: Doubleday, 1997).

Mayoux, Linda. "Beyond Naivety: Women, Gender Inequality and Participatory Development." In *Development and Change* 26 (1995), 235–258.

Mayur, Rashmi, and Bennett Daviss. "How *Not* to Develop an Emerging Nation," *The Futurist* 32, 1 (1998), 27–32.

McKibben, Bill. *Hope, Human, and Wild* (Boston: Little, Brown, 1995).

McNeely, Jeffrey A. *Conserving the World's Biological Diversity* (Gland, Switzerland: IUCN, 1990).

Mensah, Chris. "The United Nations Commission on Sustainable Development." In *Greening International Institutions,* ed. Jacob Werskman (London: Earthscan, 1996).

Miranda, Liliana, and Michaela Hordijk. "Let Us Build Cities for Life: The National Campaign of Local Agenda 21s in Peru." *Environment and Urbanization* 10, 2, (October 1998).

Mitchell, Donald O., and Merlinda C. Ingco. "Global and Regional Food Demand and Supply Prospects." In *Population and Food in the Early Twenty-First Century: Meeting Future Food Demand of an Increasing Population,* ed. Nurul Islam (Washington, D.C.: International Food Policy Research Institute, 1995).

Mokhiber, Russell, and Robert Weissman. *Corporate Predators: The Hunt for MegaProfits and the Attack on Democracy* (Monroe, Maine: Common Courage Press, 1999).

Moomaw, William, and Mark Tullis. "Charting Development Paths: A Multicountry Comparison of Carbon Dioxide Emissions." In R. Socolow et al. (1994), 157–172.

Moomaw, W.R., and G.C. Unruh. "Are Environmental Kuznets Curves Misleading Us?

The Case of CO_2 Emissions." *Environment and Development Economics* 2, 4 (1997), 451–463.

Moomaw, William, et al. "The Kyoto Protocol: A Blueprint for Sustainable Development." In *Journal of Environment and Development* 8, 1 (March 1999), 82–90.

Mosley, Paul, and David Hulme. "Microenterprise Finance: Is There a Conflict Between Growth and Poverty Alleviation?" *World Development* 26, 5 (1998), 783–790.

Munasinghe, Mohan. *Sustainable Energy Development (SED): Issues and Policy* (Washington, D.C.: World Bank Environment Department Paper No. 016, 1995).

Munasinghe, Mohan. "Countrywide Environmental Policies and Sustainable Development: Are the Linkages Perverse?" *International Yearbook of Environmental and Resource Economics,* 1998–1999, ed. Tom Tietenberg (Northampton: Elgar, 1998).

Nadal, Alejandro. *The Environmental and Social Impacts of Economic Liberalization on Corn Production in Mexico* (London: World Wildlife Fund and Oxfam UK, 2000).

Nanda, Ved P. "The Right to Development: An Appraisal." Chapter 4 in *World Debt and the Human Condition: Structural Adjustment and the Right to Development,* eds. Ved P. Nanda, George W. Shepherd, Jr., and Eileen McCarthy-Arnolds (Westport, Conn.: Greenwood Press, 1993), 41–61.

National Environmental Engineering Research Institute. *Water Resource Management in India: Present Status and Solution Paradigm* (Nagpur, India, circa 1997).

National Research Council (NRC). *Alternative Agriculture* (Washington, D.C.: National Academy Press, 1989).

Nordhaus, William, and James Tobin. *Is Growth Obsolete?* (National Bureau of Economic Research. New York: Columbia University Press, 1972).

Norgaard, Richard B. "The Case for Methodological Pluralism." In *Ecological Economics* 1 (1989), 37–57.

Norgaard, Richard B. *Development Betrayed* (London: Routledge, 1994).

Norgaard, Richard B., and Richard B. Howarth. "Sustainability and the Problem of Valuation." Chapter 7 in *Ecological Economics: The Science and Management of Sustainability,* ed. Robert Costanza (New York: Columbia University Press, 1991).

Norgaard, Richard B., et al. "Next, the Value of God, and Other Reactions." *Ecological Economics* 25 (1998), 37–39.

Norse, David. *Population and Global Climate Change: Science, Impacts, and Policy.* Proceedings from the Second World Climate Conference (1990).

Norton, Bryan G. "Sustainability, Human Welfare, and Ecosystem Health." *Environmental Values* 1 (Summer 1992), 97–111.

Norton, Bryan G. "Environmental Ethics and the Rights of Future Generations." *Environmental Ethics* 4 (Winter 1982), 319–330.

Otero, Maria, and Elisabeth Rhyne. *The New World of Microenterprise Finance: Building Healthy Financial Institutions for the Poor* (West Hartford, Conn.: Kumarian Press, 1994).

Owen, Dave. "The Emerging Green Agenda: A Role for Accounting?" In *Business and the Environment: Implications of the New Environmentalism,* ed. Denis Smith (New York: St. Martin's Press, 1993).

Paarlberg, Robert. "The Global Food Fight." *Foreign Affairs* 79, 3 (May/June 2000), 24–38.

Page, Talbot. "Intergenerational Justice as Opportunity." *Energy and the Future,* eds. Douglas MacLean and Peter G. Brown (Totowa, N.J.: Rowman and Littlefield, 1983).

Page, Talbot. "Sustainability and the Problem of Valuation." In *Ecological Economics: The Science and Management of Sustainability,* ed. Robert Costanza (New York: Columbia University Press, 1991).

Palmer-Jones, Richard, and Cecile Jackson. "Work Intensity, Gender and Sustainable Development." In *Food Policy,* 22, 1 (1997), 39–62.

Panayotou, Theodore. "Environmental Impacts of Structural Adjustment Programs: Synthesis and Recommendations." In *Environmental Impacts of Macroeconomic and Sectoral Polices,* ed. M. Munasinghe (Washington, D.C.: World Bank, 1996).

Panayotou, Theodore. *Instruments of Change: Motivating and Financing Sustainable Development* (London: Earthscan Publications, 1998).

Parikh, Jyoti. "Consumption Patterns: The Driving Force of Environmental Stress." Chapter 2 in *Pricing the Planet: Economic Analysis for Sustainable Development,* eds. Peter H. May and Ronaldo Seroa da Motta (New York: Columbia University Press, 1996), 39–48.

Pastor, Manuel. "Cuba: The Blocked Transition." *Economic Policy in Transitional Economies* 8, 1 (1998), 109–129.

Pearce, David W., and G. Atkinson. "Capital Theory and the Measurement of Sustainable Development: An Indicator of Weak Sustainability." *Ecological Economics* 8, 2 (1993), 103–108.

Pearce, David W., and Dominic Moran. *The Economic Value of Biodiversity* (London: Earthscan Publications, 1997).

Pearce, David W., and K.R. Turner. *Economics of Natural Resources and the Environment* (New York: Harvester Wheatsheaf, 1990).

Pearce, David W., and Jeremy J. Warford. *World Without End: Economics, Environment, and Sustainable Development* (New York and Oxford: Oxford University Press, 1993).

Perrings, Charles. "Reserved Rationality and the Precautionary Principle: Technological Change, Time, and Uncertainty in Environmental Decision Making." Chapter 11 in *Ecological Economics: The Science and Management of Sustainability,* ed. Robert Costanza (New York: Columbia University Press, 1991). Also summarized in Krishnan et al., 1995, 158–162.

Peters, Irene, Frank Ackerman, and Steve Bernow. "Economic Theory and Climate Change Policy." *Energy Policy* 27, 12 (1999), 501–504.

Philibert, Cédric. "The Economics of Climate Change and the Theory of Discounting." *Energy Policy* 27, 15 (1999), 913–927.

Pimentel, David, et al., eds. "Economic and Environmental Benefits of Biodiversity." *BioScience* 47, 11 (1997), 747–757.

Ponting, Clive. *A Green History of the World: The Environment and the Collapse of Great Civilizations* (New York: Penguin Books, 1993).

Population Reference Bureau. *World Population Data Sheet* (Washington, D.C.: Population Reference Bureau, 1999).

Porter, Michael E. "Capital Choices: National Systems of Investment." In *As If The Future Mattered: Translating Social and Economic Theory into Human Behavior,* ed. Neva R. Goodwin (Ann Arbor: University of Michigan Press, 1996).

Portney, Paul. "The Contingent Valuation Debate: Why Economists Should Care." *Journal of Economic Perspectives* 8, 4 (Fall 1994), 3–17.

Portney, Paul. "Economic Impact of Green Technology Questioned." *Environmental Science and Technology,* 29, 1 (January 1995), 19A.

Postel, Sandra. *Pillar of Sand: Can the Irrigation Miracle Last?* (New York: W.W. Norton, 1999).

President's Council on Sustainable Development. *Towards a Sustainable America* (Washington, D.C.: President's Council on Sustainable Development Publications, 1996).

Pretty, Jules, et al. "Regenerating Agriculture: The Agroecology of Low-External Input and Community-Based Development." In *The Earthscan Reader in Sustainable Development,* eds. John Kirby et al. (London: Earthscan Publications, 1995).

Public Citizen/RMALC. NAFTA's Broken Promises: The Border Betrayed (Washington, D.C.: Public Citizen, 1996).

Putnam, Robert D. "Democracy, Development, and the Civic Community: Evidence from an Italian Experiment." In *Culture and Development in Africa,* Ismail Serageldin and June Taboroff (Washington, D.C.: World Bank, 1994), 33–73.

Putzel, James. "Accounting for the 'Dark Side' of Social Capital: Reading Robert Putnam on Democracy." *Journal of International Development* 9, 7 (1997), 939–949.

Rabinovitch, Jonas, and Josef Leitmann. "Environmental Innovation and Management in Curitiba, Brazil." Urban Management Program Working Paper Series No. 1. (Washington, D.C.: The World Bank, 1993).

Rahman, Aminur. "Micro-credit Initiatives for Equitable and Sustainable Development: Who Pays?" *World Development* 27, 1 (1999), 67–82.

Randall, A. "Human Preferences, Economics, and the Preservation of Species." In *The Preservation of Species,* ed. Bryan G. Norton (Princeton, N.J.: Princeton University Press, 1986).

Redefining Progress. http://www.rprogress.org/resources/resources.html, 2000.

Reed, David. "The Environmental Legacy of Bretton Woods: The World Bank." In *Global Governance: Drawing Insights from the Environmental Experience,* ed. Oran Young (Cambridge, Mass.: MIT Press, 1997), 227–245.

Reed, David, ed. *Structural Adjustment, the Environment and Sustainable Development.* (London: Earthscan Publications, 1997).

Rees, William E. "How Should a Parasite Value its Host?" *Ecological Economics* 25 (1998), 49–52.

Repetto, Robert. *Population, Resources, Environment: An Uncertain Future* (Washington, D.C.: Population Reference Bureau, 1991).

Rifkin, Jeremy. *The Biotech Century* (New York: Jeremy P. Tarcher/Putnam, 1998).

Robinson, Nicholas A., ed. *Agenda 21: Earth's Action Plan* (New York: Oceana Publications, 1993).

Rodrik, Dani. *Why Do More Open Economies Have Bigger Governments?* (Cambridge, Mass.: National Bureau of Economic Research, 1996).

Rodrik, Dani. "The Global Fix." *New Republic* 2 (November 1998).

Rogoff, Kenneth. "International Institutions for Reducing Global Financial Instability." *Journal of Economic Perspectives* 13, 4 (Fall 1999), 21–42.

Roodman, David Malin. *Paying the Piper: Subsidies, Politics, and the Environment.* Worldwatch Paper 133 (Washington, D.C.: Worldwatch Institute, 1996).

Rosegrant et al. *Global Food Projections to 2020: Implications for Investment.* Food, Agriculture and the Environment Discussion Paper No. 5 (Washington, D.C.: International Food Policy Research Institute, 1995).

Roseland, Mark, Maureen Cureton, and Heather Wornell. *Toward Sustainable Communities: Resources for Citizens and Their Governments* (Stony Creek, Conn.: New Society Publishers, 1998).

Rostow, W.W. *The Stages of Economic Growth: A Non-Communist Manifesto* (Cambridge: Cambridge University Press, 1960).

Rothman, Dale S. "Environmental Kuznets Curves—Real Progress or Passing the Buck? A Case for Consumption-Based Approaches." *Ecological Economics* 25 (1998), 177–194.

Ruggerio, Renalto. Address to the World Trade Organization (March 1999).

Runge, C. Ford. *Freer Trade, Protected Environment* (New York: Council on Foreign Relations Press, 1994).

Ruth, Matthias. "Dematerialization in Five U.S. Metals Sectors: Implications for Energy Use and CO_2 Emissions." *Resources Policy* 24, 1 (1998), 1–18.

Sachs, J.D., and A.M. Warner. "Natural Resource Abundance and Economic Growth." Development Discussion paper 517a (Cambridge, Mass.: Harvard Institute for International Development, 1995).

Samuel, Warren J., and Arthur S. Miller, eds. *Corporations and Society* (Westport, Conn.: Greenwood Press, 1987).

Schmidheiny, Stephan, and Bradford Gentry. "Privately Financed Sustainable Development." In *Thinking Ecologically: The Next Generation of Environmental Policy*, eds. Marian Chertow and Daniel Esty (New Haven, Conn.: Yale University Press, 1997).

Schor, Juliet B. "Global Equity and Environmental Crisis: An Argument for Reducing Working Hours in the North." *World Development* 19, 1 (January 1991), 73–84.

Scully, Nan Dawkins, and Daphne Wysham. *The World Bank's Consultative Group to Assist the Poorest: Opportunity or Liability for the World's Poorest Women?* (Washington, D.C. : Institute for Policy Studies, 1997).

Selden, T.M., and D. Song. "Environmental Quality and Development: Is There a Kuznets Curve for Air Pollution? *Journal of Environmental Economics and Management* 27 (1994), 162–168.

Selman, P., and J. Parker. "Tales of local sustainability," *Local Environment* 4, 1 (1999), 47–60.

Sen, Amartya. *Poverty and Famines* (Oxford: Oxford University Press, 1981).

Sen, Amartya. *Inequality Reexamined* (Cambridge, Mass.: Harvard University Press, 1992).

Sen, Amartya. *Development as Freedom* (New York: Alfred A. Knopf, 1999).

Sen, Gita. "Engendering Poverty Alleviation: Challenges and Opportunities." *Development and Change* 30, 3 (July 1999), 685–692.

Seymour, Frances, and Navroz Dubasch. "World Bank's Environmental Reform Agenda." *Foreign Policy in Focus* 4, 10 (March 1999).

Shafik, N., and S. Bandyopadhyay. *Economic Growth and Environmental Quality: Time Series and Cross-Country Evidence* (Washington, D.C.: World Bank, 1992).

Sharp, Robin. "Organizing for Change: People-Power and the Role of Institutions." Chapter 2 in *Making Development Sustainable: Redefining Institutions, Policy, and Economics*, ed. Johan Holmberg (Washington, D.C.: International Institute for Environment and Development, 1992), 39–62.

Shaw, R. Paul. "The Impact of Population Growth on Environment: The Debate Heats Up." *Environmental Impact Assessment Review* 12 (1992), 11–36.

Singh, Naresh, and Perpetua Kalala, eds. *Adaptive Strategies and Sustainable Livelihoods: Integrated Summary of Community and Policy Issues* (Winnipeg, Manitoba: International Institute for Sustainable Development, 1995).

Singh, Naresh, and Vangile Titi. *Empowerment Towards Sustainable Development* (London: Zed Books, 1995).

Socolow, R., C. Andrews, F. Berkhout, and V. Thomas, eds. *Industrial Ecology and Global Change* (New York: Cambridge University Press, 1994).

Solow, R.M. "On the Intertemporal Allocation of Natural Resources." *Scandinavian Journal of Economics* 88 (1986), 141–149.

Steiner, George Albert, and John F. Steiner. *Business, Government, and Society: A Managerial Perspective* (Boston: Irwin/MacGraw-Hill, 2000).

Stern, David. "Progress on the Environmental Kuznets Curve?" *Environment and Development Economics* 3 (1998), 173–196.

Stern, D.I., M.S. Common, and E.B. Barbier. "Economic Growth and Environmental Degradation: the Environmental Kuznets Curve and Sustainable Development." *World Development* 24 (1996), 1151–1160.

Sterner, Thomas. "Environmental Tax Reform: Theory, Industrialized Country Experience and Relevance in LDCs." *New Directions in Development Economics,* eds. Mats Lundah and Benno J. Ndulu (New York: Routledge, 1996), 224–248.

Stiglitz, Joseph. "An Agenda for Development for the Twenty-First Century," presented at the World Bank Ninth Annual Conference on Development Economics, 1997.

Stiglitz, Joseph. "More Instruments and Broader Goals: Moving Toward the Post Washington Consensus." WIDER Annual Lectures No. 2, Helsinki, 1998.

Streeten, Paul. *Thinking About Development* (Cambridge and Milano, Italy: Cambridge University Press, 1995).

Streeten, Paul, with Shahid Javed Burki, Mahbub ul Haq, Norman Hicks, and Frances Stewart. *First Things First: Meeting Basic Human Needs in the Developing Countries.* Published for the World Bank. (New York and Oxford: Oxford University Press, 1981).

Stumberg, Robert. "Sovereignty By Subtraction: The Multilateral Agreement on Investment." *Cornell International Law Journal* 31, (1998) 492–595.

Suri, Vivek, and Duane Chapman. "Economic Growth, Trade, and Energy: Implications for the Environmental Kuznets Curve." *Ecological Economics* 25 (1998), 195–208.

Susskind, Lawrence. *Environmental Diplomacy* (Oxford: Oxford University Press, 1994).

Sutcliffe, Bob. "Development after Ecology." Chapter 12 in *The North, the South, and the Environment,* eds. V. Bhaskar and Andrew Glyn (New York: St. Martin's Press, 1995).

Tharamangalam, Joseph. "The Perils of Social Development without Economic Growth: The Development Debacle of Kerala, India." *Bulletin of Concerned Asian Scholars* 30, 1 (1998), 23–34.

Thrupp, Lori Ann, ed. *New Partnerships for Sustainable Agriculture* (Washington, D.C., World Resources Institute, 1996).

Tisdell, C. "Economics and the Debate about Preservation of Species, Crop Varieties, and Genetic Diversity." *Ecological Economics* 2 (1990), 77–90.

Toman, Michael A. "The Difficulty in Defining Sustainability." *Resources* 106 (1992), 3–6.

Toman, Michael A. "Research Frontiers in the Economics of Climate Change." *Environmental and Resource Economics* 11, 3–4 (1998a), 603–621.

Toman, Michael A. "Why Not to Calculate the Value of the World's Ecosystem Services and Natural Capital." *Ecological Economics* 25 (1998b), 57–60.

Torras, Mariano, and James K. Boyce. "Income, Inequality, and Pollution: A Reassessment of the Environmental Kuznets Curve." *Ecological Economics* 25 (1998), 147–160.

Union of Concerned Scientists and Tellus Institute. *A Small Price to Pay: U.S. Action to Curb Global Warming Is Feasible and Affordable* (Cambridge Mass.: Union of Concerned Scientists, 1998).

United Nations Commission on Sustainable Development (UNCSD). *Assessment of Progress in the Implementation of Agenda 21 at the National Level: Report of the Secretary General,* U.N. Doc. E/CN/.17/1997/5, 1997.

United Nations Commission on Sustainable Development (UNCSD) web address http://www.un.org/esa/sustdev/cds.htm.

United Nations Conference on Environment and Development (UNCED). *The Global Partnership for Environment and Development: A Guide to Agenda 21* (Geneva: UNCED, 1992).

United Nations Conference on Trade and Development (UNCTAD). *World Investment Report 1999: Foreign Direct Investment and the Challenge of Development* (New York: UNCTAD, 1999).

United Nations Conference on Trade and Development (UNCTAD). *World Investment Report 2000* (New York: UNCTAD, 2000).

United Nations Department for Economic and Social Information and Policy Analysis, Statistical Division. *Integrated Economic and Environmental Accounting* (Studies in Methods Series F, No. 61: Handbook of National Accounting, 1993).

United Nations Development Programme (UNDP). *Human Development Report* (New York and Oxford: Oxford University Press, various years 1990–1999).

United Nations Development Programme web address http://www.undp.org.

United Nations Economic and Social Council. *World Population Prospects: The 1998 Revision* (United Nations, 1998).

United Nations Environment Programme (UNEP). *UNEP/World Bank Workshop on the Environmental Impacts of Structural Adjustment Programs* (United Nations Environment Programme, Environmental Economics Series Paper No. 18, 1996).

United Nations Environment Programme (UNEP). *Global Environment Outlook 2000* (London: Earthscan Publications, 1999).

United Nations Environment Programme web address http://www.unep.org.

United States Environmental Protection Agency (EPA). *Searching for the Profit in Pollution Prevention: Case Studies in the Corporate Evaluation of Environmental Opportunities* (Washington, D.C.: U.S. Environmental Protection Agency 742-R-98-005, April 1998).

United States Environmental Protection Agency. *U.S. Methane Emissions 1990–2020: Inventories, Projections, and Opportunities for Reductions* (Washington: EPA Office of Air and Radiation, Document EPA 430-R-99-013, 1999).

van den Bergh, Jeroen, and Harman Verbruggen. "Spatial Sustainability, Trade, and Indicators: An Evaluation of the 'Ecological Footprint'" *Ecological Economics* 29, 1 (1999), 61–72.

Vatn, Arild, and Daniel W. Bromley. "Choices without Prices without Apologies." *Journal of Environmental Economics and Management* 26 (March 1994), 129–148.

Vincent, Jeffrey R., and Theodore Panayotou. "Consumption: Challenge to Sustainable Development . . . or Distraction?" *Science* 276 (April 1997), 53–55.

Vitousek, P.M., P.R. Ehrlich, A.H. Ehrlich, and P.A. Matson. "Human Appropriation of the Products of Photosynthesis." *BioScience* 36, 6 (1986), 368–373.

Vogel, David. *Trading Up* (Cambridge: Harvard University Press, 1996).

Wackernagel, Mathis, and Alex Long, eds. "Ecological Economics Forum." *Ecological Economics* 29 (1999), 13–60.

Wackernagel, Mathis, and William Rees. *Our Ecological Footprint: Reducing Human Impact on Earth* (Stony Creek, Conn.: New Society Publishers, 1996).

Wackernagel, Mathis, et al. *Ecological Footprints of Nations: How Much Nature Do They Use? How Much Nature Do They Have?* (Toronto, Canada: International Council for Local Environmental Initiatives, 1997).

Wackernagel, Mathis, et al. "National Natural Capital Accounting with the Ecological Footprint Concept." *Ecological Economics* 29, 3 (1999), 375–390.

Wallach, Lori. "Lori's War." *Foreign Policy* (Spring 2000), 28–58.

Walley, Noah, and Bradley Whitehead. "It's Not Easy Being Green." *Harvard Business Review* (Sept.-Oct. 1994), 120–134.

Wilson, E.O. *The Diversity of Life* (New York: W.W. Norton, 1992).

Wilson, E.O., ed. *Biodiversity* (Washington, D.C.: National Academy Press, 1988).

Wolfe, Marshall. "Globalization and Social Exclusion: Some Paradoxes." Chapter 4 in *Social Exclusion: Rhetoric, Reality, Responses,* Gerry Rodgers, Charles Gore, and Jose B. Figueiredo (Geneva: International Institute for Labour Studies, 1995), 81–101.

World Bank. *World Development Report: Development and the Environment* (New York and Oxford: Oxford University Press, 1992).

World Bank. *Mainstreaming the Environment* (Washington, D.C.: World Bank, 1995).

World Bank. *World Development Report 1996: From Plan to Market* (New York: Oxford University Press, 1996).

World Bank. *Expanding the Measure of Wealth: Indicators of Environmentally Sustainable Development* (Washington, D.C.: The World Bank, 1997a).

World Bank. *World Development Report 1997: The State in a Changing World* (New York and Oxford: Oxford University Press, 1997b).

World Bank. *Entering the 21st Century: World Development Report 1999/2000* (New York: Oxford University Press, 2000).

World Commission on Environment and Development (WCED). *Our Common Future* (New York: Oxford University Press, 1987).

World Resources Institute (WRI). *World Resources 1990–91: A Guide to the Global Environment* (New York: Oxford University Press, 1990).

World Resources Institute (WRI). *World Resources 1996–97: The Urban Environment* (New York: Oxford University Press, 1996).

World Resources Institute (WRI). *World Resources 1998–99: A Guide to the Global Environment* (New York and Oxford: Oxford University Press, 1998).

Worldwatch Institute. *Vital Signs 1998* (New York: Norton, 1998).

Young, Oran. "Hitting the Mark." *Environment* (October 1999), 20–29.

Zlotnik, Hanna. "International Migration 1965–96: An Overview." *Population and Development Review* 24, 3 (1998), 429–468.

Subject Index

Name Index

Island Press Board of Directors